# NANOCOMPOSITE MATERIALS
## Synthesis, Properties and Applications

# NANOCOMPOSITE MATERIALS
## Synthesis, Properties and Applications

EDITED BY

Jyotishkumar Parameswaranpillai | Nishar Hameed
Thomas Kurian | Yingfeng Yu

CRC Press is an imprint of the
Taylor & Francis Group, an **informa** business

CRC Press
Taylor & Francis Group
6000 Broken Sound Parkway NW, Suite 300
Boca Raton, FL 33487-2742

First issued in paperback 2019

ISBN-13: 978-1-4822-5807-3 (hbk)
ISBN-13: 978-0-367-87080-5 (pbk)

This book contains information obtained from authentic and highly regarded sources. Reasonable efforts have been made to publish reliable data and information, but the author and publisher cannot assume responsibility for the validity of all materials or the consequences of their use. The authors and publishers have attempted to trace the copyright holders of all material reproduced in this publication and apologize to copyright holders if permission to publish in this form has not been obtained. If any copyright material has not been acknowledged please write and let us know so we may rectify in any future reprint.

**Visit the Taylor & Francis Web site at**
**http://www.taylorandfrancis.com**

**and the CRC Press Web site at**
**http://www.crcpress.com**

# Contents

# Acknowledgments

We thank Dr. George Z. Kyzas, Dr. Anna Biedunkiewicz, Dr. Ningzhong Bao, Dr. Peng Tao, Dr. Gheorghe Rusu, Dr. Cassagnau Philippe, Dr. Ke-Ke Yang, Dr. Francesco Ciardelli, Dr. Andrea Pucci, Dr. Dimitris Achilias, and Dr. Seno Jose for their time and efforts in evaluating the chapters.

# Editors

**Jyotishkumar Parameswaranpillai** is an INSPIRE Faculty member in the Department of Polymer Science and Rubber Technology of Cochin University of Science and Technology, India. He earned his PhD in polymer science and technology (chemistry) from Mahatma Gandhi University, Kerala, India. He has published approximately 40 papers in high-quality international peer-reviewed journals on polymer nanocomposites, polymer blends and alloys, and biopolymer, and has edited two books. Dr. Parameswaranpillai's biographical profile is listed in the *Marquis Who's Who in the World 2015* (32nd edition) in reference to his outstanding achievement in the fields of polymer science and technology. In addition, he received the Venus International Foundation award for the best young faculty (science/chemistry) in the country for the year 2015.

**Nishar Hameed** earned his PhD from Deakin University in 2011. He received the Alfred Deakin Best Doctoral thesis award and Vice Chancellors Award for outstanding academic achievement from Deakin University. Dr. Hameed's research has been mainly focused on the preparation, processing, and characterization of nanostructured polymer materials. He has extensive expertise in the development of self-assembled block copolymers and bio-renewable plastics from natural resources. His current research is mainly focused on processing and characterization of nanostructured precursors for high-performance carbon fibers and thermoformable composite materials. As an early career researcher, Dr. Hameed has an excellent track record in polymer processing and characterization with nearly 40 high-quality journal papers, more than 23 conference papers, and one patent. In recent years, he has won a number of prestigious awards including the Victoria Fellowship and the Endeavour Research Fellowship.

**Thomas Kurian** is a professor of polymer technology and former head of the Department of Polymer Science and Rubber Technology, Cochin University of Science and Technology, Kochi, Kerala, India. He has authored several journal articles and book chapters. Dr. Kurian earned his PhD from the Indian Institute of Technology (IIT) Kharagpur; BSc in chemistry from Kerala University; BTech, MTech, MBA, and PG Cert. (Germ.) from Cochin University of Science and Technology; and PGDTRM (Post-Graduate Diploma in Teaching and Research in Management) from IGNOU (Indira Gandhi National Open University), India. He was JSPS (Japan Society for the Promotion of Science) postdoctoral fellow and JSPS bridge fellow at Yamagata University, Japan.

**Yingfeng Yu** is a professor of polymer science and engineering and head of the laboratory for polymer electronic packaging materials located at Fudan University, China. His research interests include structure and property relationship study of polymer materials in microelectronics and composites, thermosetting resins and high-performance polymers, polymer blends, and nanocomposites. He has published 60 papers in international journals devoted to polymers and materials.

# Contributors

**Hrushikesh A. Abhyankar**
Centre of Automotive Technology
Cranfield University
Cranfield, Bedfordshire, United Kingdom

**Farbod Alimohammadi**
Department of Chemistry
Temple University
Philadelphia, Pennsylvania

and

Young Researchers and Elite Club
South Tehran Branch
Islamic Azad University
Tehran, Iran

**Nektaria-Marianthi Barkoula**
Department of Materials Engineering
University of Ioannina
Ioannina, Greece

**Danilo J. Carastan**
Center for Engineering, Modeling and
    Applied Social Sciences
Federal University of ABC (UFABC)
Santo André, Brazil

**Jerzy J. Chruściel**
Textile Research Institute
Łódź, Poland

**Aliasghar Davoodi**
Chemical, Polymeric and Petrochemical
    Technology Development Research Division
Research Institute of Petroleum Industry
Tehran, Iran

**Nicole R. Demarquette**
Metallurgical and Materials Engineering
    Department
University of São Paulo
São Paulo, Brazil

**Enis S. Džunuzović**
Faculty of Technology and Metallurgy
University of Belgrade
Karnegijeva, Belgrade, Serbia

**Jasna V. Džunuzović**
Institute of Chemistry, Technology
    and Metallurgy (ICTM)-Center
    of Chemistry
University of Belgrade
Karnegijeva, Belgrade, Serbia

**William J. Eldridge**
Department of Biomedical Engineering
Duke University
Durham, North Carolina

**Mazeyar Parvinzadeh Gashti**
Department of Textile, College
    of Engineering
Islamic Azad University
Tehran, Iran

**K. G. Gatos**
Megaplast S.A., Research and Development
    Center
Athens, Greece

**Nishar Hameed**
Carbon Nexus, Institute for Frontier
    Materials
Deakin University
Geelong, Victoria, Australia

**Linxiang He**
Department of Physics and Materials
    Science
City University of Hong Kong
Kowloon, Hong Kong

**Amir Kiumarsi**
Chang School of Continuing Education
Ryerson University
Toronto, Canada

**Thomas Kurian**
Department of Polymer Science and Rubber
    Technology
Cochin University of Science and
    Technology
Cochin, Kerala, India

**Athanasios K. Ladavos**
Department of Business Administration
    of Food and Agricultural Enterprises
University of Patras
Agrinio, Greece

**Y. W. Leong**
Institute of Materials Research
    and Engineering
Research Link
Singapore

**Ghasem Naderi**
Department of Rubber
Iran Polymer and Petrochemical Institute
Tehran, Iran

**Wojciech Nogala**
Institute of Physical Chemistry
Polish Academy of Sciences (PAS)
Warsaw, Poland

**Jyotishkumar Parameswaranpillai**
Department of Polymer Science and Rubber
    Technology
Cochin University of Science and
    Technology
Cochin, Kerala, India

**A. Ramazani S.A.**
Institute for Nanoscience
    and Nanotechnology
and
Department of Chemical and Petroleum
    Engineering
Sharif University of Technology
Tehran, Iran

**Veronica Marchante Rodriguez**
Centre of Automotive Technology
Cranfield University
Cranfield, Bedfordshire, United Kingdom

**M. Shaban**
Institute for Nanoscience and Nanotechnology
Sharif University of Technology
Tehran, Iran

**Shirin Shokoohi**
Chemical, Polymeric and Petrochemical
    Technology Development Research Division
Research Institute of Petroleum Industry
Tehran, Iran

**Y. Tamsilian**
Institute for Nanoscience and Nanotechnology
Sharif University of Technology
Tehran, Iran

**Sie Chin Tjong**
Department of Physics and Materials Science
City University of Hong Kong
Kowloon, Hong Kong

**Adam Wax**
Department of Biomedical Engineering
Duke University
Durham, North Carolina

**Zhun Xu**
MicroPhotoAcoustics (MPA) Inc
Ronkonkoma, New York

**Yingfeng Yu**
Department of Macromolecular Science
Fudan University
Shanghai, China

# 1 Introduction to Nanomaterials and Nanocomposites

*Jyotishkumar Parameswaranpillai, Nishar Hameed, Thomas Kurian, and Yingfeng Yu*

## CONTENTS

## 1.1 INTRODUCTION TO NANOMATERIALS

Recently, scientists are more interested in nanomaterials and nanocomposites because they form the foundation of nanotechnology. Nanomaterials and nanocomposites have a wide range of applications from toys to aircraft. Nanomaterials are diminutive and their size is measured in nanometers. By definition, they must have at least one dimension at the nanoscale. Nanoscale materials are too small to be seen with the naked eye and even with conventional optical microscopes. Therefore, we need highly sophisticated techniques for the characterization of nanomaterials. Because of their high surface area, most of the techniques are designed to analyze the surface. The techniques such as transmission electron microscopy (TEM), cryogenic transmission electron microscopy (cryo-TEM), high resolution scanning electron microscopy (HRSEM), atomic force microscopy (AFM), scanning tunneling microscopy (STM), laser scanning confocal microscopy (LSCM), scanning electrochemical microscopy (SECM), x-ray scattering, x-ray diffraction (XRD), x-ray fluorescence spectroscopy (XRF), x-ray photoelectron spectroscopy (XPS) rheometry, and electrical conductivity are used to identify size, composition, morphology, crystal structure, and orientation of nanoparticles. Nanomaterials should be carefully analyzed. Moreover, we should take care of the following: (1) specimen handling and (2) environmental conditions. It is important to mention that a multitechnique approach is required for the better characterization of nanomaterials.

Nanomaterials can be either natural or synthetic (engineered nanomaterials). Examples for naturally occurring nanomaterials are volcanic ash, soot from forest fires, etc. Engineered nanomaterials are produced with specific shape, size, and surface properties. Depending on the size and shape, nanomaterials are classified into 0-D (quantam dots, nanoparticles), 1-D (carbon nanotubes, nanorods, and nanowires), 2-D (nanofilms), and 3-D nanomaterials, where 0-D, 1-D, and 2-D nanomaterials are in close contact with each other to form interfaces (powders, fibrous, multilayer, and polycrystalline materials). Usually the performance of the nanomaterials depends more on the surface area than the material composition. The recent advances in nanotechnology enable us to develop nanomaterials to improve the quality of life. Examples for engineered nanomaterials are carbon nanotubes (CNTs), carbon nanofibers (CNFs), graphene, fullerenes, silica, clay, metal, and metal oxide nanomaterials.

## 1.2 ADVANTAGES OF NANOMATERIALS

Nanomaterials exhibit unique optical, magnetic, electrical, chemical, and other properties. These properties have great impact in electronics, sensors, energy devices, medicine, cosmetics, catalysis, and many other fields [1–9]. For example, a wide range of electronic products such as nanotransistors, organic light emitting diodes (OLED), and plasma displays are due to the invention of nanomaterials. High performance portable batteries, fuel cells, and solar cells are examples of the impact of nanomaterials in energy. Smart drugs can target specific organs or cells in the body and are very effective in healing. Nanomaterials can be added to polymers to make them stronger and lighter to develop smart uniforms, nonwetting fabrics, fire-retardant fabrics, self-cleaning fabrics, self-healing fabrics, decontaminating fabrics, lightweight high performance military aircrafts, automobiles, etc. [10,11].

Because of their large surface area, metallic-based nanomaterials are highly reactive with gases. This property of metallic-based nanomaterials can be used for applications such as gas sensors and hydrogen storage devices [12,13]. Nanopesticides, nanofertilizers, and nanoherbicides may have positive impact on crops [14]. The field of nanocatalysis is rapidly growing. The large surface area and effective surface activity make nanomaterials attractive candidates for use as catalysts. For example, for the production of biodiesel from waste cooking oil aluminum dodeca-tungstophosphate ($Al_{0.9}H_{0.3}PW_{12}O_{40}$) nanotubes are used as solid catalyst [15]. In cosmetics, titanium dioxide and zinc oxide nanoparticles are widely used [8].

## 1.3 CHALLENGES AND OPPORTUNITIES

Even though nanomaterials possess a wide range of applications, there are several issues to be resolved. The effect of nanoparticles on the human/animal body is still a concern. Since much about this regime is still unclear, careful analysis must be done to unravel the safety problems of nanoparticles. For example, inhaling of nanoparticles may cause irritation in the lungs and may lead to lung damage and cancer. The tendency of nanoparticles to agglomerate because of weak van der Waals forces is a concern. Technology should be developed to keep the nanoparticles apart so that their characteristic properties can be retained. Nanomaterials and products made of them are very expensive compared to traditional materials. It has been suggested that by increasing the use of nanomaterials in various new applications, their production rate can be increased and thereby their cost can be decreased. It is important to recognize that atomic weapons are possible to develop, which may be very destructive. Therefore, nanomaterials should be used for constructive applications. Another increasingly worrying fact is that widespread use of nanoparticles in agriculture may spread nanoparticles into the environment and may adversely affect the soil fertility, plant growth, and animals. Another prodigious task is to control the structure and chemistry of metallic-based nanomaterials, before using them in gas sensors and for the storage of hydrogen. The requirements of new materials, performance improvement, extended product life, and cost of the product are still challenges. All of these facts point toward the need for new improved technologies to enhance the performance of many products with reduced cost.

## 1.4 NANOCOMPOSITES

Since nanomaterials possess exceptional properties, they are widely used to mix with the bulk polymeric material to improve their properties. By definition, nanocomposites are materials that are reinforced with nanoparticles. Based on the matrix material, nanocomposites are classified into polymer matrix composites, metal matrix composites, and ceramic matrix composites. In polymer matrix composites, the most important topic to be considered is the dispersion of the nanofillers in bulk polymer matrix. Homogeneous distribution of nanomaterials results in improved properties. But the tendency of particle agglomeration due to the weak van der Waals forces between the nanomaterials results in deterioration in properties. For example, homogeneous dispersion of CNTs, graphene, CNFs, and clay in the polymer matrix improved mechanical, thermal, electrical, optical, gas barrier, and flame retardancy properties of nanocomposites [16–21]. It is now well established that for better dispersion of the nanomaterials in a polymer matrix, the nanomaterials can be surface modified or functionalized [22–24]. Recently, it has been shown that addition of compatibilizer also improved dispersion of the nanomaterials in the polymer matrix [17]. Surface modification and functionalization of nanomaterials improve the interfacial interaction or compatibility between the filler and matrix, which results in better dispersion which in turn facilitates effective stress transfer of the matrix and filler to develop high performance lightweight composites for advanced applications. A number of techniques such as TEM, SEM, AFM, STM, XRD, and FTIR can be used to find the size and distribution of filler in polymer matrix.

## 1.5 TOPICS COVERED BY THE BOOK

The book outlines a comprehensive overview on nanomaterials and nanocomposites and their fundamental properties using a wide range of state-of-the-art techniques. Chapter 2 deals with the fundamentals of nanomaterials and their application in nanocomposites. An overview of 0-D, 1-D, and 2-D nanomaterials is given in this chapter. Further applications of 0-D, 1-D, and 2-D nanomaterials in thermoplastic, thermoset, and elastomeric composites are discussed in detail. In the last part of the chapter, the current trends in nanocomposites are described. Chapter 3 provides a comprehensive description on the synthesis of 0-D, 1-D, 2-D, and 3-D nanomaterials. A number of routes including microwave-assisted synthesis, sonochemical synthesis, chemical reduction method, biological method, sol–gel method, aerosol method, emulsion method, microfluidic method, laser ablation method, arc discharge method, and chemical bath technique for the synthesis of different nanomaterials are presented. In Chapter 4, optical properties of nanomaterials are discussed. Different techniques such as electronic absorption (UV–vis), photoluminescence (PL), infrared (IR) absorption, Raman scattering, dynamic light scattering, and x-ray-based techniques and their potential applications in the optical properties of nanomaterials are discussed. Chapter 5 focuses on one of the most important and powerful research techniques—"microscopy"—which provides valuable information regarding size, shape, and morphology of nanomaterials. The purpose, advantages, and limitations of different microscopic techniques such as SEM, TEM, scanning electrochemical microscopy (SECM), photoacoustic microscopy, and hyperspectral microscopy have been elaborated and pointed out the importance of using multimicroscopic techniques to open new horizons in the characterization of nanomaterials. In Chapter 6, the strength, hardness, toughness, fatigue, and creep of nanomaterials are presented. Chapter 7 discusses the various processing techniques such as solution mixing, melt compounding, and *in situ* polymerization for the preparation of different polymer-based nanocomposites. The preparation of CNTs and graphene sheets-based polymer nanocomposites are considered. The advantages and disadvantages of different processing techniques are discussed in detail. Chapter 8 focuses on the significance of thermal stability of polymer composites in many demanding practical applications. The applications of techniques like DSC, TGA, DTA, DMA, and TMA to analyze the thermal stability of polymers is presented in this chapter. Chapter 9 provides a comprehensive overview of the recent findings and advances on the investigation of linear and nonlinear optical properties of polymer nanocomposites. Chapter 10 presents the role of rheology in identifying the structure and processing characteristics of polymer nanocomposites. A comprehensive overview of the progress in mechanical and thermomechanical performance of polymer nanocomposites is given in Chapter 11.

## REFERENCES

1. Z. Liu, J. Xu, D. Chen, and G. Shen, Flexible electronics based on inorganic nanowires, *Chem. Soc. Rev.*, 2015, 44, 161–192.

2. W. Yang, P. Wan, H. Meng, J. Hu, and L. Feng, Supersaturation-controlled synthesis of diverse $In_2O_3$ morphologies and their shape-dependent sensing performance, *CrystEngComm*, 2015, 17, 2989–2995.
3. V. Chabot, D. Higgins, A. Yu, X. Xiao, Z. Chen, and J. Zhang, A review of graphene and graphene oxide sponge: Material synthesis and applications to energy and the environment, *Energy Environ. Sci.*, 2014, 7, 1564–1596.
4. X. Lai, J. E. Halpert, and D. Wang, Recent advances in micro-/nano-structured hollow spheres for energy applications: From simple to complex systems, *Energy Environ. Sci.*, 2012, 5, 5604–5618.
5. R. K. Koninti, A. Sengupta, K. Gavvala, N. Ballav, and P. Hazra, Loading of an anti-cancer drug onto graphene oxide and subsequent release to DNA/RNA: A direct optical detection, *Nanoscale*, 2014, 6, 2937–2944.
6. S. Xu, J. Shi, D. Feng, L. Yang, and S. Cao, Hollow hierarchical hydroxyapatite/Au/polyelectrolyte hybrid microparticles for multi-responsive drug delivery, *J. Mater. Chem. B*, 2014, 2, 6500–6507.
7. Z. Zhao, D. Huang, Z. Yin, X. Chi, X. Wang, and J. Gao, Magnetite nanoparticles as smart carriers to manipulate the cytotoxicity of anticancer drugs: Magnetic control and pH-responsive release, *J. Mater. Chem.*, 2012, 22, 15717–15725.
8. K. Schilling, B. Bradford, D. Castelli, E. Dufour, J. F. Nash, W. Pape, S. Schulte, I. Tooley, J. Bosch, and F. Schellauf, Human safety review of "nano" titanium dioxide and zinc oxide, *Photochem. Photobiol. Sci.*, 2010, 9, 495–509.
9. V. Polshettiwar and R. S. Varma, Green chemistry by nano-catalysis, *Green Chem.*, 2010, 12, 743–754.
10. G. Thilagavathi1, A.S.M. Raja1, and T. Kannaian, Nanotechnology and protective clothing for defence personnel, *Defence Sci. J.*, 2008, 58, 451–459.
11. R. Bogue, Nanocomposites: A review of technology and applications, *Assembly Autom.*, 2011, 31, 106–112.
12. S. K. Gupta, A. Joshi, and M. Kaur. Development of gas sensors using ZnO nanostructures, *J. Chem. Sci.*, 2010, 122, 57–62.
13. S. S. Mao, S. Shen, and L. Guo, Nanomaterials for renewable hydrogen production, storage and utilization, *Prog. in Nat. Sci.: Mater. Int.*, 2012, 22, 522–534.
14. C. R. Chinnamuthu and P. M. Boopathi, Nanotechnology and agroecosystem, *Madras Agric. J.*, 2009, 96, 17–31.
15. J. Wang, Y. Chen, X. Wang, and F. Cao. Aluminium dodecatungstophosphate ($Al_{0.9}H_{0.3}PW_{12}O_{40}$) nanotube as soild acid catalyst one-pot production of biodiesel from waste cooking oil, *BioResources*, 2009, 4, 1477–1486.
16. N. Domun, H. Hadavinia, T. Zhang, T. Sainsbury, G. H. Liaghat, and S. Vahid, Improving the fracture toughness and the strength of epoxy using nanomaterials—A review of the current status. *Nanoscale*, 2015, DOI: 10.1039/C5NR01354B.
17. J. Parameswaranpillai, G. Joseph, K. P. Shinu, S. Jose, N. V. Salim, and N. Hameed, Development of hybrid composites for automotive applications: Effect of addition of SEBS on the morphology, mechanical, viscoelastic, crystallization and thermal degradation properties of PP/PS–x GnP composites, *RSC Adv.*, 2015, 5, 25634–25641.
18. P. Jyotishkumar, E. Logakis, S. M. George, J. Pionteck, L. Häussler, R. Haßler, P. Pissis, and S. Thomas, Preparation and properties of multiwalled carbon nanotube/epoxy-amine composites, *J. Appl. Polym. Sci.*, 2013, 127, 3063–3073.
19. L. Liu, M. Yu, J. Zhang, B. Wang, W. Liu, and Y. Tang, Facile fabrication of color-tunable and white light emitting nano-composite films based on layered rare-earth hydroxides, *J. Mater. Chem. C*, 2015, 3, 2326–2333.
20. H. Liu, T. Kuila, N. H. Kim, B. Ku, and J. H. Lee, *In situ* synthesis of the reduced graphene oxide–polyethyleneimine composite and its gas barrier properties, *J. Mater. Chem. A*, 2013, 1, 3739–3746.
21. K. Zhou, S. Jiang, C. Bao, L. Song, B. Wang, G. Tang, Y. Hu, and Z. Gui. Preparation of poly(vinyl alcohol) nanocomposites with molybdenum disulfide ($MoS_2$): Structural characteristics and markedly enhanced properties, *RSC Adv.*, 2012, 2, 11695–11703.
22. B. Faure, G. Salazar-Alvarez, A. Ahniyaz, I. Villaluenga, G. Berriozabal, Y. R. De Miguel, and L. Bergstrom, Dispersion and surface functionalization of oxide nanoparticles for transparent photo-catalytic and UV-protecting coatings and sunscreens, *Sci. Technol. Adv. Mater.* 2013, 14, 023001.
23. E. Glogowski, R. Tangirala, T. P. Russell, and T. Emrick, Functionalization of nanoparticles for dispersion in polymers and assembly in fluids. *J. Polym. Sci. A: Polym. Chem.*, 2006, 44, 5076–5086.
24. W. R. Glomm, Functionalized gold nanoparticles for application in biotechnology, *J. Disp. Sci. Technol.*, 2005, 26, 389–414.

# 2 Classification of Nanomaterials and Nanocomposites

*K. G. Gatos and Y. W. Leong*

## CONTENTS

## 2.1 INTRODUCTION TO NANOMATERIALS AND NANOCOMPOSITES

During the last three decades, a significant interest in nanomaterials by both industry and academia has emerged. The reason behind was the need for solutions in real-life applications, where existing approaches were facing limitations (e.g., in electronics, catalysis, medicine, etc.). The development of upgraded characterization and processing techniques assisted the efforts of the scientists to evaluate the theoretical studies on nanomaterials. This progress supplied more advanced tools for the investigation and manipulation of the nanoscale world (Nabok 2005).

In general, nanomaterials are materials that are characterized by at least one dimension in the nanometer (1 nm = $10^{-9}$m) range (Rao and Cheetham 2006). This length scale, which lies within the atomic and the microscale, brings new physical and chemical properties to a material (i.e., size effects). This becomes especially apparent when the nanomaterial has at least one dimension between 1 and 100 nm (Murphy et al. 2004). Thus, it is now widely accepted that this latter specific range covers the term of nanomaterial.

There are several types of nanomaterials classified according to their composition and shape. A practical way for their classification is according to the structural features of their elementary units and the number of dimensions, which are outside the nanoscale. Thus, they can be zero-dimensional (0-D) like nanoparticles, one-dimensional (1-D) like nanorods, and two-dimensional (2-D) like nanolayers (Vollath 2008). The various combinations of nanomaterials may configure a broad spectrum of nanostructures characterized by their discernible form and dimensionality. Accordingly, for example, a linear or a planar bonding of nanoparticles, of a defined composition, creates a one- or a two-dimensional nanostructured material, respectively. Prerequisite for the latter classification in addition is to exhibit size effects (Pokropivny and Skorokhod 2006). Further, the construction of ordered arrays of nanostructures can provide strategies for the manufacturing of nanodevices (Rao and Cheetham 2001).

Besides, the combination of two or more components or phases with different physical or chemical properties, wherein at least one of these components is in the nanometer scale, and results to differentiated properties compared to the individual constituents, introduces the class of nanocomposites. These present enhanced properties as compared to their respective microcomposites mainly due to the size effects of the nanomaterials involved (Pokropivny et al. 2007). The distinguished category of nanocomposites in which at least one discontinuous nanosized phase is dispersed in a continuous polymeric medium (i.e., matrix) relates to polymeric nanocomposites and it is the subject of the present chapter. The limited distance between the nanofillers, along with the huge surface area per mass, imparts significant property enhancement (Manias 2007).

## 2.2 NANOMATERIALS

There is a plethora of nanomaterials examined in the literature. Their significant role in several applications makes their research a hot topic within materials science.

### 2.2.1 0-D NANOMATERIALS

Particulate nanomaterials constitute the most common category with a vast spectrum of applications. It is well established that the confinement of the electron motion in three dimensions induces new optical, electronic, magnetic, and chemical properties, which are not observed in their bulk (Alivisatos 2004). This is related with the quantum effects and the high surface to volume ratio of the nanoparticles. Figure 2.1 shows the effect of the reduction of particle size on surface area and on melting temperature of gold (Buzea et al. 2007). Inert metals like gold and platinum become catalysts and opaque copper substances become transparent under nanodimensions (Parab et al. 2009). Common metal nanoparticles used in materials science involve silver, nickel, gold, and palladium, while widely exploited metal oxides relate to titanium, copper, and iron (Basu and Pal 2011). For example, nanosilver is famous for its antimicrobial properties (Sotiriou and Pratsinis 2010), palladium nanoparticles have been extensively studied in relation to their catalytic action (Cookson 2012), and titanium dioxide nanoparticles are well known for their high photocatalytic activation (Pelaez et al. 2009) and use in cosmetics (Tyner et al. 2011).

Besides, nanoparticles such as silica (Rosso et al. 2006) or alumina (Siengshin et al. 2008) are well known for their prominent role as fillers in polymeric matrices. Depending on the application, the nanoparticles are capped with functional groups, which provide solubility and stability (Katz et al. 2003, Serrano-Ruiz et al. 2010). Grafting macromolecules onto nanoparticles possess advantages over modification by low molecular weight surfactants or coupling agents (Zhang et al. 2009). Especially for biomedical applications dendritic structures have been proposed for the functionalization of the nanoparticles (Nazemi and Gillies 2013). Bimetallic core–shell nanostructures have also been proposed for their unique catalytic and biological sensing properties such as the gold–platinum core–shell nanostructure (Mo and Du 2009).

A significant group within the 0-D nanomaterials relates to carbonic nanomaterials. Fullerenes have been extensively examined, presenting interesting properties. Among this group, the

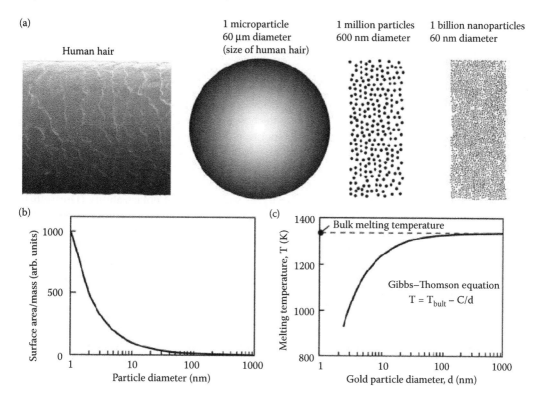

**FIGURE 2.1** (a) Schematics illustrating a microparticle of 60 μm diameter about the size of a human hair shown in the left at scale, and the number of nanoparticles with a diameter of 600 and 60 nm having the same mass as one microparticle of 60 μm diameter. (b) Surface area normalized to mass versus particle diameter. (c) Gold melting temperature as a function of particle diameter, according to Gibbs–Thomson equation, shown inset, the gold bulk melting temperature is 1336 K. (With kind permission from Springer Science+Business Media: *Biointerphases*, Nanomaterials and nanoparticles: Sources and toxicity, 2, 2007, 17–71, Buzea, C., Pacheco, I. I., Robbie, K.)

buckminsterfullerene ($C_{60}$) with 60 carbon atoms is the most popular one. The high degree of symmetry in its structure is mainly responsible for its characteristic physicochemical, electronic, and magnetic properties (Mateo-Alonso et al. 2006). A rather common particulate carbonic material which can be seen in nanodimensions is carbon black, mainly exploited for its electrical, coloring, and reinforcing properties (Psarras and Gatos 2011).

Biodegradability is desirable for several applications and nanofillers like starch have been proposed for polymer reinforcement (Dufresne et al. 1996). Such types of fillers are mainly extracted from maize, potatoes, corn, and rice and their hydrophilicity can be resolved by suitable modification (Valodkar and Thakore 2011).

Polymeric materials have been used also for the fabrication of nanoparticles (Serrano et al. 2006). In this case, controlling of the self-assembled block copolymer morphologies is a facile way to create various nanoobjects like spheres and hollow spheres (Lazzari and López-Quintela 2003).

## 2.2.2 1-D Nanomaterials

Highly emerging categories of nanomaterials relate to nanoscale-rods, -fibers, -wires, and -tubes. Boron nitride nanotubes have been proposed for applications in electronic devices as nanoscale insulators. For such applications, tubes of more than 1 mm in length have been manufactured (Chen et al. 2008). In general, the template method has been extensively exploited in the production of metal (e.g., Ni, Ag, and Fe), chalcogenide (e.g., $MoS_2$ and $NdS_2$), and oxide (e.g., $TiO_2$ and $SiO_2$)

nanotubes (Rao and Govindaraj 2011). The well-known chemical vapor deposition (CVD) technique has been extensively used for the production of zinc oxide nanomaterials (Schmidt-Mende and MacManus-Driscol 2007). Semiconducting ZnO nanobelts, which were obtained by the CVD method, serve applications in optoelectronics and piezoelectricity (Wang 2004).

Glass nanofibers fabricated by a high temperature taper-drawing process are promising building blocks of nanoscale photonic devices due to their low-loss optical wave guiding (Tong and Mazur 2008). Additionally, bioactive glass nanofibers, which were manufactured by electrospinning using a glass sol–gel precursor with a bioactive composition, have been shown to possess osteogenic potential in vitro (Kim et al. 2006).

Carbon nanotubes have attracted a great scientific interest within the 1-D nanomaterials. Depending on their structure, they can be metallic or semiconducting, presenting properties of ballistic transport, very high thermal conductivity, as well as optical polarizability (Dresselhaus and Dai 2004). Apart from their use in nanodevices (Kinoshita et al. 2010), for several other applications covalent (Bottari et al. 2013) or noncovalent functionalization should take place (Tasis et al. 2009). The first functionalization method yields higher stability while the latter route preserves the electronic structure of the nanotube (Bottari et al. 2013). Carbon nanotubes have been very promising for their reinforcing efficiency in polymeric matrices (Schadler et al. 1998, Moniruzzaman and Winey 2006). Notwithstanding, within the carbonic 1-D nanomaterials, the carbon nanofiber appears as a valuable candidate for the cost-effective manufacturing of polymer nanocomposites due to its exceptional thermal, electrical, and reinforcing properties (Zhang et al. 2010).

The naturally occurring clay nanotube halloysite has been used for its capability to bear chemically active agents for drug delivery (Abdullayev and Lvov 2013). Additionally, its reinforcing action in polymers has been examined (Hedicke-Höchstötter et al. 2009). Bio-nanomaterials are also an important group within the 1-D nanomaterials. For example, cellulose is a linear polysaccharide macromolecule whose aggregates form fibrils of nanodimensions. This material has been proposed for the manufacturing of polymer nanocomposites with applications in the packaging sector mainly due to its biodegradability (Siró and Plackett 2010).

### 2.2.3   2-D NANOMATERIALS

The nanoconfinement in one dimension of a material creates a 2-D structure, which relates to nanoscale layers, disks, platelets, and flakes. Iron oxide nanoflakes have been proposed as anode material for lithium batteries with better performance in capacity fading on cycling compared to their nanoparticle or nanotube counterparts (Reddy et al. 2007). Patterned gold nanodisk arrays have been used to enhance the efficiency of organic photovoltaic devices (Diukman et al. 2011), while silver nanoplates were found to improve the bacterial killing capacity compared to silver nanoparticles or nanorods (Pal et al. 2007).

Regarding the carbonic 2-D nanomaterials, graphene is the most heavily investigated one during the last decade. Its 1-atom thick planar honeycomb structure imparts high mechanical, thermal, electrical, and gas barrier properties (Seppälä and Luong 2014). It can find application in chemical and biological sensors (Jiang 2011), photovoltaics (Jariwala et al. 2013), and the interest in the electron-device community is tremendously growing (Schwierz 2010). For example, palladium–graphene junctions have been proposed for the development of high performance transistors due to their low contact resistance (Xia et al. 2011). Besides, graphene appears to be a potential candidate for solar energy applications as transparent conductors and photoactive components in solar cells (Jariwala et al. 2013). Additionally, the superior tribological properties of graphene make it suitable to serve as an additive in lubricant oils or as a self-lubricating material (Berman et al. 2014). For dispersing graphene in polymeric matrices, functionalization is usually required (Kuilla et al. 2010). Apart from single graphite layers (i.e., graphene), exfoliated graphitic nanoplatelets can find various applications as a rather inexpensive 2-D nanofiller (Stankovich et al. 2006b).

Silicate platelets have been extensively investigated for manufacturing of polymer nanocomposites. In nature, these platelets are abundantly found in phyllosilicate minerals in the form of aggregated stacks (Theng 1979). The forms of clay (of natural origin) or layered silicates (of synthetic origin), which are able to expand their interlayer spacing and disperse their single silicate platelets in a medium are potential candidates for mixing with polymers. It is very common to attach organic cations on the silicate layers in order to render the latter organophilic and assist their exfoliation in the polymer matrix (Bhattacharya et al. 2008). The dispersion of a small amount of such platy nanofillers in a polymer yields significant improvement in mechanical, gas barrier, and flame retardant properties (Gatos and Karger-Kocsis 2011). In contrast to layered silicates, layered double hydroxides (LDH) have positively charged layers stacked one on another with small anions and water molecules sandwiched in the interlayer region (Pradhan et al. 2008).

### 2.2.4 Nanomaterials of Complex Structure

In practice, various combinations of the above-described nanomaterials may create complicated structures with special form and dimensionality. For example, zinc oxide can be formed in a plethora of structures and symmetries such as tripods, tetrapods, and nanocombs (Banerjee et al. 2006). Nanoporous materials, which are assigned as 3-D nanostructures (Foa Torres et al. 2014), have potential applications in separation, catalytic, biomedical, and heat transfer applications (Plawsky et al. 2009). Besides, researchers have created numerous complex structures like flower-like or urchin-like, which actually belong to the group of 3-D nanostructured materials. Nanomaterials or nanostructured materials, which are not found dispersed in the polymer matrix, for example, polymer confined in a nanoporous inorganic matrix (Duran et al. 2011), will not be a subject of the present chapter.

## 2.3 THERMOPLASTIC NANOCOMPOSITES

The development of multifunctional thermoplastic materials for a particular application often requires the incorporation of one or more types of nanofillers. Some of the functions that can be attained with the presence of nanofillers in polymer matrices include (but are not limited to) enhancements in mechanical performance, chemical resistance, barrier to oxygen or moisture, compatibility between different phases in polymer blends, and electrical or thermal conductivity. The dimensionality (size, shape, and morphology) of the nanofiller is one of the main considerations when choosing suitable nanofillers to be incorporated into a thermoplastic resin. Types of nanofillers with different dimensionalities but of the same chemical composition may result in nanocomposites with significant differences in properties such as viscosity, mechanical performance, crystallinity, thermal stability, and conductivity. This is especially apparent for systems containing 1-D and 2-D nanofillers whereby the aspect ratio and orientation of the fillers are important in determining the final properties of the composite.

### 2.3.1 Thermoplastics Filled with 0-D Nanomaterials

The incorporation of nanoparticles into commodity thermoplastics such as polyolefins, as well as engineering thermoplastics such as polyamide, poly(ethylene terephthalate), and polycarbonate were mainly targeted at influencing impact response, structural rigidity, thermal stability, thermal and electrical conductivity, and wear resistance.

The initial intention of incorporating low cost minerals such as $CaCO_3$ in polyolefin was merely to reduce the overall material's cost. Later, it was discovered that $CaCO_3$ was able to enhance both the stiffness and the impact performance of the polyolefin especially when the particles were homogeneously dispersed. Zhu et al. have emphasized the importance of $CaCO_3$ nanofiller dispersion on the crystallization, morphology, and mechanical performance of PP (Zhu et al. 2014). These authors recommended an effective method to obtain quantitative particle size distribution curves from microscopy images, which can be correlated to the mechanical performance of the composites.

More recently, $CaCO_3$ nanoparticles have been used to impart further functionalities with the creation of different polymorphic crystalline forms of $CaCO_3$, that is, calcite, aragonite, and vaterite. However, not all polymorphs can be easily produced since calcite is thermodynamically more stable than all other forms. Aragonite can be found naturally in shells of cockles and snails. It has been reported that using diethylenetriaminepentaacetic acid (DTPA), the polymorphism of $CaCO_3$, with the application of various temperatures can be controlled (Gopi et al. 2013).

The physical properties of $CaCO_3$ depend on the type of polymorphic composition and this in turn influences the properties of the composites (Karamipour et al. 2011, Thumsorn et al. 2011, Yao et al. 2014). Thumsorn et al. reported that the incorporation of aragonite-$CaCO_3$ derived from cockle shells into PP was able to impart higher stiffness to the composites as compared with those filled with conventional calcite-$CaCO_3$ (Thumsorn et al. 2011). Yao et al. have used aragonite clam shell-derived $CaCO_3$ (CS) and compared its function with commercially available calcite nano-$CaCO_3$ (Yao et al. 2014). CS was treated with furfural (FCS) and hydrochloric acid (ACS) in order to improve its hydrophobicity, enhancing the impact performance of the system. It was reported that the FCS treatment was particularly effective for the dispersion of FCS in PP even at high filler contents. $CaCO_3$ acted as an effective nucleating agent for PP decreasing the crystallization rate and size of the spherulites. Karamipour et al. (2011) have found that the presence of nano-$CaCO_3$ enabled the PP matrix to sustain high modulus at elevated temperatures especially when the nanofiller contents increased above 5 wt%. Many of these reports, however, have not indicated remarkable increments in the composite strength. The reason behind is the bonding between $CaCO_3$ particles and the matrix resin, which consist of weak van der Waals forces insufficient to cause "bridging interactions" between the polymer chains mediated by the filler particles (Jancar et al. 2010). Therefore, $CaCO_3$ is not reported to cause notable shifts in glass transition temperature ($T_g$) of polymers although it affects their crystallization kinetics.

The usage of fullerene-like nanoparticles such as tungsten disulfide ($WS_2$) and molybdenum disulfide ($MoS_2$) as reinforcements for polymers is an emerging strategy that can provide similar or enhanced performance (e.g., wear) when compared to nanocomposites reinforced with 1-D or 2-D nanofillers. Further to that, this strategy is expected to be cost-effective and environmentally friendly (Naffakh et al. 2013). These fullerene-like nanoparticles (FNP) are inherently unstable in the sheet form due to their abundant ring atomic structure and tend to curl into hollow nanoparticles.

Due to their higher surface to volume ratio, nanoparticles have significantly higher surface energy than micron-sized particles. Therefore, they are prone to agglomeration during melt blending with thermoplastics. Various approaches are used to disperse 0-D nanoparticles including vigorous mechanical shearing, solution blending, surface treatment, polymer grafting onto nanofiller, *in situ* polymerization, and *in situ* growth of nanoparticles. Nanosilica, for example, possesses surface hydroxyl groups and tends to agglomerate via hydrogen bonding. In this case, use of organic surfactants or grafting of polymers to the surface of the nanoparticle is a common approach to increase affinity between the nanoparticles and the thermoplastic matrix.

Aso et al. (2007) successfully improved the dispersion of nanosilica in thermoplastic elastomer copolyetherester through surface modification with dimethyl dichlorosilane (DDS). They reported that the interaction between the methylene groups in the silane and the methylenic chains in the elastomer was much stronger as compared to the hydrogen bonds created between the hydroxyl groups in untreated nanosilica and the ester groups in the elastomer. Through this interaction, the elongation at break as well as the creep resistance of the nanocomposite was significantly improved. In this system, it was found a good state of dispersion even at filler loadings up to 6 wt%. On the contrary, agglomeration was detected for 3 wt% of untreated nanosilica.

In the case of nano-$CaCO_3$, surface treatment with stearic acid or a certain type of silanes (aluminate or titanate) has been proposed to promote dispersion in PP (Wan et al. 2006; Liu et al. 2007; Meng and Dou 2009). Use of compatibilizers such as PP-g-MAH combined with silane treated filler is another approach for a fine dispersion of nanoparticles in this matrix (Palza et al. 2011). In another case, PP-methyl-POSS has been exploited to disperse the untreated nanosilica (Lin et al. 2009). For $TiO_2$ nanoparticles, trisilanol isobutyl polyhedral oligomeric silsesquioxane (TSIB-POSS) has been reported to act as a compatibilizer for promoting dispersion in PP. The polar silanol groups in TSIB-POSS were able to interact with the $TiO_2$ particles while the nonpolar groups interacted with PP (Wheeler et al. 2008).

Leong et al. synthesized toughening agents based on a block copolymer of polyhedral oligo-meric silsesquioxane (MA-POSS) and enantiopure poly(D-lactic acid) (PDLA) or poly(L-lactic acid) (PLLA) [PDLA-b-P(MA-POSS) or PLLA-b-P(MA-POSS)], and incorporated them as filler in a PLA matrix via melt compounding (Leong et al. 2014). This approach exploited the formation of a stereocomplex to enhance the filler-matrix interfacial adhesion and subsequently to promote filler dispersion. The resulting PLA nanocomposite exhibited synergistic enhancement in stiffness and toughness even at very low nanofiller concentrations.

### 2.3.2 THERMOPLASTICS FILLED WITH 1-D NANOMATERIALS

Due to their considerable length in one dimension, nanofibers, nanotubes, or nanowhiskers tend to entangle each other. Unlike 0-D nanoparticles where agglomeration can often be overcome by mechanical agitation or surface treatment, entanglements between 1-D nanofillers require signifi-cantly more energy to separate and disperse in thermoplastic matrices. Therefore, it is more impor-tant to prevent 1-D nanofillers from entangling in the first place in order to promote dispersion during the mixing process with the thermoplastics.

Carbon nanotube (CNT) is one of the most promising 1-D nanofillers since its inception in 1991 (Ijima 1991). It has been incorporated into thermoplastics to achieve properties such as electri-cal and thermal conductivity, stiffness, strength, and fracture toughness. Various strategies have been used to enhance the dispersion of CNTs in thermoplastics, for example, electropolymerization, electrodeposition, esterification, ring-opening metathesis polymerization (ROMP), atomic trans-fer radical polymerization (ATRP), radical-addition fragmentation chain transfer polymerization (RAFT), and amidation (Chen et al. 1998; Gómez et al. 2003; Cui et al. 2004, Kong et al. 2004; Qin et al. 2004). Wu et al. (2010) dispersed multiwalled CNT (MWCNT) in polyamide by creating a MWCNT–polyamic acid colloidal suspension, which under an electric field deposited coherent MWCNT–polyimide film onto the anode through an imidization reaction. This method was effec-tive in creating a uniform dispersion of MWCNT throughout the deposited polyimide film. The schematic representation of the electrophoretic deposition (EPD) process is illustrated in Figure 2.2 while the excellent state of MWCNT dispersion in polyimide is shown in Figure 2.3. Chen et al.

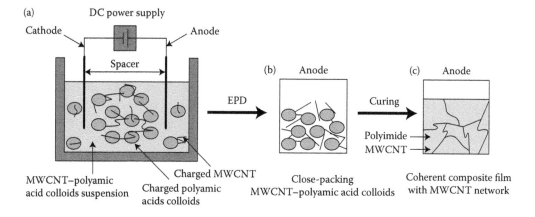

**FIGURE 2.2** **(See color insert.)** Schematic description of the fabrication of MWCNT/polyimide composite films through EPD. (Reprinted from *Polymer*, 51, Wu, D. C. et al., Multi-walled carbon nanotube/polyimide composite film fabricated through electrophoretic deposition, 2155–60, Copyright 2010, with permission from Elsevier.)

**FIGURE 2.3** TEM photograph depicting the state of MWCNT dispersion in polyimide films fabricated through the EPD process. MWCNT content is 0.8 wt%. (Reprinted from *Polymer*, 51, Wu, D. C. et al., Multi-walled carbon nanotube/polyimide composite film fabricated through electrophoretic deposition, 2155–60, Copyright 2010, with permission from Elsevier.)

(2000) used electrolysis to uniformly coat polypyrrole (PPy) on individual CNTs in order to create a doped conducting polymer.

Besides electrochemical methods, polymer grafting and surface modification have also been widely used to prepare nanocomposites with highly dispersed 1-D nanofillers. Tang and Xu prepared CNT/phenylacetylene via *in situ* polymerization of phenylacetylene (PA) in the presence of CNT. This resulted in the helical wrapping of PA onto the CNT surface and enhanced its solubility in organic solvents (Tang and Xu 1999).

Creation of covalent bonding between CNT and polymers has been reported through chemical functionalization of the CNT surface. Philip et al. (2005) functionalized the surface of CNT with *p*-phenylenediamine so that the phenylamine functional groups formed covalent bonds with polyaniline (PANI). Additionally, grafting of polymeric surfactants onto the surface of pristine, oxidized or surface functionalized CNT is another approach to promote dispersion. Hill et al. (2002) grafted polystyrene copolymer, poly(styrene-*co*-p(4-(4′-vinylphenyl)-3-oxabutanol)) to acyl-chloride activated single-wall and multiple-walled CNT in order to make CNT soluble in organic solvents such as tetrahydrofuran (THF). Therefore, CNTs were mixed successfully with polystyrene due to their increased solubility, followed by film casting.

In terms of electrical conductivity, 1-D nanofillers like CNTs permit efficient charge transport along certain directions due to their high aspect ratio. Nanofillers including metals, metal oxides, functional ceramics, and conjugated polymers could also be exploited as elements in many kinds of electronic nanodevices, as well as EMI shielding materials (Hu et al. 1999; Xia et al. 2003). Relevant potential applications and the required electrical conductivity ranges are depicted in Figure 2.4. In these nanocomposites, the percolation or the connectivity between the conducting nanofillers plays an important role in determining the extent of conductivity in the macroscale (Foygel et al. 2005). These networks can be formed in two-, quasi-two-, or three-dimensions depending on the processing methods used and the desired application (Mutiso and Winey 2013). Factors in view of the nanofiller that influence electrical percolation are mainly related to aspect ratio, intrinsic electrical conductivity, chemical purity, physical and crystalline structure, size distribution, dispersion, and orientation within the polymer matrix.

### 2.3.3 THERMOPLASTICS FILLED WITH 2-D NANOMATERIALS

2-D nanomaterials often appear with a structure of stacked sheets and galleries between sheets that are bound together by van der Waals forces. These stacked sheets may appear in a certain order and

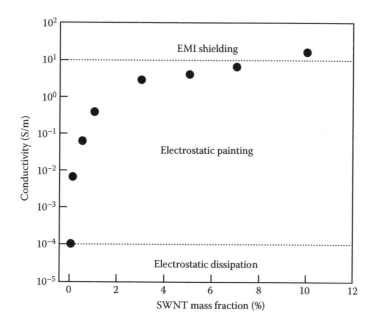

**FIGURE 2.4** Room temperature electrical conductivity of poly(phenyleneethynylene)–SWNTs/polystyrene composites versus the SWNT weight loading. Dashed lines represent the approximate conductivity lower bound required for several electrical applications. (Reprinted with permission from Ramasubramaniam, R., Chen, J., Liu, H. *Applied Physics Letters* 83:2928–30. Copyright 2003, American Institute of Physics.)

therefore the distance between these sheets (height of the galleries) can be ascertained by means of X-ray diffraction (XRD). However, when the stacking of the sheets is not ordered, XRD might not be able to determine accurately the size of the galleries. The determination of the distance between the sheets is important and often used to indicate the success or failure of the techniques employed to promote exfoliation and dispersion of the 2-D nanofiller.

The interest in using 2-D nanofillers such as graphene to impart electrical conductivity to polymers has grown especially when Geim and Novoselov won the Nobel Prize in physics in 2010 for successfully extracting and characterizing the electrical properties of single-layered graphene. This was evident from the exponential growth in the number of publications related to graphene and its nanocomposites since then (Randviir et al. 2014). Graphene is defined as monolayer carbon atoms that are tightly packed into a 2-dimensional sheet and it is the fundamental building block of graphite. Geim and Novoselov had successfully isolated single-layer graphene sheets by using adhesive tape (Novoselov et al. 2004). Later, more complex methods such as CVD synthesis and mechanical exfoliation of graphite enabled a better yield. Commercially available "graphene," which is used for mixing with thermoplastics, typically consists of stacked sheets that are bonded together by van der Waals forces. An optimum approach to produce single-layered graphene at a commercial scale is still illusive. Therefore, graphene oxide (GO) is frequently used instead due to its good dispersibility and processability in aqueous environments (Hu et al. 2014). The sheet-thickness of GO is typically double than that of graphene due to the presence of bulky epoxide and hydroxyl functional groups on the surface and carbonyl and carboxyl groups at the edges of the graphene sheet, respectively. These groups are formed during the oxidation process (Kulkarni et al. 2010; Hu et al. 2013; Hofmann et al. 2014) and provide hydrophilicity to GO sheets. Thus, GO can be exfoliated and dispersed in aqueous solutions, although incompatibility with organic polymers can be induced. Moreover, these functional groups are insulators and cause deterioration in electrical conductivity. Stankovich et al. provided an approach to disperse GO in polystyrene by first completely exfoliating graphite through Hummers' method (Hummers and Offeman 1958) followed by treating GO with phenyl-isocyanate.

This treatment reduced the hydrophilicity of GO allowing mixing with polystyrene. Further chemical reduction of GO into graphene was possible by adding dimethylhydrazine/DMF solution, which enhanced the electrical properties of the former. The key in this process, in order to prevent reagglomeration of graphene, is to perform the reduction in the presence of the polymer. Through this approach, it was managed to create highly electrically conductive ($\approx$0.1 S m$^{-1}$) polystyrene nanocomposites with a percolation threshold of 0.1 vol% in graphene content (Stankovich et al. 2006a). The compounding method that is followed to mix the exfoliated graphite nanoplatelets in a thermoplastic matrix affects both the percolation threshold value and the mechanical properties of the nanocomposite as it was demonstrated for a polypropylene matrix (Kalaitzidou et al. 2007).

Layered silicates or clays are another class of 2-D nanofillers that have been featured since the Toyota research group synthesized polyamide 6 (PA6)/clay nanocomposites with significantly enhanced thermal and mechanical properties (Kojima et al. 1990). Clays are also considered as inexpensive among the various commercial nanofillers.

The compatibility of clay with thermoplastic resins is usually achieved through treatment of the former with organic hosts. This provides a way for macromolecules to intercalate into the clay galleries. The interlayer spacing increase can be detected via shifts in x-ray diffraction (XRD) peaks to lower angles according to Bragg's law. The disappearance of the relevant peaks indicates that clay sheets are exfoliated, that is, the repetitive order of the clay stacking is lost. The state of exfoliation and dispersion of the clay sheets can be further examined by transmission electron microscopy (TEM). Figure 2.5 provides TEM images along with the corresponding XRD spectra for different types of polymer/clay nanocomposites according to the state of intercalation, flocculation, and exfoliation (Sinha Ray and Okamoto 2003). Initial applications of clay-thermoplastic nanocomposites were focused on enhancement of stiffness, strength, heat distortion temperature, and water and gas barrier properties (Balazs et al. 1999). Uskov (1960) first noted an increment in the glass transition temperature of poly(methyl methacrylate) (PMMA) with the incorporation of octadecylammonium modified montmorillonite clay. Later, Nahin and Backlund (1963) patented a method to reinforce polyolefin with organoclay through irradiation crosslinking, which provided significant enhancements in tensile strength and chemical resistance. Nevertheless, they had not discussed the reinforcement mechanism provided by the clay on the polymer matrix. The reinforcement mechanism emerged when the Toyota researchers conducted a detailed analysis in utilizing *in situ* polymerization to attain intercalation and exfoliation of clay in polymer matrices. Initially, the investigated polymers involved polyamide (Kojima et al. 1990), polystyrene, elastomers, and polyimide (Okada and Usuki 1995, Okada et al. 1987).

Recent advancements in clay–polymer nanocomposites consider clay as a halogen-free nontoxic flame retardant filler. The mechanisms of flame retardancy in clay–polymer nanocomposites have been derived through systematic cone calorimetry experiments as well as XRD and TEM examinations (Kiliaris and Papaspyrides 2010). The heat release rate (HRR) and mass loss rate (MLR) are among the most important parameters that are affected by the presence of clay during combustion of the nanocomposites. It has been shown that the presence of clay reduced HRR, which in turn reduced MLR due to the development of a multilayered carbonaceous silicate structure on the surface of the nanocomposite during combustion. The formation of this carbonaceous structure is contributed by the pyrolysis of organic compounds present on the surface of the clay particles. The degradation and subsequent disintegration of organic compounds may cause rearrangement of the clay particles into stacks. This latter prevents oxygen and heat from penetrating into the bulk hindering further combustion (Gilman 1999; Alexandre and Dubois 2000; Wang et al. 2002; Zanetti and Costa 2004). Lewin (2003, 2006) proposed that the migration and accumulation of silicate layers on the surface was caused by the lower surface free energy of the filler as compared to the polymer. The temperature and viscosity gradients provided a convective motion, which transported the clay particles toward the surface. Moreover, the formation of bubbles caused by the volatile compounds released from the degrading polymer could further carry the clay particles to the surface. Nevertheless, the bursting of these bubbles may cause a porous or discontinuous surface. This bubbling effect was found to depend on the molecular weight of the polymer, the clay concentration

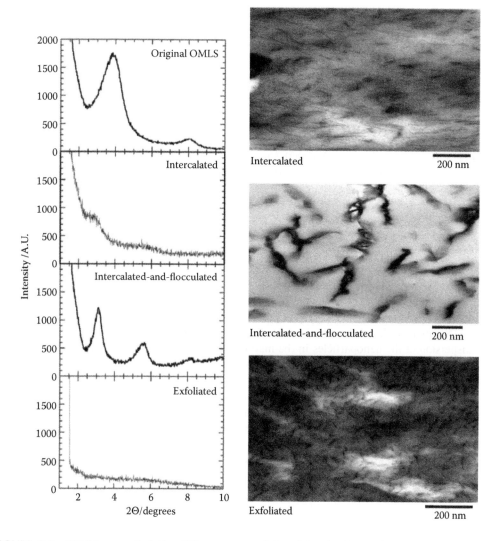

**FIGURE 2.5** TEM images depicting different states of clay dispersion in a polymer matrix and the corresponding XRD spectra. (Reprinted from *Progress in Polymer Science*, 28, Sinha Ray, S., Okamoto, M., Polymer/layered silicate nanocomposites: A review from preparation to processing, 1539–641, Copyright 2003, with permission from Elsevier.)

and the aspect ratio of clay particles (Kashiwagi et al. 2004). On the contrary, the time required for ignition (TTI) can be worsened by the presence of clay in thermoplastics. The initial quick ignition of the clay nanocomposites is attributed to the release of volatile compounds from the decomposition of clay surfactants, which catalyzes the combustion (Tang et al. 2002; Zanetti et al. 2004; Zong et al. 2007). Another factor to be considered is the increase in melt viscosity with the incorporation of clay. This would not only affect processability but also nullify the dripping effect that would otherwise prevent burning to a large area.

## 2.4 THERMOSET NANOCOMPOSITES

Thermoset polymers can be found in everyday as well as in hi-tech applications including, among others, adhesives, coatings, and aircraft fuselages. Their low processing viscosity along with their low creep and high stiffness when cured has been proven to be beneficial for composite structures

(Srivastava et al. 2013). Nevertheless, in many cases where stiffness and toughness enhancement are required at the same time, nanomaterials are suitable to serve as fillers (Kotsilkova 2007). Additionally, thermosets' wear, fracture toughness, and electrical and thermal performance have been significantly improved with nanomaterials.

### 2.4.1 THERMOSETS FILLED WITH 0-D NANOMATERIALS

Incorporation of nanoparticles may affect the curing reaction of the thermoset matrix with parallel side effects. In such an example, $Fe_2O_3$ of a diameter less than 50 nm in 10% filler content increased the temperature of 5% weight loss during thermogravimetric analysis (TGA) from 227°C to 292°C. This effect was related with the ability of the nanoparticles to catalyze the curing reaction. This performance was accompanied by increase of the glass transition temperature (Zabihi et al. 2012).

Adding lanthanide anions doped $LaF_3$ nanoparticles of 5 nm in diameter in epoxy, the $T_g$ was decreased from 190°C to 110°C for 10 wt% filler amount. Nevertheless, in both filled and unfilled matrix, the conversion was corresponded to about 70% of epoxy groups, while the thermal stability during TGA for the nanocomposite was slightly improved. This reduction in the $T_g$ was charged to the presence of oleic groups, which were attached on the nanoparticles' surfaces to avoid agglomeration during processing (Sangermano et al. 2009).

The thermal conductivity of epoxy has been improved by incorporating aluminum nitride (AlN) nanoparticles of 50 nm in diameter. This improvement was further enhanced by modifying the surface of AlN nanoparticles with gamma-aminopropyl triethoxysilane. This effect was related to the better dispersion of the nanoparticles in the matrix along with the decrease of both the interfacial phonon scattering and interface heat resistance (Peng et al. 2010).

The filler that has been extensively exploited in thermoset matrices is silica. Nanosilica particles of about 20 nm in diameter have been reported to increase both modulus and toughness of an epoxy, while the $T_g$ remained rather unaffected. The fracture toughness increase, which was measured compared to the neat matrix, was linked to the debonding of the nanoparticles and the subsequent plastic void growth (Johnsen et al. 2007). For *in situ* generated silica particles of about 25 nm in diameter, the well-dispersed nanosilica particles (see Figure 2.6) were responsible for the enhanced mechanical performance of the nanocomposite up to 80°C hindering the fast crack-growth (Zhang et al. 2008). In order to improve the fatigue performance of epoxy filled with nanosilica, the synergy

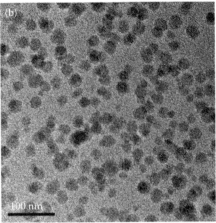

**FIGURE 2.6** Transmission electron microscopy (TEM) micrographs taken from epoxy-based nanocomposites with 8 vol% silica nanoparticles: (a) lower magnification and (b) higher magnification. (Reprinted from *Polymer*, 49, Zhang, H. et al. Fracture behaviours of in situ silica nanoparticle-filled epoxy at different temperatures, 3816–25, Copyright 2008, with permission from Elsevier.)

by rubber particles of about 100 nm in diameter has been proposed. The rubber particles had a rubber core of 90 nm covered by an epoxy-compatible random copolymer shell of 10–20 nm thickness (Liu et al. 2012). Additionally, in an effort to extend the action of silica in a thermoset matrix, phosphorous flame retardant was immobilized on the surface of silica nanoparticles. These nanocomposites presented high values of char yield and limited oxygen index (Kawahara et al. 2013).

Although surface modified silica particles of 10 nm in diameter have been presented to improve the tribological properties of an epoxy coating, the $Al_2O_3$ particles of 17 nm in diameter performed better (Scholz et al. 2014). The wear behavior of $Al_2O_3$ can be further improved with the appropriate surface treatment. It was shown that 0.24 vol% of $Al_2O_3$ nanoparticles of about 4 nm in diameter, when grafted with polyacrylamide, decreased by 97% the specific wear rate of the neat matrix. Among the various pretreatments, the graft polymerization was superior to silane treatment in improving the tribological performance of epoxy (Shi et al. 2004).

The scratch resistance of thermosets is of great importance especially for gel-coats. $TiO_2$ particles of 32 nm in diameter were found to improve the scratch resistance of an epoxy matrix at 10 wt%. It has to be mentioned that the scratch resistance of the same epoxy matrix when filed with micron-size $TiO_2$ particles of 0.24 μm at 10 wt% remained actually unaffected (Ng et al. 1999). Another candidate for improving the scratch resistance of thermosets is nanodiamond (Ayatollahi et al. 2012). Nanodiamonds of 5 nm in diameter at 25 vol% enhanced the mechanical and tribological properties of the epoxy matrix. At the same time, the thermal conductivity of the composite can be improved (Neitzel et al. 2011).

Taking advantage of the optical properties of ZnO, this nanomaterial has been proposed to be mixed with epoxy for ultraviolet light-based white light emitting dioxide applications (Li et al. 2006). Epoxy filled with ZnO nanoparticles at 0.06 wt% improved the UV light shielding efficiency maintaining the epoxy visible light transparency (Ding et al. 2012). In another case, for providing an encapsulation material for light emitting dioxides applications which requires low thermal resistivity and high refractive index, an epoxy filled with silane-modified $ZrO_2$ nanoparticles has been proposed (Chung et al. 2012).

In order to improve the electrical conductivity of thermosets, carbon black (CB) appears to be an inexpensive nanomaterial. CB with particle size of 25–75 nm has been mixed with epoxy presenting an increase in electrical conductivity at a filler content of 1 wt%. Above this filler percentage, the dielectric permittivity constant was increased in the frequency region of $10^{-2}$ to $10^{-6}$ (Kosmidou et al. 2008). When fullerene ($C_{60}$) was served as filler in an epoxy matrix, the increase in dielectric permittivity was detected at the amount of 0.08 wt% (Pikhurov and Zuev 2014).

## 2.4.2 THERMOSETS FILLED WITH 1-D NANOMATERIALS

The most popular 1-D nanofiller, which has been used in thermoset matrices, is carbon nanotube (CNT). Issues of high concern relate to the dispersion of the CNTs in the thermoset matrix, as well as the interfacial adhesion between filler and matrix. Sonication technique (Gkikas et al. 2012) and roll-mill mixing (Tuğrul Seyhan et al. 2007) have been usually in action for homogeneous dispersion of CNTs. Besides, functionalized CNTs, which assist dispersion, are able to interact with the thermoset matrix to yield enhanced thermo-mechanical properties (Gojny and Schulte 2004). An optimized dispersion of surface-treated CNTs would have mainly a notable effect on fracture toughness followed by any improvement on strength or stiffness (Gojny et al. 2005). In such an example, an amino-functionalized double-wall CNT (DWCNT-$NH_2$) reinforced epoxy at 0.5 wt% filler content created a rough fracture surface compared to the neat matrix as depicted in Figure 2.7. Grafting of well-defined siloxane brushes on the nanotube surface has been reported to alter the relaxation dynamics of the network, increasing the glass transition temperature of the nanocomposite, as well as its strength and toughness (Vennerberg et al. 2014).

CNTs are usually exploited to induce electrical conductivity in a thermoset matrix (Vavouliotis et al. 2010). The dispersion quality of the CNTs due to different processing parameters affects this

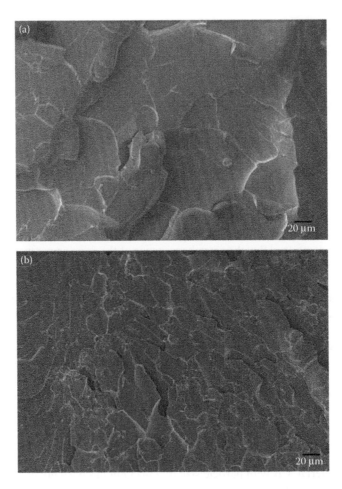

**FIGURE 2.7**   SEM-micrographs of a fracture surface at low magnifications (×1000), showing (a) the epoxy and (b) a DWCNT-NH₂/epoxy composite. The composite containing CNTs exhibits a significantly rougher fracture surface compared to the neat epoxy, indicating a toughening effect of the nanoparticles. (Reprinted from *Composites Science and Technology*, 65, Gojny, F. H. et al., Influence of different carbon nanotubes on the mechanical properties of epoxy matrix composites – A comparative study, 2300–13, Copyright 2005, with permission from Elsevier.)

property as it alters the induced aspect ratio, the created conductive network, and the percolation threshold (Li et al. 2007). Notwithstanding, the initial aspect ratio of the CNTs for the preparation of the nanocomposite affects also the electrical conductivity. In such an example, an epoxy was mixed with two kinds of multiwall CNT grades at 1 wt% filler content without functionalization. The first presented tube dimensions of about 140 nm in diameter and 7 μm in length while the second included tube dimensions of about 35 nm in diameter and 30 μm in length. It was found that the nanotubes with the high aspect ratio presented 10 orders of magnitude greater conductivity compared to the conductivity induced by the CNTs of low aspect ratio (Hernádez-Pérez et al. 2008). It appears that the former system was above percolation threshold while the latter below, for the same filler content. Noteworthy, optimized processing parameters that yielded well-dispersed CNTs of a high aspect ratio have presented a percolation threshold at about 0.1 wt% in a vinyl ester matrix (Thostenson et al. 2009). Additionally, well-dispersed CNTs were able to improve significantly the thermal conductivity of the related nanocomposites (Song and Young 2005).

Other 1-D nanomaterials that have been used to enhance the properties of thermosets include attapulgite, sepiolite, titanium dioxide nanotubes, and ceria. The first is a rod-like silicate, which after functionalization it was reported to increase the modulus of the respective epoxy nanocomposite with a parallel decrease of the coefficient of thermal expansion (Lu et al. 2005). Organophilic sepiolite was found to improve the toughness of epoxy resin (Nohales et al. 2011), while titanium dioxide nanotubes presented to enhance the corrosion resistance in salt of an epoxy coating (El Saeed et al. 2015). Finally, ceria nanorods were found to increase the impact strength of an epoxy matrix (He et al. 2011).

### 2.4.3 THERMOSETS FILLED WITH 2-D NANOMATERIALS

Platy fillers seem to be attractive for several thermoset's applications. Pioneering research by Pinnavaia (Lan and Pinnavaia 1994) and Giannelis (Messersmith and Giannelis 1994) showed that organically modified layered silicates can be effectively dispersed in epoxy matrices. There are several factors that affect the formation of intercalated or exfoliated organoclay/thermoset nanocomposites such as clay characteristics (Kornmann et al. 2001), intercalant type (Zilg et al. 2000), crosslinking process (Park and Jana 2003), and further compatibilizers (Fröhlich et al. 2004). Organoclay/epoxy nanocomposites are usually met to increase stiffness (Wang et al. 2005) and decrease the gas permeation (Osman et al. 2004).

Another 2-D nanomaterial with significant impact on the manufacturing of thermoset nanocomposites is graphene. Due to processing parameters, usually stacks of graphite layers having thickness in nanodimensions are seen. Graphite nanoplatelets (GNP) increased the electrical conductivity of an epoxy matrix presenting a percolation threshold at 0.3 wt% (Chandrasekaran et al. 2013). As shown in Figure 2.8, this performance was detected for nanocomposites prepared by three-roll mill (3RM) rather than those prepared by sonication combined with high-speed shear mixing (Soni_hsm). Although the percolation threshold obtained by CNTs is revealed at lower filler fraction values than by graphene sheets, the overall mechanical performance yielded by the latter is better (Martin-Callego et al. 2013). The fracture mechanism of epoxy filled with GNP, which involved crack deflection, crack pinning, secondary cracks, and separation between graphitic

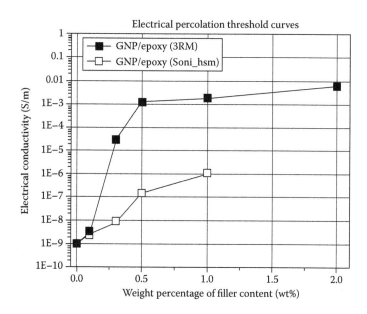

**FIGURE 2.8** Electrical conductivity of GNP/epoxy nanocomposite as a function of weight percentage of the filler. (Reprinted from *European Polymer Journal*, 49, Chandrasekaran, S., Seidel, C., Schulte, K., Preparation and characterization of graphite nano-platelet (GNP)/epoxy nano-composite: Mechanical, electrical and thermal properties, 3878–88, Copyright 2013, with permission from Elsevier.)

layers, was responsible for the significant improvement of fracture toughness (Chandrasekaran et al. 2014). For applications of epoxy/graphene nanocomposites, apart from mechanical and electrical properties, thermal conductivity is important. Silane functionalized graphitic nanoplatelets, which presented higher thermal conductivity in an epoxy matrix than the unmodified version, appeared the electrical resistivity drop at a higher filler fraction (Ganguli et al. 2008). Other properties that have been reported to be improved by using graphene in a thermoset matrix relate to dimensional stability (Park and Kim 2014), electromagnetic interference shielding (Yousefi et al. 2014), hydrothermal ageing (Starkova et al. 2013), and wear resistance (Shen et al. 2013).

## 2.5 ELASTOMERIC NANOCOMPOSITES

Elastomeric nanocomposites possess a special category of nanocomposites, which find numerous applications in everyday life. These nanomaterials, which are mixed with elastomers, usually target to enhance the mechanical, the electrical, as well as the wear performance. They include inorganic or organic fillers of metallic, ceramic, or polymeric structure.

### 2.5.1 ELASTOMERS FILLED WITH 0-D NANOMATERIALS

Metallic nanoparticles such as iron or nickel have been used to bring dielectric and magnetic properties to elastomers. In such an example, natural rubber (NR) filled with nickel nanoparticles of 25–40 nm in diameter was found to improve the magnetic and dielectric properties of the vulcanizate. This property improvement was more apparent for filler loadings up to 100 phr (Jamal et al. 2009). Magnetite ($Fe_3O_4$) has been used to yield electromagnetic interference (EMI) shielding in rubber matrices. Acrylonitrile butadiene rubber (NBR), which was filled with magnetite of 6–8 nm in diameter, maximized its EMI shielding at the frequency range of 1–12 GHz at a filler percentage of 40 wt% (Al-Ghamdi et al. 2012). NiZn ferrite nanoparticles of 7 nm in diameter have also been exploited successfully for similar purposes (Flaifel et al. 2014). Magnetite has been also served as filler in epoxidized natural rubber (ENR) aiming at oil spill recovery applications. This nanocomposite, on the one hand, was reused in several cycles without mass loss and on the other hand, was recovered using an appropriate magnetic field (Venkatanarasimhan and Raghavachari 2013).

Nevertheless, the poor interface of such nanoparticles with the rubber matrix affects the ultimate mechanical properties of the rubber nanocomposites. This property deterioration becomes obvious as the filler percentage imparts the desirable electromagnetic properties to the vulcanizate. In an attempt to maintain and improve the mechanical performance of rubber nanocomposites, $TiO_2$ nanoparticles were modified with silane coupling agent when mixed with silicon rubber (Dang et al. 2011).

Moreover, the role of $TiO_2$ nanoparticles of 5 nm in diameter has been investigated in UV photodegradation. It was found that the styrene butadiene rubber (SBR) nanocomposites with 0.2 wt% nanoparticles degraded four times faster than the pure matrix. A similar photodegradation effect to this matrix, however, at a lower intensity, was delivered by 1 wt% zirconium dioxide ($ZrO_2$) nanoparticles of 5–15 nm in diameter (Arantes et al. 2013).

Alumina nanoparticles mixed with NR in 3–5 phr imparted to the nanocomposites optimum acid and alkaline resistance as evaluated by their mechanical response (Fu et al. 2012). The reinforcing efficiency of alumina in rubber stocks has been investigated in the case of polyurethane rubber (PUR). Decreasing the boehmite alumina nanoparticle size from 90 nm to 25 nm the modulus, hardness, and dynamic mechanical properties were significantly improved for the respective nanocomposites (Gatos et al. 2007b). As shown in Figure 2.9, the tanδ peak value of the nanocomposite was decreased by lowering alumina's particle size. Accordingly, the relevant TEM images indicated a better dispersion for the alumina of 25 nm compared with that of 90 nm in the rubber matrix. In the case of PUR, the use of a suitable polyhedral oligomeric silsesquioxane (POSS) as crosslinking

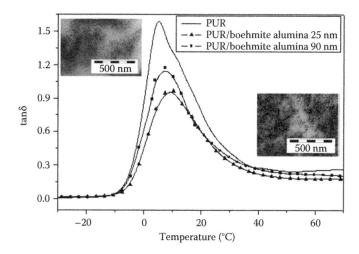

**FIGURE 2.9** Mechanical loss factor (tanδ) of PUR and PUR/boehmite alumina nanocomposites in 10 phr filler loading versus temperature (in tension mode at 10 Hz frequency). The TEM images of PUR/alumina 25 nm and PUR/alumina 90 nm are depicted in the left and right side of the curve-peak, respectively.

agent for producing PUR hybrid networks has been presented to enhance the dynamic mechanical performance of the matrix (Liu and Zheng 2005).

Another nanoparticle, which has been extensively exploited for its reinforcing efficiency in rubber matrices, is silica. In addition, its usage in tire applications due to its good abrasion properties, low rolling resistance, and low heat build-up, rendered its research to receive special attention among rubber technologists. For lowering the tendency of silica to agglomerate and improving its dispersion in a rubber matrix, silane coupling agents are usually used (Sae-oui et al. 2005). This weakening of the silica–silica interaction in the presence of coupling agents within the rubber matrix and the induced filler dispersion was able to be visualized with three-dimensional transmission electron microscopy (3-D TEM) images (Kohjiya et al. 2008). The decrease of the storage modulus when the strain amplitude is increasing (i.e., Payne effect) has been investigated for several coupling agents in an SBR matrix. It was found that with the same level of silica dispersion and similar interaggregate distance, the longer the alkylsilanes the more effective the reduction of the Payne effect was (Guy et al. 2005).

For controlling the particle dispersion in a rubber matrix, an *in situ* silica synthesis by the sol–gel technique can be exploited. In such an example, silica particles of 10 nm in diameter have been generated in a crosslinked NR at a filler volume percentage up to 11% and the system was examined with differential scanning calorimetry (DSC) and dielectric spectroscopy. It was found in the vicinity of the silica particles the existence of a polymer fraction with slower segmental relaxation times (2–3 orders of magnitude) and reduced heat capacity increment at glass transition temperature compared to bulk behavior. This interfacial layer was estimated to correspond to 2–4 nm (Fragiadakis et al. 2011).

Among the particulate nanomaterials, carbon black (CB) belongs to the most traditional one. Pioneering work by Guth proposed models for rubber reinforcement based on the various types of CB dispersion within a rubber matrix. These models demonstrated colloidal CB spheres in a continuous rubber medium (up to 10% filler loading), aggregated CB and chain-creation (up to 30% filler loading) and CB diluted by the rubber (above 30% filler loading) (Guth 1945). Much before the introduction of the term of rubber nanocomposites, Mullins and Tobin compared results of two NR stocks filled with CB having particle size of 400 nm or 40 nm. They found a more complex performance in the case of the 40 nm CB, which was attributed to the higher tendency for agglomeration into chain-like clusters (Mullins and Tobin 1965). Later, it has been suggested that the reinforcing efficiency of CB is related to its structure (Vilgis et al. 2009) and pH value (Lawandy et al. 2009).

There are several types of CB depending on their manufacturing method (e.g., by channel or furnace processes). The CB particles present differences on the number of active sites, which can be estimated by the amount of bound (solvent-insoluble) rubber measured in a curative-free compound (Medalia 1978). Using an atomic force microscopy (AFM), the bound rubber was measured in an HNBR matrix filled with 5 vol% CB of 13–15 nm particle radius. It was visualized that the bound rubber had added about 10 nm in aggregate's radius, while the respective elastic properties had been affected (stiffening) by one order of amplitude. This latter effect which was measured also via torsional harmonic AFM indentation at room temperature indicated an interphase of up to 20 nm in thickness beyond the aggregate's surface (Qu et al. 2011).

In the case of vulcanizates, the stronger rubber–CB interaction results in a thicker rubber shell around the CB particle, which increases the effective volume of filler loading (Wang 1998). In this respect, graphitization of CB (which induces deactivation of the filler's surface) reduces the mechanical properties of the vulcanizate, but increases the conductivity. This latter, which was investigated in 40 phr CB loading, is due to the shortening of the gap distance between adjacent particles (i.e., polymer bridges between the filler) and said gap was estimated using dielectric spectroscopy to be below 4 nm (Geberth and Klüppel 2012). Furthermore, potential piezoresistive applications of silicon rubber may be realized by taking advantage of the conductive role of CB (Luheng et al. 2009).

A nanomaterial that is abundant in nature and has been proposed for substituting CB in some rubber applications is lignin. Compared to CB, this filler is less dense, nonconductive, and allows light-colored rubber mixtures. Lignin of 60 nm particle size was nicely dispersed in an NR vulcanizate, shifting the thermal degradation of the matrix at higher temperatures (Jiang et al. 2013). Another abundant in nature and biodegradable material that has been proposed to substitute CB is starch. NR/starch films improved their modulus and tensile strength values for filler loadings up to 20 wt%, however, at the cost of their elongation at break (Rajisha et al. 2014).

Rubber technologists also examined polymeric nanoparticles for obtaining tailor-made rubber nanocomposites. In such an attempt, core–shell nanoparticles composed of crosslinked polystyrene (PS) and polyisoprene (PI) shell were synthesized with a particle diameter of 50–70 nm. These nanoparticles, being in latex form, were mixed with SBR latex. The increase in PI thickness improved rubber–filler interactions yielding enhanced mechanical performance for the vulcanizate (Lu et al. 2012).

### 2.5.2 Elastomers Filled with 1-D Nanomaterials

Halloysite is a naturally occurring aluminosilicate nanotube, which has been added in rubber matrices inducing increased crosslink density and improved mechanical and thermal properties (Ismail et al. 2008). It has been found that the degree of dispersion in the case of solution mixing is better than that of the mechanical method and the former process may produce vulcanizates with slightly better mechanical performance than the latter one (Ismail et al. 2013). For optimizing both the dispersion within the polymeric matrix and the relevant properties, sorbic acid (Guo et al. 2009) and bifunctional organosilanes are usually proposed. NR, which had been mixed with halloysite nanotubes (HNT) in 10 phr filler loading, improved significantly the thermal degradation temperature of the vulcanizate by about 64°C. Nevertheless, the same degree of thermal stability with and without silane coupling agent was measured for these mixtures (Rooj et al. 2010).

Sepiolite is another silicate with fibrous morphology, which improves rubber properties at low filler contents. It is a 2:1 phyllosilicate; however, its structure allows the creation of channels running along the length of the fiber. In an example of hydrogenated acrylonitrile butadiene rubber (HNBR), a sepiolite modified with quaternary ammonium salt improved the tensile and dynamic mechanical properties along with pronounced thermal stability even at 4 phr filler loading (Choudhury et al. 2010). Advantages of fibrous fillers over the traditional particulates have been shown in the case of silica. Various aspect ratios of silica produced by the sol–gel method have been synthesized within

SBR matrix at 35 phr filler content. It was found that the dynamic mechanical performance was improved in case of rod-like silica (Scotti et al. 2014).

Regarding the carbonic nanomaterials, carbon nanofibers have improved the mechanical performance of NBR matrix at filler loadings below 10 wt%. The same nanofiller was identified also to reduce the flammability of a NBR vulcanizate (Felhös et al. 2008). A nanotube, which has been extensively investigated in elastomers is CNT and mostly the MWCNT version. A low amount of CNT is adequate for producing a conductive rubber nanocomposite with enhanced mechanical performance. Since the dispersion of CNTs is of vital importance for the above characteristics, several ways have been exploited for tuning this; however, with care not to improve the former at the cost of the latter property (Jiang et al. 2008). A SBR/BR (50/50) blend was mixed on a two-roll mill with a suspension of hydroxyl-modified MWCNTs in ethanol containing a nonionic surfactant. The conductive character of the nanocomposite was already revealed at 2 wt% filler content, while the reinforcing efficiency was sound at 5 wt% (Das et al. 2008). A further improvement of the mechanical properties at even lower filler loading, while maintaining the electrical performance of such nanocomposites, may be obtained by using ionic liquids. In such an example, MWCNTs were mixed with 1-butyl 3-methyl imidazolium bis(trifluoromethylsulphonyl)imide (BMI) and the obtained black paste was compounded with polychloroprene rubber (CR) on an open mill (Subramaniam et al. 2011). The effect of BMI on the electrical performance of the vulcanizate can be attributed to the smaller gap distance between CNTs, which promoted the quantum mechanical tunneling at low temperatures and the thermal activated hopping above room temperature (Steinhauser et al. 2012).

Recently, the piezoresistive behavior at low strains of EPDM/MWCNT nanocomposite (linear relation up to 10%) (Ciselli et al. 2010) pointed rubber/CNT valcanizates as possible candidates for pressure sensor and strain gauge applications. Nevertheless, at higher strain levels, the situation is rather different. Figure 2.10 depicts an NR filled with 3 phr of MWCNTs (solution mixed) stretched at three cycles increasing every time the strain level. More specifically, the nanocomposite was stretched up to 100% (curve 1) and released, stretched again at 200% (curves 2 and 3) and released, and stretched for a third time until failure (curves 4 and 5). It becomes apparent that the gradual

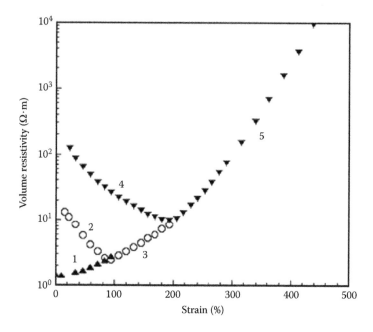

**FIGURE 2.10** Strain dependence of the electrical resistivity for NR filled with 3 phr of MWNTs. (Reprinted from Bokobza, L. 2012. *Express Polymer Letters* 6:213–23. With permission from Budapest University of Technology and Economics, Department of Polymer Engineering/BME-PT and GTE.)

decrease of resistivity relates to the increasing distances between the conductive inclusions and breakdown of contacts. After unloading the sample, the resistivity is higher than initially measured, indicating that the contacts are not reformed after the removal of the stress (Bokobza 2012).

Another potential advantage of CNTs versus traditional particulate fillers (like CB or silica) is their tribological properties. Therein, the filler loading may increase above usual percentages (which is lower than 10 wt%) and reach even up to 30 wt% (Felhös et al. 2008).

A nanomaterial that has been mixed with elastomers to enhance their mechanical performance and solvent uptake properties is cellulose. NR, which was mixed via the latex route with cellulose extracted from the rachis of the palm of the date palm tree, decreased significantly the toluene uptake even at 1 wt% content compared to the neat matrix. The high filler-matrix adhesion was responsible for the increased storage modulus and decreased tanδ values during dynamic mechanical tests (Bendahou et al. 2010). The remarkable reduction of solvent uptake was repeated for other solvents like benzene and p-xylene, where cellulose nanofibers (23–42 nm in diameter and length of some microns) were more effective than cellulose nanowhiskers (5–14 nm in diameter and 300–400 nm length) (Visakh et al. 2012). In case of PUR, cellulose nanowhiskers were found to improve both the mechanical performance and the thermal stability of the matrix (Park et al. 2013).

### 2.5.3 ELASTOMERS FILLED WITH 2-D NANOMATERIALS

During the last two decades, a lot of research has been conducted on elastomers filled with platy nanomaterials (Gatos and Karger-Kocsis 2010). Following the findings at the Toyota Central Research Laboratories (Japan) on the barrier properties of rubber/layered silicate nanocomposites (Kojima et al. 1993), various 2:1 phyllosilicates have served as fillers in a plethora of rubber matrices (Potts et al. 2012). In the example of EPDM, it was presented that the modification of montmorillonite affects the intercalation and exfoliation of the silicate layers within the rubber matrix (Gatos and Karger-Kocsis 2005). It has been presented that a low amount of silicate content (<10 phr) with the appropriate modification is sufficient to improve the mechanical performance of the vulcanizate without compensating the elongation at break values (Gatos et al. 2004). The effect of the aspect ratio of the layered silicate on the barrier properties of the rubber matrix was evident in the case of HNBR. As depicted in Figure 2.11,

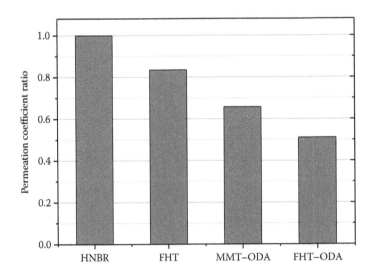

**FIGURE 2.11** Oxygen permeability results for HNBR vulcanizates as a function of the aspect ratio of the filler in dry conditions and 10 phr filler loading. Note that the oxygen permeation value for HNBR was 87 cm$^3$ mm/m$^2$ day atm and the relevant ratio was evaluated by dividing each permeation coefficient by that of the neat matrix. Octadecylamine (ODA) was used to modify both fluorohectorite (FHT) and montmorillonite (MMT).

organically modified fluorohectorite (higher aspect ratio) reduced the oxygen permeability of HNBR in a greater extent than montmorillonite (lower aspect ratio) modified with the same intercalant (Gatos and Karger-Kocsis 2007). The dispersion of the silicate layers within the rubber matrix can also affect and improve other properties like wear (Gatos et al. 2007a), thermal stability in terms of flammability (Wu et al. 2008), and ageing resistance (Choudhury et al. 2010).

Another type of layered silicate, which has received some attention by rubber technologists, is layered double hydroxide (LDH). In such an example, LDH modified with the anionic surfactant sodium 1-decanesulfonate was mixed with a nonpolar (EPDM) and a polar (XNBR) elastomer. It was found that mechanical property enhancement due to nanoreinforcement took place only for the polar matrix. It was speculated that the modified LDH likely also affected the curing of XNBR (Pradhan et al. 2008).

However, the filler, which has become currently the hot-topic of rubber nanocomposites, relates to graphene. An inexpensive way to produce graphene is through oxidation of graphite followed by exfoliation. This process creates numerous oxygen-containing functional groups on the graphene surfaces such as carboxyl, epoxide, and hydroxyl groups. On one hand, these groups make graphene oxide hydrophilic and on the other hand, deteriorate the superior electric properties of graphene. Thus, graphene oxide may further be reduced to regain structure and property (Yang et al. 2010). It has been proposed that the tendency of ZnO to react with the oxidized surface of graphene alters the vulcanization of EPDM/graphene nanocomposites even at 1 phr filler content (Allahbakhsh et al. 2013). Note that ZnO is a rather typical component of a curing recipe. A shorter induction period along with increased crosslink density has also been presented for NR/graphene nanocomposites (Wu et al. 2013). In the case of an NBR matrix, a self-crosslinking reaction has been detected to improve the performance of the vulcanizates (Yang et al. 2007). The effect of graphene to boost the mechanical performance of a rubber matrix usually appears at low filler loadings, as shown in the case of NR where 0.5 phr of graphene increased 48% the tensile strength and 80% the modulus without sacrificing the elongation at break values (Xing et al. 2014). Additionally, due to the platy nature of graphene and its high aspect ratio, significant barrier properties are expected. It has been presented that the addition of 1.9 vol% of graphene oxide reduced the nitrogen permeability coefficient of XNBR by 55% (Kang et al. 2014).

Moreover, a property of graphene, which is expected to encourage valuable applications in relevant rubber nanocomposites, is electrical conductivity. NR reinforced by reduced graphene oxide presented electrical conductivity of 7.31 S/m at a filler content of 4.16 vol%. At the same time, the percolation threshold of these nanocomposites was detected at 0.21 vol% filler fraction (Luo et al. 2014). In another example, poly(isobutylene-*co*-isoprene) filled with reduced graphene oxide at 5 wt% presented linear changes of relative resistance under uniaxial pressure (piezo-resistance), suggesting potential uses as flexible force sensors (Kumar et al. 2013). It has to be mentioned that the solution assisted techniques for mixing graphene oxide with rubber matrices is beneficial in revealing the desirable properties of the vulcanizates at low filler loadings as compared to dry mixing (e.g., roll milling) (Potts et al. 2012).

## 2.6 OUTLOOK

### 2.6.1 Trends in Thermoplastic Nanocomposites

In view of the growing applications involving additive manufacturing and 3D printing, there is an increasing demand for more functionalized materials as opposed to conventional neat thermoplastic resins to be used as the filament feeding material. The prospects of 3D printed articles are tremendous. At home, 3D printing could provide opportunities for vast customization of wearable devices that are otherwise unavailable or too expensive to manufacture. In the industry, rapid prototyping could quickly turn ideas into 3D articles that provide customers and buyers the opportunity to view, touch, and play with a prototype. This provides the added dimension and confidence that could not be achieved by mere graphics or drawings. Since 3D printing is a layer-by-layer melt deposition process, the typical requirements for the feed material would be fast melting and cooling, low warpage and shrinkage, good interlayer bonding, and excellent flow properties. The usage of conventional reinforcements such as micron-sized particulates or fibers in the thermoplastic compound is not feasible because these materials will typically cause clogging of the printer nozzles. Therefore, the usage of nanofillers is feasible because only very low filler concentrations are needed to effectively modify the properties of the thermoplastic resin. By this way, additional functionalities can be provided such as mechanical property enhancement, thermal and electrical conductivity, thermal stability, dimensional stability, and stable flow properties. Potential products that can benefit from such new 3D printing materials include, among others, structures for unmanned aerial vehicles (UAV), repair parts for aircraft, marine and building structures, wearable devices, jewelry, customized sports equipment, and assistive devices.

### 2.6.2 TRENDS IN THERMOSET NANOCOMPOSITES

Lately, combined action of nanomaterials has been proven to be beneficial for the final properties of a thermoset matrix. CNTs along with graphene nanosheets in a ratio 8:2 were found to improve the flexural properties and decrease the electrical percolation threshold for the respective epoxy nanocomposites (Yue et al. 2014). Exploitation of wider graphene nanoplatelets has been proposed to yield better synergy effects in such hybrid systems (Chatterjee et al. 2012). Moreover, silica nanoparticle-coated graphite nanoplatelets at 2 wt% content were served as thermally conductive and electrical insulating filler in an epoxy matrix (Choi et al. 2014). In another example, matching the properties of silver nanoparticles and organoclay in an epoxy matrix, an infection-resistant and strong implantable scaffold film for skin tissue regeneration has been produced (Barua et al. 2014).

Conventional thermoset composites can also benefit from the addition of nanomaterials. In such examples, 10 wt% of nanosilica increased three to four times the fatigue life of glass-reinforced plastic composite laminates (Manjunatha et al. 2010), while a traditional glass fabric reinforced epoxy composite increased its flexural modulus and interlaminar shear strength only with 0.5 wt% CNT–$Al_2O_3$ hybrid nanomaterial (Li et al. 2014).

Besides, nanomaterials are able to offer new insights in conventional thermoset composites. In this view, taking advantage of the conductive CNT network, the impact damage evolution of a woven glass fiber/epoxy composite was monitored by *in situ* electrical resistance measurements (Gao et al. 2011).

### 2.6.3 TRENDS IN ELASTOMERIC NANOCOMPOSITES

Demanding rubber applications make imperative the quest of more sophisticated fillers. In this view, rubber technologists combine nanofillers in order to tune the performance of rubber nanocomposites. Combination of 40 phr CB with less than 5 phr oxidized graphite delivered a stable wearing process on an NBR vulcanizate presenting a low coefficient of friction (COF) and low wear rate even at high loads and high sliding velocities (Wang et al. 2012). On an example of SBR/BR blend, a combination of 45 phr CB with 3 phr of organoclay improved significantly the abrasion resistance of the relevant vulcanizates (Shan et al. 2011).

Cellulose at filler loading of 25 phr combined with 5 phr silica had a great effect on improving heat build-up and compression set at dynamic compression in NR compared with NR/silica vulcanizates of 30 phr filler loading. At the same time, the tanδ values at 0°C and 60°C remained almost unaffected or reduced, respectively, having in mind recipes for tire applications (Xu et al. 2012). Likewise, NR filled with 84 phr of silica modified with silane coupling agent in combination with 6 phr of MWCNTs yielded electrical conductivity of about $10^{-3}$ S/cm maintaining a moderate Payne effect and a high tanδ value at 0°C (Fritzsche et al. 2009).

In another case, silicon rubber presented enhanced electrical properties when filled with a combination of 2.5 phr CB and 1.0 phr CNT. Compared with the single filler mixtures, the electrical resistance for the dual system had five times lower value than CB and nearly 3 orders of magnitude lower than CNT for the same filler content (Witt et al. 2013). It becomes clear from this chapter that nanomaterials within polymer matrices succeed to achieve more (improved properties) with less (filler content).

## REFERENCES

Abdullayev, E., Lvov, Y. 2013. Halloysite clay nanotubes as a ceramic "skeleton" for functional biopolymer composites with sustained drug release. *Journal of Materials Chemistry B*, 1:2894–903.

Alexandre, M., Dubois. P. 2000. Polymer-layered silicate nanocomposites: Preparation, properties and uses of a new class of materials. *Materials Science and Engineering* 28:1–63.

Al-Ghamdi, A. A., Al-Hartomy, O. A., Al-Salamy, F. et al. 2012. Novel electromagnetic interference shielding effectiveness in the microwave band of magnetic nitrile butadiene rubber/magnetite nanocomposites. *Journal of Applied Polymer Science* 125:2604–13.

Alivisatos, P. 2004. The use of nanocrystals in biological detection. *Nature Biotechnology* 22:47–52.

Allahbakhsh, A., Mazinani, S., Kalaee, M. R., Sharif, F. 2013. Cure kinetics and chemorheology of EPDM/graphene oxide nanocomposites. *Thermochimica Acta* 563:22–32.

Arantes, T. M., Sala, R. L., Leite, E. R., Longo, E., Camargo, E. R. 2013. Comparison of the nanoparticles performance in the photocatalytic degradation of a styrene-butadiene rubber nanocomposite. *Journal of Applied Polymer Science* 128:2368–74.

Aso, O., Eguiazábal, J. I., Nazábal, J. 2007. The influence of surface modification on the structure and properties of a nanosilica filled thermoplastic elastomer. *Composites Science and Technology* 67:2854–63.

Ayatollahi, M. R., Alishahi, E., Doagou-R, S., Shadlou, S. 2012. Tribological and mechanical properties of low content nanodiamond/epoxy nanocomposites. *Composites: Part B* 43:3425–30.

Balazs, A. C., Singh, C., Zhulina, E., Lyatskaya, Y. 1999. Modeling the phase behavior of polymer/clay nanocomposites. *Accounts of Chemical Research* 32:651–57.

Banerjee, D., Lao, J., Ren Z. 2006. Design of nanostructured materials. In: Schulz, M. J., Kelkar, A. D., Sundaresan, M. J. (eds.), *Nanoengineering of Structural, Functional and Smart Materials*, Boca Raton, FL: CRC Press/Taylor & Francis, pp. 15–56.

Barua, S., Chattopadhyay, P., Aidew, L., Buragohain, A. K., Karak, N. 2014. Infection-resistant hyperbranched epoxy nanocomposite as a scaffold for skin tissue regeneration. *Polymer International* 64:303–11.

Basu, M., Pal, T. 2011. Metal and metal oxide nanostructure on resin support. In: Chauhan, B. P. S. (ed.), *Hybrid Nanomaterials*, Hoboken, NJ: Wiley, pp. 23–63.

Bendahou, A., Kaddami, H., Dufresne, A. 2010. Investigation on the effect of cellulosic nanoparticles' morphology on the properties of natural rubber based nanocomposites. *European Polymer Journal* 46:609–20.

Berman, D., Erdemir, A., Sumant, A. V. 2014. Graphene: A new emerging lubricant. *Materials Today* 17:31–42.

Bhattacharya, S. N., Kamal, M. R., Gupta, R. K. 2008. *Polymeric Nanocomposites*. Munich: Hanser.

Bokobza, L. 2012. Multiwall carbon nanotube-filled natural rubber: Electrical and mechanical properties. *Express Polymer Letters* 6:213–23.

Bottari, G., Urbani, M., Torres T. 2013. Covalent, donor–acceptor ensembles based on phthalocyanines and carbon nanostructures. In: T. Torres, G. Bottari (eds.), *Organic Nanomaterials*, Hoboken, NJ: Wiley, pp. 163–86.

Buzea, C., Pacheco, I. I., Robbie, K. 2007. Nanomaterials and nanoparticles: Sources and toxicity. *Biointerphases* 2:17–71.

Chandrasekaran, S., Seidel, C., Schulte, K. 2013. Preparation and characterization of graphite nano-platelet (GNP)/epoxy nano-composite: Mechanical, electrical and thermal properties. *European Polymer Journal* 49:3878–88.

Chandrasekaran, S., Sato, N., Tölle, F., Mülhaupt, R., Fiedler, B., Schulte, K. 2014. Fracture toughness and failure mechanism of graphene based epoxy composites. *Composites Science and Technology* 97:90–9.

Chatterjee, S., Nafezarefi, F., Tai, N. H., Schlagenhauf, L., Nüesch, F. A., Chu, B. T. T. 2012. Size and synergy effects of nanofiller hybrids including graphene nanoplatelets and carbon nanotubes in mechanical properties of epoxy composites. *Carbon* 50:5380–6.

Chen, H., Chen, Y., Liu, Y., Fu, L., Huang, C., Llewellyn, D. 2008. Over 1.0 mm-long boron nitride nanotubes. *Chemical Physics Letters* 463:130–3.

Chen, J., Hamon, M. A., Hu, H. et al. 1998. Solution properties of single-walled carbon nanotubes. *Science* 282:95–8.

Chen, G. Z., Shaffer, M. S. P., Coleby, D. et al. 2000. Carbon nanotube and polypyrrole composites: Coating and doping. *Advanced Materials* 12:522–6.

Choi, S., Yang, J., Kim, Y., Nam, J., Kim, K., Shim, S. E. 2014. Microwave-accelerated silica nanoparticle-coated graphite nanoplatelets and properties of their epoxy composites. *Composites Science and Technology* 103:8–15.

Choudhury, A., Bhowmick, A. K., Ong, C. 2010. Effect of different nanoparticles on thermal, mechanical and dynamic mechanical properties of hydrogenated nitrile butadiene rubber nanocomposites. *Journal of Applied Polymer Science* 116:1428–41.

Choudhury, A., Bhowmick, A. K., Soddemman, M. 2010. Effect of organo-modified clay on accelerated aging resistance of hydrogenated nitrile rubber nanocomposites and their life time prediction. *Polymer Degradation and Stability* 95:2555–62.

Chung, P. T., Yang, C. T., Wang, S. H., Chen C. W., Chiang, A. S. T., Liu, C.-Y. 2012. ZrO$_2$/epoxy nanocomposite for LED encapsulation. *Materials Chemistry and Physics* 136:868–76.

Ciselli, P., Lu, L., Busfield, J. J. C., Peijs, T. 2010. Piezoresistive polymer composites based on EPDM and MWNTs for strain sensing applications. *e-Polymers* 10:125–37.

Cookson, J. 2012. The preparation of palladium nanoparticles. *Platinum Metals Review* 56:83–98.

Cui, J., Wang, W., You, Y., Liu, C., Wang, P. 2004. Functionalization of multiwalled carbon nanotubes by reversible addition fragmentation chain-transfer polymerization. *Polymer* 45:8717–21.

Dang, Z.-M., Xia, Y.-J., Zha, J.-W., Juan, J.-K., Bai, J. 2011. Preparation and dielectric properties of surface modified TiO$_2$/silicone rubber nanocomposites. *Materials Letters* 65:3430–2.

Das, A., Stöckelhuber, K. W., Jurk, R. et al. 2008. Modified and unmodified multiwalled carbon nanotubes in high performance solution-styrene-butadiene and butadiene rubber blends. *Polymer* 49:5276–83.

Ding, K. E., Wang, G. L., Zhang, M. 2012. Preparation and optical properties of transparent epoxy composites containing ZnO nanoparticles. *Journal of Applied Polymer Science* 126:734–9.

Diukman, I., Tzabari, L., Berkovitch, N., Tessler, N., Orenstein, M. 2011. Controlling absorption enhancement in organic photovoltaic cells by patterning Au nano disks within the active layer. *Optics Express* 19:64–71.

Dresselhaus, M. S., Dai, H. 2004. Carbon nanotubes: Continued innovations and challenges. *MRS Bulletin* 29:237–43.

Dufresne, A., Cavaille, J. Y., Helbert, W. 1996. New nanocomposite materials: Microcrystalline starch reinforced thermoplastic. *Macromolecules* 29:7624–6.

Duran, H., Steinhart, M., Butt, H.-J., Floudas, G. 2011. From heterogeneous to homogeneous nucleation of isotactic poly(propylene) confined to nanoporous alumina. *Nano Letters* 11:1671–5.

El Saeed, A. M., El-Fattah, M. A., Dardir, M. M. 2015. Synthesis and characterization of titanium oxide nanotubes and its performance in epoxy nanocomposite coatings. *Progress in Organic Coatings* 78:83–9.

Felhös, D., Karger-Kocsis, J., Xu, D. 2008. Tribological testing of peroxide cured HNBR with different MWCNT and silica contents under dry sliding and rolling conditions against steel. *Journal of Applied Polymer Science* 108:2840–51.

Flaifel, M. H., Ahmad, S. H., Abdullah, M. H. et al. 2014. Preparation, thermal, magnetic and microwave absorption properties of thermoplastic natural rubber matrix impregnated with NiZn ferrite nanoparticles. *Composites Science and Technology* 96:103–8.

Foa Torres, L. E. F., Roche, S., Charlier, J.-C. 2014. *Introduction to Graphene-Based Nanomaterials*. New York: Cambridge University Press.

Foygel, M., Morris, R., Anez, D., French, S., Sobolev, V. 2005. Theoretical and computational studies of carbon nanotube composites and suspensions: Electrical and thermal conductivity. *Physical Review B* 71:104201.

Fragiadakis, D., Bokobza, L., Pissis, P. 2011. Dynamics near the filler surface in natural rubber-silica nanocomposites. *Polymer* 52:3175–82.

Fritzsche, J., Lorenz, H., Klüppel, M. 2009. CNT based elastomer-hybrid-nanocomposites with promising mechanical and electrical properties. *Macromolecular Materials and Engineering* 294:551–60.

Fröhlich, J., Thomann, R., Gryshchuk, O., Karger-Kocsis, J., Mülhaupt, R. 2004. High-performance epoxy hybrid nanocomposites containing organophilic layered silicates and compatibilized liquid rubber. *Journal of Applied Polymer Science* 92:3088–96.

Fu, J.-F., Chen, L.-Y., Yang, H. et al. 2012. Mechanical properties, chemical and aging resistance of natural rubber filled with nano-Al$_2$O$_3$. *Polymer Composites* 33:404–11.

Ganguli, S., Roy, A. K., Anderson, D. P. 2008. Improved thermal conductivity for chemically functionalized exfoliated graphite/epoxy composites. *Carbon* 46:806–17.

Gao, L., Chou, T.-W., Thostenson, E. T., Zhang, Z., Coulaud, M. 2011. In-situ impact damage in epoxy/glass fiber composites using percolating carbon nanotube networks. *Carbon* 49:3382–5.

Gatos, K. G., Sawanis, N. S., Apostolov, A. A., Thomann, R., Karger-Kocsis, J. 2004. Nanocomposite formation in hydrogenated nitrile rubber (HNBR)/organo-montmorillonite as a function of the intercalant type. *Macromolecular Materials and Engineering* 289:1079–86.

Gatos, K. G., Karger-Kocsis, J. 2005. Effects of primary and quaternary amine intercalants on the organoclay dispersion in a sulfur-cured EPDM rubber. *Polymer* 46:3069–76.

Gatos, K. G., Karger-Kocsis, J. 2007. Effect of the aspect ratio of silicate platelets on the mechanical and barrier properties of hydrogenated acrylonitrile butadiene rubber (HNBR)/layered silicate nanocomposites. *European Polymer Journal* 43:1097–104.

Gatos, K. G., Kameo, K., Karger-Kocsis, J. 2007a. On the friction and sliding wear of rubber/layered silicate nanocomposites. *Express Polymer Letters* 1:27–31.

Gatos, K. G., Martínez Alcázar, J. G., Psarras, G. C., Thomann, R., Karger-Kocsis, J. 2007b. Polyurethane latex/water dispersible boehmite alumina nanocomposites: Thermal, mechanical and dielectrical properties. *Composites Science and Technology* 67:157–67.

Gatos, K. G., Karger-Kocsis, J. 2010. Rubber/clay nanocomposites: Preparation, properties and applications. In: Thomas, S., Stephen, R. (eds.), *Rubber Nanocomposites*, Singapore: Wiley, pp. 169–95.

Gatos, K. G., Karger-Kocsis, J. 2011. Rubber–clay nanocomposites based on nitrile rubber. In: Galimberti, M. (ed.), *Rubber–clay nanocomposites*, Hoboken, NJ: Wiley, pp. 409–30.

Geberth, E., Klüppel, M. 2012. Effect of carbon black deactivation on the mechanical and electrical properties of elastomers. *Macromolecular Materials and Engineering* 297:914–22.

Gilman, J. W. 1999. Flammability and thermal stability studies of polymer layered-silicate (clay) nanocomposites. *Applied Clay Science* 15:31–49.

Gkikas, G., Barkoula, N.-M., Paipetis, A. S. 2012. Effect of dispersion conditions on the thermo-mechanical and toughness properties of multi walled carbon nanotubes-reinforced epoxy. *Composites: Part B* 43:2697–705.

Gojny, F. H., Schulte, K. 2004. Functionalisation effect on the thermo-mechanical behaviour of multi-wall carbon nanotube/epoxy composites. *Composites Science and Technology* 64:2303–8.

Gojny, F. H., Wichmann, M. H. G., Fiedler, B., Schulte, K. 2005. Influence of different carbon nanotubes on the mechanical properties of epoxy matrix composites – A comparative study. *Composites Science and Technology* 65:2300–13.

Gómez, F. J., Chen, R. J., Wang, D., Waymouth, R. M., Dai, H. 2003. Ring opening metathesis polymerization on non-covalently functionalized single-walled carbon nanotubes. *Chemical Communication* 190:190–1.

Gopi, S., Subramanian, V. K., Palanisamy, K. 2013. Aragonite–calcite–vaterite: A temperature influenced sequential polymorphic transformation of CaCO3 in the presence of DTPA. *Materials Research Bulletin* 48:1906–12.

Guo, B., Chen, F., Lei, Y., Liu, X., Wan, J. Jia, D. 2009. Styrene-butadiene rubber/halloysite nanotubes nanocomposites modified by sorbic acid. *Applied Surface Science* 255:7329–36.

Guth, E. 1945. Theory of filler reinforcement. *Journal of Applied Physics* 16:596–604.

Guy, L., Bomal. Y., Ladouce-Stelandre, L. 2005. Elastomers reinforcement by precipitated silicas. *Kautschuk Gummi Kunststoffe* 58:43–9.

He, X., Zhang, D., Li, H., Fang, J., Shi, L. 2011. Shape and size effects of seria nanoparticles n the impact strength of ceria/epoxy resin composites. *Particuology* 9:80–5.

Hedicke-Höchstötter, K., Lim, G. T., Altstädt, V. 2009. Novel polyamide nanocomposites based on silicate nanotubes of the mineral halloysite. *Composites Science and Technology* 69:330–4.

Hernádez-Pérez, A., Avilés, F., May-Pat, A., Valadez-González, A., Herrera-Franco, P. J., Bartolo-Pérez, P. 2008. Effective properties of multiwalled carbon nanotube/epoxy composites using two different tubes. *Composites Science and Technology* 68:1422–31.

Hill, D. E., Lin, Y., Rao, A. M., Allard, L. F., Sun, Y. P. 2002. Functionalization of carbon nanotubes with polystyrene. *Macromolecules* 35:9466–71.

Hofmann, D., Keinath, M., Thomann, R., Mülhaupt, R. 2014. Thermoplastic carbon/polyamide 12 composites containing functionalized graphene, expanded graphite, and carbon nanofillers. *Macromolecular Materials and Engineering* 229:1329–42.

Hu, J., Odom, T. W., Lieber, C. M. 1999. Chemistry and physics in one dimension: Synthesis and properties of nanowires and nanotubes. *Accounts of Chemical Research* 32:435–45.

Hu, K., Gupta, M. K., Kulkarni, D. D., Tsukruk, V. V. 2013. Ultra-robust graphene oxide-silk fibroin nanocomposite membranes. *Advanced Materials* 25:2301–7.

Hu, K., Kulkarni, D. D., Choi, I., Tsukruk, V. V. 2014. Graphene–polymer nanocomposites for structural and functional applications. *Progress in Polymer Science* 39:1934–72.

Hummers, W. S., Offeman, R. E. 1958. Preparation of graphitic oxide. *Journal of the American Chemical Society* 80:1339.

Ijima, S. 1991. Helical microtubules of graphitic carbon. *Nature* 354:56–8.

Ismail, H., Pasbakhsh, P., Ahmad Fauzi, M. N., Abu Bakar, A. 2008. Morphological, thermal and tensile properties of halloysite nanotubes filled ethylene propylene diene monomer (EPDM) nanocomposites. *Polymer Testing* 27:841–50.

Ismail, H., Salleh, S. Z., Ahmad, Z. 2013. Properties of halloysite nanotubes-filled natural rubber prepared using different mixing methods. *Materials and Design* 50:790–7.

Jamal, E. M. A., Joy, P. A., Kurian, P., Anantharaman, M. R. 2009. Synthesis of nickel-rubber nanocomposites and evaluation of their dielectric properties. *Materials Science and Engineering B* 156:24–31.

Jancar, J., Douglas, J. F., Starr, F. W. et al. 2010. Current issues in research on structure property relationships in polymer nanocomposites. *Polymer* 51:3321–43.

Jariwala, D., Sangwan, V. K., Lauhon, L. J., Marks, T. J., Hersam, M. C. 2013. Carbon nanomaterials for electronics, optoelectronics, photovoltaics, and sensing. *Chemical Society Reviews* 42:2824–60.

Jiang, C., He, H., Jiang, H., Ma, L., Jia, D. M. 2013. Nano-lignin filled natural rubber composites: Preparation and characterization. *Express Polymer Letters* 7:480–93.

Jiang, H. 2011. Chemical preparation of graphene-based nanomaterials and their applications in chemical and biological sensors. *Small* 7:2413–27.

Jiang, M.-J., Dang, Z.-M., Yao, S.-H., Bai, J. 2008. Effect of surface modification of carbon nanotubes on the microstructure and electrical properties of carbon nanotubes/rubber nanocomposites. *Chemical Physics Letters* 457:352–6.

Johnsen, B. B., Kinloch, A. J., Mohammed, R. D., Taylor, A. C., Sprenger, S. 2007. Toughening mechanisms of nanoparticle-modified epoxy polymers. *Polymer* 48:530–41.

Kalaitzidou, K., Fukushima, H., Drzal, L. T. 2007. A new compounding method for exfoliated graphite–polypropylene nanocomposites with enhanced flexural properties and lower percolation threshold. *Composites Science and Technology* 67:2045–51.

Kang, H., Zuo, K., Wang, Z., Zhang, L., Liu, L., Guo, B. 2014. Using a green method to provide graphene oxide/elastomers nanocomposites with combination of high barrier and mechanical performance. *Composites Science and Technology* 92:1–8.

Karamipour, S., Ebadi-Dehaghani, H., Ashouri, D., Mousavian, S. 2011. Effect of nano-$CaCO_3$ on rheological and dynamic mechanical properties of polypropylene: Experiments and models. *Polymer Testing* 30:110–7.

Kashiwagi, T., Harris Jr., R. H., Zhang, X., Briber, R. M., Cipriano, B. H., Ragharan, S. R. 2004. Flame retardant mechanism of polyamide 6-clay nanocomposites. *Polymer* 45:881–91.

Katz, E., Shipway, A. N., Willner, I. 2003. Chemically functionalized metal nanoparticles. In: Liz-Marzan, L. M., Kamat, P. (eds.), *Nanoscale Materials*, Boston: Kluwer, pp. 5–78.

Kawahara, T., Yuuki, A., Hashimoto, K., Fujiki, K., Yamauchi, T., Tsubokawa, N. 2013. Immobilization of flame-retardant onto silica nanoparticle surface and properties of epoxy resin filled with the flame-retardant-immobilized silica (2). *Reactive & Functional Polymers* 73:613–8.

Kiliaris, P., Papaspyrides, C. D. 2010. Polymer/layered silicate (clay) nanocomposites: An overview of flame retardancy. *Progress in Polymer Science* 35:902–58.

Kim, H.-W., Kim, H.-E., Knowles, J. C. 2006. Production and potential of bioactive glass nanofibers as a next-generation biomaterial. *Advanced Functional Materials* 16:1529–35.

Kinoshita, M., Steiner, M., Engel, M. et al. 2010. The polarized carbon nanotube thin film LED. *Optics Express* 18:25738–45.

Kohjiya, S., Kato, A., Ikeda, Y. 2008. Visualization of nanostructure of soft matter by 3D-TEM: Nanoparticles in a natural rubber matrix. *Progress in Polymer Science* 33:979–97.

Kojima, Y., Fukumori, F., Usuki, A., Okada, A., Kurauchi, T. 1993. Gas permeabilities in rubber–clay hybrid. *Journal of Materials Science Letters* 12:889–90.

Kojima, Y., Usuki, A., Kawasumi, M., Okada, A., Fukushima, Y., Kurauchi, T., Kamigaito, O. 1990. Mechanical properties of nylon-6/clay hybrids. *Journal of Materials Research* 8:1185–9.

Kong, H., Gao, C., Yan, D. Y. 2004. Controlled functionalization of multiwalled carbon nanotubes by in-situ atom transfer radical polymerization. *Journal of the American Chemical Society* 126:412–3.

Kornmann, X., Lindberg, H., Berglund, L. A. 2001. Synthesis of epoxy-clay nanocomposites: Influence of the nature of the clay on structure. *Polymer* 42:1303–10.

Kosmidou, T. V., Vatalis, A. S., Delides, C. G., Logakis, E., Pissis, P., Papanicolaou, G. C. 2008. Structural, mechanical and electrical characterization of epoxy-amine/carbon black nanocomposites. *Express Polymer Letters* 2:364–72.

Kotsilkova R. 2007. Performance of thermoset nanocomposites. In: Kotsilkova, R. (ed.), *Thermoset Nanocomposites for Engineering Applications*, Shawbury: Smithers Rapra Technology Limited, pp. 207–276.

Kuilla, T., Bhadra, S., Yao, D., Kim, N. H., Bose, S., Lee, J. H. 2010. Recent advances in graphene based polymer composites. *Progress in Polymer Science* 35:1350–75.

Kulkarni, D. D., Choi, I., Singamaneni, S. S., Tsukruk, V. V. 2010. Graphene oxide-polyelectrolyte nanomembranes. *ACS Nano* 4:4667–76.

Kumar, S. K., Castro, M., Saiter, A. et al. 2013. Development of poly(isobutylene-co-isoprene)/reduced graphene oxide nanocomposites for barrier, dielectric and sensing applications. *Materials Letters* 96:109–12.

Lan, T., Pinnavaia, T. J. 1994. Clay-reinforced epoxy nanocomposites. *Chemistry of Materials* 6:2216–9.

Lawandy, S. N., Halim, S. F., Darwish, N. A. 2009. Structure aggregation of carbon black in ethylene-propylene diene polymer. *Express Polymer Letters* 3:152–8.

Lazzari, M., López-Quintela, M. A. 2003. Block copolymers as a tool for nanomaterial fabrication. *Advanced Materials* 15:1–12.

Leong, Y. W., Tan, M. B. H., Lin, T. T. et al. 2014. *Hybrid Polymers*. United States Patent 8,853,330 B2.

Lewin, M. 2006. Reflections on migration of clay and structural changes in nanocomposites. *Polymers for Advanced Technologies* 17:758–63.

Lewin, M. 2003. Some comments on the modes of action of nanocomposites in the flame retardancy of polymers. *Fire and Materials* 27:1–7.

Li, Y.-Q., Fu, S.-Y., Mai, Y.-W. 2006. Preparation and characterization of transparent ZnO/epoxy nanocomposites with high UV-shielding efficiency. *Polymer* 47:2127–32.

Li, W., Dichiara, A., Zha, J., Su, Z., Bai, J. 2014. On the improvement of mechanical and thermo-mechanical properties of glass fabric epoxy composites by incorporating CNT–Al$_2$O$_3$ hybrids. *Composites Science and Technology* 103:36–43.

Li, J., Ma, P. C., Chow, W. S., To, C. K., Tang, B. Z., Kim, J.-K. 2007. Correlations between percolation threshold, dispersion state, and aspect ratio of carbon nanotubes. *Advanced Functional Materials* 17:3207–15.

Lin, O. H., Akil, H. M., Ishak, Z. A. M. 2009. Comparative study between PP-g-MAH and PP-methyl-POSS as compatibilizer for nanoSiO2/PP composites: Mechanical and morphological properties. *Advanced Composite Letters* 18:183–90.

Liu, H., Zheng, S. 2005. Polyurethane networks nanoreinforced by polyhedral oligomeric silsesquioxane. *Macromolecular Rapid Communications* 26:196–200.

Liu, Z. Y., Yu, R. Z., Yang, M. B., Feng, J. M., Yang, W., Yin, B. 2007. An approach to preparation of polymer nano CaCO$_3$ composites. *Acta Polymerica Sinica* 1:115–22.

Liu, H.-Y., Wang, G., Mai, Y.-W. 2012. Cyclic fatigue crack propagation of nanoparticle modified epoxy. *Composites Science and Technology* 72:1530–8.

Lu, M., He, B., Wang, L. et al. 2012. Preparation of polystyrene–polyisoprene core–shell nanoparticles for reinforcement of elastomers. *Composites: Part B* 43:50–6.

Lu, H., Shen, H., Song, Z., Shing, K. S., Tao, W., Nutt, S. 2005. Rod-like silicate epoxy nanocomposites. *Macromolecular Rapid Communications* 26:1445–50.

Luheng, W., Tianhuai, D., Peng, W. 2009. Influence of carbon black concentration on piezoresistivity for carbon-black-filled silicone rubber composite. *Carbon* 47:3151–7.

Luo, Y., Zhao, P., Yang, Q., He, D., Kong, L., Peng, Z. 2014. Fabrication of conductive elastic nanocomposites via framing intact interconnected graphene networks. *Composites Science and Technology* 100:143–51.

Manjunatha, C. M., Taylor, A. C., Kinloch, A. J., Sprenger, S. 2010. The tensile fatigue behaviour of a silica nanoparticle-modified glass fibre reinforced epoxy composite. *Composites Science and Technology* 70:193–9.

Manias, E. 2007. Nanocomposites: Stiffer by design. *Nature Materials* 6:9–11.

Martin-Callego, M., Bernal, M. M., Hernadez, M., Verdejo, R., Lopez-Mnchado, L. A. 2013. Comparison of filler percolation and mechanical properties in graphene and carbon nanotubes filled epoxy nanocomposites. *European Polymer Journal* 49:1347–53.

Mateo-Alonso, A., Tagmatarchis, N., Prato, M. 2006. Fullerenes and their derivatives. In: Gogotsi, Y. (ed.), *Carbon Nanomaterials*, Boca Raton, FK: CRC Press/Taylor & Francis, pp. 1–39.

Medalia, A. I. 1978. Effect of carbon black on dynamic properties of rubber vulcanizates. *Rubber Chemistry and Technology* 51:437–523.

Meng, M. R., Dou, Q. 2009. Effect of filler treatment on crystallization, morphology and mechanical properties of polypropylene/calcium carbonate composites. *Journal of Macromolecular Science Part B Physics* 48:213–25.

Messersmith, P. B., Giannelis, E. P. 1994. Synthesis and characterization of layered silicate-epoxy nanocomposites. *Chemistry of Materials* 6:1719–26.

Mo, M.-S., Du, X.-S. 2009. Building nonmagnetic metal@oxide and bimetallic nanostructures: Potential applications in the life sciences. In: Kumar, C. S. S. R. (ed.), *Mixed Metal Nanomaterials*, Weinheim: Wiley, pp. 161–96.

Moniruzzaman, M., Winey, K. I. 2006. Polymer nanocomposites containing carbon nanotubes. *Macromolecules* 39:5194–205.

Mullins, L., Tobin, N. R. 1965. Stress softening in rubber vulcanizates. Part I. Use of a strain amplification factor to describe the elastic behavior of filler-reinforced vulcanized rubber. *Journal of Applied Polymer Science* 9:2993–3009.

Murphy, C. J., Jana, N. R., Gearheart, L. A. et al. 2004. Synthesis, assembly and reactivity of metallic nanorods. In: Rao, C. N. R., Müller, A., Cheetham, A. K. (eds.), *The Chemistry of Nanomaterials*, Weinheim: Wiley, pp. 285–307.

Mutiso, R. M., Winey, K. I. 2013. Electrical percolation in quasi-two-dimensional metal nanowire networks for transparent conductors. *Physical Review E* 88:032134.

Mutiso, R. M., Sherrott, M. C., Rathmell, A. R., Wiley, B. J., Winey, K. I. 2013. Integrating simulations and experiments to predict sheet resistance and optical transmittance in nanowire films for transparent conductors. *ACS Nano* 7:7654–63.

Nabok, A. 2005. *Organic and Inorganic Nanostructures*. Norwood: Artech House Inc.

Naffakh, M., Díez-Pascual, A. M., Marco, C., Ellis, G. J., Gómez-Fatou, M. A. 2013. Opportunities and challenges in the use of inorganic fullerene-like nanoparticles to produce advanced polymer nanocomposites. *Progress in Polymer Science* 38:1163–231.

Nahin, P. G., Backlund, P. S. 1963. Organoclay–polyolefin compositions. United States Patent 3,084,117.

Nazemi, A., Gillies, E. R. 2013. Dendritic surface functionalization of nanomaterials: Controlling properties and functions of biomedical applications. *Brazilian Journal of Pharmaceutical Sciences* 49:15–32.

Neitzel, I., Mochalin, V., Knoke, I., Palmese, G. R., Gogotsi, Y. 2011. Mechanical properties of epoxy composites with high contents of nanodiamonds. *Composites Science and Technology* 71:710–6.

Ng, C. B., Schadler, L. S., Siegel, R. W. 1999. Synthesis and mechanical properties of $TiO_2$-epoxy nanocomposites. *NanoStructured Materials* 12:507–10.

Nohales, A., Muñoz-Espí, E., Félix, P., Gómez, C. M. 2011. Sepiolite-reinforced epoxy nanocomposites: Thermal, mechanical and morphological behavior. *Journal of Applied Polymer Science* 119:539–47.

Novoselov, K. S., Geim, A. K., Morozov, S. V., Jiang, D., Zhang, Y., Dubonos, S. V., Grigorieva, I. V., Firsov, A. A. 2004. Electric field effect in atomically thin carbon films. *Science* 306:666–9.

Okada, A., Fukushima, Y., Kawasumi, M. et al. 1987. Composite material and its preparation. United States Patent 4, 739,007.

Okada, A., Usuki. A. 1995. The chemistry of polymer–clay hybrids. *Materials Science and Engineering C* 3:109–15.

Osman, M. A., Mittal, V., Morbidelli, M., Suter, U. W. 2004. Epoxy-layered silicate nanocomposites and their gas permeation properties. *Macromolecules* 37:7250–7.

Pal, S., Tak, Y. K., Song, J. M. 2007. Does the antibacterial activity of silver nanoparticles depend on the shape of the nanoparticle? A study of the gram-negative bacterium *Escherichia coli*. *Applied and Environmental Microbiology* 73:1712–20.

Palza, H., Vergara, V., Zapata, P. 2011. Composites of polypropylene melt blended with synthesized silica nanoparticles. *Composites Science and Technology* 71:535–40.

Parab, H. J., Chen, H. M., Bagkar, N. C., Liu, R.-S., Hwu, Y.-K., Tsai, D. P. 2009. Approaches to the synthesis and characterization of spherical and anisotropic noble metal nanomaterials. In: Kumar, C. (ed.), *Metalic Nanomaterials*, Weinheim: Wiley, pp. 405–460.

Park, S. H., Oh, K. W., Kim, S. H. 2013. Reinforcement effect of cellulose nanowhisker on bio-based polyurethane. *Composites Science and Technology* 86:82–8.

Park, J. H., Jana, S. C. 2003. Mechanism of exfoliation of nanoclay particles in epoxy-clay nanocomposites. *Macromolecules* 36:2758–68.

Park, S., Kim D. S. 2014. Preparation and physical properties of an epoxy nanocomposite with amine-functionalized graphenes. *Polymer Engineering and Science* 54:985–91.

Pelaez, M., de la Cruz, A. A., Stathatos, E., Falaras, P., Dionysiou, D. D. 2009. Visible light-activated N-F-codoped TiO$_2$ nanoparticles for the photocatalytic degradation of microcystin-LR in water. *Catalysis Today* 144:19–25.

Peng, W., Huang, X., Yu, J., Jiang, P., Liu, W. 2010. Electrical and thermophysical properties of epoxy aluminum/nitride nanocomposites: Effects of nanoparticle surface modification. *Composites: Part A* 41:1201–09.

Philip, B., Xie, J., Abraham, J. K., Varada, V. K. 2005. Polyaniline/carbon nanotube composites: Starting with phenylamino functionalized carbon nanotubes. *Polymer Bulletin* 53:127–38.

Pikhurov, D. V., Zuev, V. V. 2014. The effect of fullerene C$_{60}$ on the dielectric behaviour of epoxy resin at low nanofiller loading. *Chemical Physics Letters* 601:13–5.

Plawsky, J. L., Kim, J. K., Schubert, E. F. 2009. Engineered nanoporous and nanostructured films. *Materials Today* 12:36–45.

Pokropivny, V., Lohmus, R., Hussainova, I., Pokropivny, A., Vlassov, S. 2007. *Introduction to Nanomaterials and Nanotechnology.* Tartu: Tartu University Press.

Pokropivny, V. V., Skorokhod, V. V. 2006. Classification of nanostructures by dimensionality and concept of surface forms engineering in nanomaterial science. *Materials Science and Engineering C* 27:990–3.

Potts, J. R., Shankar, O., Du, L., Ruoff, R. S. 2012. Processing–morphology–property relationships and composite theory analysis of reduced graphene oxide/natural rubber nanocomposites. *Macromolecules* 45:6045–55.

Pradhan, S., Costa, F. R., Wagenknecht, U., Jehnichen, D., Bhowmick, A. K., Heinrich, G. 2008. Elastomer LDH/nanocomposites: Synthesis and studies on nanoparticle dispersion, interfacial properties and interfacial adhesion. *European Polymer Journal* 44:3122–32.

Psarras, G. C., Gatos, K. G. 2011. Relaxation phenomena in elastomeric nanocomposites. In: Mittal, V., Kim, J. K., Pal, K. (eds.), *Recent Advances in Elastomeric Nanocomposites*, Heidelberg: Springer, pp. 89–118.

Qin, S., Qin, D., Ford, W. T., Resasco, D. E., Herrera, J. E. 2004. Polymer brushes on single-walled carbon nanotubes by atom transfer radical polymerization of n-butyl methacrylate. *Journal of the American Chemical Society* 126:170–6.

Qu, M., Deng, F., Kalkhoran, S. M., Gouldstone, A., Robisson, A., van Vliet, K. J. 2011. Nanoscale visualization and multiscale mechanical implications of bound rubber interphases in rubber-carbon black nanocomposites. *Soft Matter* 7:1066–77.

Rajisha, K. R., Maria, H. J., Pothan, L. A., Ahmad, Z., Thomas, S. 2014. Preparation and characterization of potato starch nanocrystal reinforced natural rubber nanocomposite. *International Journal of Biological Macromolecules* 67:147–53.

Ramasubramaniam, R., Chen, J., Liu, H. 2003. Homogeneous carbon nanotube/polymer composites for electrical applications. *Applied Physics Letters* 83:2928–30.

Randviir, E. P., Brownson, D. A. C., Banks, C. E. 2014. A decade of graphene research: Production, applications and outlook. *Materials Today* 17:426–32.

Rao, C. N. R., Cheetham, A. K. 2001. Science and technology of nanomaterials: Current status and future prospects. *Journal of Materials Chemistry* 11:2887–94.

Rao, C. N. R., Cheetham, A. K. 2006. Materials science at the nanoscale. In: Gogotsi, Y. (ed.), *Nanomaterials Handbook*, Boca Raton, FL: CRC Press/Taylor & Francis, pp. 1–12.

Rao, C. N. R. and Govindaraj, A. 2011. *Nanotubes and Nanowires.* Cambridge: Royal Society of Chemistry.

Reddy, M. V., Yu, T., Sow, C.-H. et al. 2007. α-Fe$_2$O$_3$ nanoflakes as a anode material for Li-ion batteries. *Advanced Functional Materials* 17:2792–9.

Rooj, S., Das, A., Thakur, V., Mahaling, R. N., Bhowmick, A. K., Heinrich, G. 2010. Preparation and properties of natural nanocomposites based on natural rubber and naturally occurring halloysite nanotubes. *Materials and Design* 31:2151–6.

Rosso, P., Ye, L., Friedrich, K., Sprenger, S. 2006. A toughened epoxy resin by silica nanoparticle reinforcement. *Journal of Applied Polymer Science* 100:1849–55.

Rybiński, P., Janowska, G. 2012. Thermal properties and flammability of nanocomposites based on nitrile rubbers and activated halloysite nanotubes and carbon nanofibers. *Thermochimica Acta* 549:6–12.

Sae-oui, P., Sirisinha, C., Hatthapanit, K., Thepsuwan, U. 2005. Comparison of reinforcing efficiency between Si-69 and Si-264 in an efficient vulcanization system. *Polymer Testing* 24:439–46.

Sangermano, M., Roppolo, I., Shan, G., Andrews, M. P. 2009. Nanocomposite epoxy coatings containing rare earths ion-doped LaF$_3$ nanoparticles: Film preparation and characterization. *Progress in Organic Coatings* 65:431–4.

Serrano-Ruiz, D., Rangou, S., Avgeropoulos, A., Zafeiropoulos, N. E., López-Cabarcos, E., Rubio-Retama, J. 2010. Synthesis and chemical modification of magnetic nanoparticles covalently bound to polystyrene-SiCl$_2$-poly(2-vinylpyridine). *Journal of Polymer Science: Part B: Polymer Physics* 48:1668–75.

Schadler, L. S., Giannaris, S. C., Ajayan, P. M. 1998. Load transfer in carbon nanotube epoxy composites. *Applied Physics Letters* 73:3842–4.

Schmidt-Mende, L., MacManus-Driscol, J. L. 2007. ZnO-nanostructures, defects and devices. *Materials Today* 10:40–8.

Scholz, S., Kroll, L., Schettler, F. 2014. Nanoparticle reinforced epoxy gelcoats for fiber-plastic composites under multiple load. *Progress in Organic Coatings* 77:1129–36.

Schwierz, F. 2010. Graphene transistors. *Nature Nanotechnology* 5:487–96.

Scotti, R., Conzatti, L., D'Arienzo, M. et al. 2014. Shape controlled spherical (0D) and rod-like (1D) silica nanoparticles in silica/styrene butadiene rubber nanocomposites: Role of the particle morphology on the filler reinforcing effect. *Polymer* 55:1497–506.

Seppälä, J., Luong, N. D. 2014. Chemical modification of graphene for functional polymer nanocomposites. *Express Polymer Letters* 8:373.

Serrano, E., Tercjak, A., Kortaberria, G. et al. 2006. Nanostructured thermosetting systems by modification with epoxidized styrene-butadiene star block copolymers. Effect of epoxidation degree. *Macromolecules* 39:2254–61.

Shan, C., Gu, Z., Wang, L. et al. 2011. Preparation, characterization, and application of NR/SBR/organoclay nanocomposites in the tire industry. *Journal of Applied Polymer Science* 119:1185–94.

Shen, X.-J., Pei, X.-Q., Fu, S.-Y., Friedrich, K. 2013. Significantly modified tribological performance of epoxy nanocomposites at very low graphene oxide content. *Polymer* 54:1234–42.

Shi, G., Zhang, M. Q., Rong, M. Z., Wetzel, B., Friedrich, K. 2004. Sliding wear behavior of epoxy containing nano-$Al_2O_3$ particles with different pretreatments. *Wear* 256:1072–81.

Siengshin, S., Karger-Kocsis, J., Psarras, G. C., Thomann, R. 2008. Polyoxymethylene/polyurethane/alumina ternary composites: Structure, mechanical, thermal and dielectric properties. *Journal of Applied Polymer Science* 110:1613–23.

Sinha Ray, S., Okamoto, M. 2003. Polymer/layered silicate nanocomposites: A review from preparation to processing. *Progress in Polymer Science* 28:1539–641.

Siró, I., Plackett, D. 2010. Microfibrillated cellulose and new nanocomposite materials: A review. *Cellulose* 17:459–94.

Song, Y. S., Youn, J. R. 2005. Influence of dispersion states of carbon nanotubes on physical properties of epoxy nanocomposites. *Carbon* 43:1378–85.

Sotiriou, G. A., Pratsinis, S. E. 2010. Antibacterial activity of nanosilver ions and particles. *Environmental Science and Technology*, 44:5649–54.

Srivastava, I., Rafiee, M. A., Yavari, F., Rafiee, J., and Koratkar, N. 2013. Epoxy nanocomposites: Graphene a promising filler. In: Mukhopadhyay, P., Gupta, R. K. (eds.), *Graphite, Graphene and their Polymer Nanocomposites*, Boca Raton, FL: CRC Press, Taylor & Francis Group, pp. 315–351.

Stankovich, S., Dikin, D. A., Dommett, G. H. B. et al. 2006a. Graphene-based composite materials. *Nature* 442:282–6.

Stankovich, S., Piner, R. D., Chen, X., Wu, N., Nguyen, S. T., Ruoff, R. S. 2006b. Stable aqueous dispersions of graphitic nanoplatelets via the reduction of exfoliated graphite oxide in the presence of poly(sodium 4-styrenesulfonate). *Journal of Materials Chemistry* 16:155–8.

Starkova, O., Chandrasekaran, S., Prado, L. A. S. A., Tölle, F., Mülhaupt, R., Schulte, K. 2013. Hydrothermally resistance thermally reduced graphene oxide and multi-wall carbon nanotube based epoxy nanocomposites. *Polymer Degradation and Stability* 98:519–26.

Steinhauser, D., Subramaniam, K., Das, A., Heinrich, G., Klüppel, M. 2012. Influence of ionic liquids on the dielectric relaxation behavior of CNT based elastomer nanocomposites. *Express Polymer Letters* 6:927–36.

Subramaniam, K., Das, A., Heinrich, G. 2011. Development of conducting polychloroprene rubber using imidazolium based ionic liquid modified multi-walled carbon nanotubes. *Composites Science and Technology* 71:1441–9.

Tang, B., Xu, H. 1999. Preparation, alignment, and optical properties of soluble poly(phenylacetylene)-wrapped carbon nanotubes. *Macromolecules* 32:2569–76.

Tang, Y., Hu, Y., Wang, S., Gui, Z., Chen, Z., Fan, W. 2002. Preparation and flammability of ethylene-vinyl acetate copolymer/montmorillonite nanocomposites. *Polymer Degradation and Stability* 78:555–9.

Tasis, D., Mikroyiannidis, J., Karoutsos, V., Galiotis, C., Papagelis, K. 2009. Single-walled carbon nanotubes decorated with a pyrene-fluorenevinylene conjugate. *Nanotechnology* 20:1–7.

Theng, B. K. G. 1979. *Formation and Properties of Clay–Polymer Complexes*. Amsterdam: Elsevier.

Thostenson, E. T., Ziaee, S., Chou, T.-W. 2009. Processing and electrical properties of carbon nanotube/vinyl ester nanocomposites. *Composites Science and Technology* 69:801–4.

Thumsorn, S., Yamada, K., Leong, Y. W., Hamada, H. 2011. Development of cockleshell-derived CaCO$_3$ for flame retardancy of recycled PET/Recycled PP blend. *Materials Sciences and Applications* 2:59–69.

Tong, L., Mazur, E. 2008. Glass nanofibers for micro- and nano-scale photonic devices. *Journal of Non-Crystalline Solids* 354:1240–4.

Tuğrul Seyhan, A., Gojny, F. H., Tanoğlu, M., Schulte, K. 2007. Critical aspects related to processing of carbon nanotube/unsaturated thermoset polyester nanocomposites. *European Polymer Journal* 43:374–9.

Tyner, K. M., Wokovich, A. M., Codar, D. E., Doub, W. H., Sadrieh, N. 2011. The state of nano-sized titanium dioxide (TiO$_2$) may affect sunscreen performance. *International Journal of Cosmetic Science* 33:234–44.

Uskov, I. A. 1960. Filled polymers. III. Polymerization of methyl methacrylate during dispersion of sodium bentonite. *Vysokomol Soed* 2:926–930.

Valodkar, M., Thakore, S. 2011. Organically modified nanosized starch derivatives as excellent reinforcing agents for bionanocomposites. *Carbohydrate Polymers* 86:1244–51.

Vavouliotis, A., Fiamegou, E., Karapappas, P., Psarras, G. C., Kostopoulos, V. 2010. DC and AC conductivity in epoxy resin/multiwall carbon nanotubes percolative system. *Polymer Composites* 31:1874–80.

Venkatanarasimhan, S., Raghavachari, D. 2013. Epoxidized natural rubber–magnetite nanocomposites for oil spill recovery. *Journal of Materials Chemistry A* 1:868–76.

Vennerberg, D., Rueger, Z., Kessler, M. R. 2014. Effect of silane structure on the properties of silanized multiwalled carbon nanotube–epoxy nanocomposites. *Polymer* 55:1854–65.

Vilgis, T. A., Heinrich, G., Klüppel, M. 2009. *Reinforcement of Polymer Nano-Composites*. New York: Cambridge University Press.

Visakh, P. M., Thomas, S., Oksman, K., Mathew, A. P. 2012. Cellulose nanofibres and cellulose nanowhiskers based natural rubber composites: Diffusion, sorption and permeation of aromatic organic solvents. *Journal of Applied Polymer Science* 124:1614–23.

Vollath, D. 2008. *Nanomaterials: An Introduction to Synthesis, Properties and Applications*. Weinheim: Wiley.

Wan, W., Yu, D., Xie, Y., Guo, X., Zhou, W., Cao, J. 2006. Effects of nanoparticle treatment on the crystallization behavior and mechanical properties of polypropylene/calcium carbonate nanocomposites. *Journal of Applied Polymer Science* 102:3480–8.

Wang, M.-J. 1998. The role of filler networking in dynamic properties of filled rubber. *Rubber Chemistry and Technology* 71:520–89.

Wang, Z. L. 2004. Nanostructures of ZnO. *Materials Today* 7:26–33.

Wang, J., Du, J., Zhu, J., Wilkie, C. A. 2002. An XPS study of the thermal degradation and flame retardant mechanism of polystyrene–clay nanocomposites. *Polymer Degradation and Stability* 77:249–52.

Wang, K., Chen, L., Wu, J., Toh, M. L., He, C., Yee, A. F. 2005. Epoxy nanocomposites with highly exfoliated clay: Mechanical properties and fracture mechanisms. *Macromolecules* 38:788–800.

Wang, L. L., Zhang, L. Q., Tiang, M. 2012. Mechanical and tribological properties of acrylonitrile-butadiene rubber filled with graphite and carbon black. *Materials and Design* 39:450–7.

Wheeler, P. A., Misra, R., Cook, R. D., Morgan, S. E. 2008. Polyhedral oligomeric silsesquioxane trisilanols as dispersants for titanium oxide nanopowder. *Journal of Applied Polymer Science* 108:2503–8.

Witt, N., Tang, Y., Ye, L., Fang, L. 2013. Silicone rubber nanocomposites containing a small amount of hybrid fillers with enhanced electrical sensitivity. *Materials and Design* 45:548–54.

Wu, D. C., Shen, L., Low, J. E. et al. 2010. Multi-walled carbon nanotube/polyimide composite film fabricated through electrophoretic deposition. *Polymer* 51:2155–60.

Wu, Y. P., Huang, H., Zhao, W., Zhang, H., Wang, Y., Zhang, L. 2008. Flame retardance of montmorillonite/rubber composites. *Journal of Applied Polymer Science* 107:3318–24.

Wu, J., Xing, W., Huang, G., Li, H., Tang, M., Wu, S., Liu, Y. 2013. Vulcanization kinetics of graphene/natural rubber nanocomposites. *Polymer* 54:3314–23.

Xia, F., Perebeinos, V., Lin, Y.-M., Wu, Y., Avouris, P. 2011. The origins and limits of metal–graphene junction resistance. *Nature Nanotechnology* 6:179–84.

Xia, Y., Yang, P., Sun, Y. et al. 2003. One-dimensional nanostructures: Synthesis, characterization, and applications. *Advanced Materials* 15:353–89.

Xing, W., Wu, J., Huang, G., Li, H., Tang, M., Fu, X. 2014. Enhanced mechanical properties of graphene/natural rubber nanocomposites at low content. *Polymer International* 63:1674–81.

Xu, S. H., Gu, J., Luo, Y. F., Jia, D. M. 2012. Effects of partial replacement of silica with surface modified nanocrystalline cellulose on properties of natural rubber nanocomposites. *Express Polymer Letters* 6:14–25.

Yang, J., Tiang, M., Jia, Q.-X. et al. 2007. Improved mechanical and functional properties of elastomer/graphite nanocomposites prepared by latex compounding. *Acta Materialia* 55:6372–82.

Yang, W., Ratinac, K. R., Ringer, S. P., Thordarson, P., Gooding, J. J., Braet, F. 2010. Carbon nanomaterials in biosensors: Should you use nanotubes or graphene? *Angewandte Chemie International Edition* 49:2114–38.

Yao, Z., Xia, M., Ge, L., Chen, T., Li, H., Ye, Y., Zheng, H. 2014. Mechanical and thermal properties of polypropylene (PP) composites filled with $CaCO_3$ and shell waste derived bio-fillers. *Fibers and Polymers* 15:1278–87.

Yousefi, N., Sun, X., Lin, X. et al. 2014. Highly aligned graphene polymer nanocomposites with excellent dielectric properties for high-performance electromagnetic interference shielding. *Advanced Materials* 26:5480–87.

Yue, L., Pircheraghi, G., Monemian, S. A., Manas-Zloczower, I. 2014. Epoxy composites with carbon nanotubes and graphene nanoplatelets – dispersion and synergy effects. *Carbon* 78:268–78.

Zabihi, O., Hooshafza, A., Moztarzadeh, F., Payravand, H., Afshar, A., Alizadeh, R. 2012. Isothermal curing behavior and thermo-physical properties of epoxy-based thermoset nanocomposites reinforced with $Fe_2O_3$ nanoparticles. *Thermochimica Acta* 527:190–8.

Zanetti, M., Costa, L. 2004. Preparation and combustion behaviour of polymer/layered silicate nanocomposites based upon PE and EVA. *Polymer* 45:4367–73.

Zanetti, M., Bracco, P., Costa. L. 2004. Thermal degradation behavior of PE/clay nanocomposites. *Polymer Degradation and Stability* 85:657–65.

Zhang, G., Karger-Kocsis, J., Zou, J. 2010. Synergetic effect of carbon nanofibers and short carbon fibers on the mechanical and fracture properties of epoxy resin. *Carbon* 48:4289–300.

Zhang, M. Q., Rong, M. Z., Ruan W. H. 2009. Nanoparticles/polymer composites: Fabrication and mechanical properties. In: Karger-Kocsis, J., Fakirov, S. (eds.), *Nano- and Micro- Mechanics of Polymer Blends and Composites*, Munich: Hanser, pp. 93–140.

Zhang, H., Tang, L.-C., Zhang, Z., Friedrich, K., Sprenger, S. 2008. Fracture behaviours of in-situ silica nanoparticle-filled epoxy at different temperatures. *Polymer* 49:3816–25.

Zhu, Y. D., Allen, G. C., Jones, P. G., Adams, J. M., Gittins, D. I., Heard, P. J., Skuse D. R. 2014. Dispersion characterisation of $CaCO_3$ particles in $PP/CaCO_3$ composites. *Composites: Part A* 60:38–43.

Zilg, C., Thomann, R., Finter, J., Mülhaupt, R. 2000. The influence of silicate modification and compatibilizers on mechanical properties and morphology of anhydride-cured epoxy nanocomposites. *Macromolecular Materials and Engineering* 280/281:41–6.

Zong, R., Hu, Y., Liu, N., Li, S., Liao, G. 2007. Investigation of thermal degradation and flammability of polyamide-6 and polyamide-6 nanocomposites. *Journal of Applied Polymer Science* 104:2297–303.

# 3 Synthesis of Nanomaterials

*A. Ramazani S.A., Y. Tamsilian, and M. Shaban*

## CONTENTS

## 3.1   INTRODUCTION

Nanomaterials, dimension size between 1 and 100 nanometers, are categorized in four sections [1]: (1) zero-dimensional (confined all dimensions in nanoscale such as clusters, quantum dots, and nanoparticles), (2) one-dimensional (confined two dimensions in nanoscale such as nanotubes, nanorods, and nanowires), (3) two-dimensional (confined one dimensions in nanoscale such as nanofilms, nanolayers, and nanocoatings), and (4) three-dimensional (confined no dimensions in nanoscale including nanocomposites, porous materials, powders, fibrous, multilayer, and polycrystalline materials) (see Figure 3.1).

   In general, several methods that have been mostly used for preparation of nanomaterials are categorized into two main approaches: top-down and bottom-up [2,3]. Top-down approach refers to reducing the dimension of the original materials using physical techniques. Bottom-up or chemical approach is a popular method in which nanomaterials could be produced from the atomic or molecular scale [4]. Figure 3.2 shows physical and chemical approaches to obtain nanomaterials.

## 3.2   ZERO-DIMENSIONAL MATERIALS

### 3.2.1   Synthesis of Metal Nanoclusters

Metal nanoclusters (NCs) containing a number of atoms are a new category of nanomaterials that cover the distance between metal atoms and nanoparticles (NPs). They have attracted a great deal of attention in recent years due to optical, electrical, chemical, and molecule-like features [5–8].

   Wang's group used a general approach for synthesis of nanocrystals such as noble metals [9]. This approach is using a general phase transfer and separation mechanism taking place at the interfaces of the liquid, solid, and solution phases existing through synthesis. In this technique, uniform

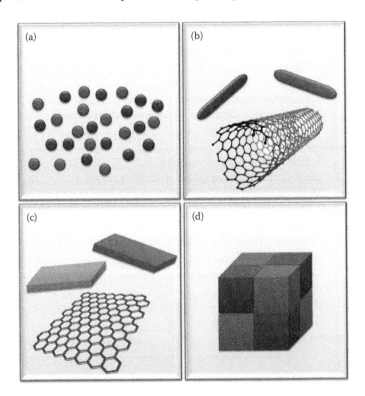

**FIGURE 3.1   (See color insert.)** Classification of nanomaterials (a) 0D spheres and clusters, (b) 1D nanofibers, wires, and rods, (c) 2D films, plates, and networks, and (d) 3D nanomaterials.

Physical method                    Chemical method

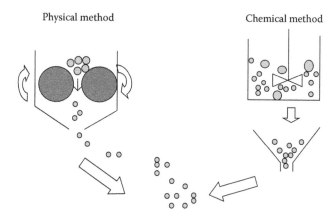

**FIGURE 3.2    (See color insert.)** Physical and chemical approach to produce nanomaterials. (Adapted from I. Rahman, V. Padavettan, *Journal of Nanomaterials*, 1, 2012, 1–15.)

noble metal quantum dots, or nanocrystals, were produced using the reduction of noble metal ions via ethanol at various temperatures of 20–200°C under hydrothermal or atmospheric situations.

Up to now, many researchers made considerable attempts at the synthesis of several noble metal nanoclusters such as Cu, Pd, Ag, Pt, and Au NCs [10–12], but between them, gold nanoclusters have received particular interest due to some good properties such as their facile synthesis, nontoxicity, outstanding photophysical features, and astonishing chemical stability. It is interesting to note that luminescent gold nanoclusters (AuNCs) are a new category of luminescent nanomaterials which have a particle size under 2 nm and contain about 100 gold atoms [13]. Various methods have been applied for preparation of uniform luminescent AuNCs with high quality.

It is well known that sodium borohydride ($NaBH_4$) is used normally for reduction of $HAuCl_4$ solution to Au nanoparticles (AuNPs) because the Au atoms have a great tendency to aggregation when a strong reducing agent such as $NaBH_4$ is utilized. Therefore, partially weak reducing agents could be applied to synthesize small luminescent AuNCs. Consequently, thiol-containing precursors are appropriate reducing agents and stabilizers for preparation of this kind of nanomaterial. Also the reaction conditions and synthesis processes could be controlled by different techniques [14].

### 3.2.1.1    Microwave-Assisted Synthesis

Microwave irradiation (MWI) is one of the most significant methods for the synthesis of nanomaterials because of several outstanding superiorities such as low energy consumption, uniform heating, environment-friendly features, and cost-effectiveness [15–17]. Furthermore, the main advantage of MWI in comparison to conventional heating is that this method is rapid and uniform.

Yue and coworkers prepared extremely fluorescent AuNCs containing 16 gold atoms using a one-step microwave-assisted technique with MWI for 6 h with power of 700 W. They utilized bovine serum albumin (BSA) as the reducing agent and the stabilizer which is reacted with $HAuCl_4$ as an Au precursor. In this method, the reaction was under particularly simple conditions (pH = 12) at 37°C via MWI as an alternative of direct heating to maintain the temperature [18].

Moreover, in this method the reaction time can be reduced from several hours to 1 h with modified processes using microwaves. For instance, Chen et al. applied eight cycles of consecutive microwave heating (5 min per cycle with the power of 90 W) to prepare the lysozyme-directed AuNCs [19].

A very rapid and strong microwave-assisted green synthesis of Ag nanoclusters with high fluorescent in the presence of a common polyelectrolyte, polymethacrylic acid sodium salt (PMAA-Na), was reported by Liu and coworkers [20]. In addition, we could reduce the reaction time to some minutes by controlling the irradiation power [21,22]. The microwave-assisted method provides fast and uniform heating; thus, it can speed up the synthesis of nanomaterials.

### 3.2.1.2  Sonochemical Synthesis

The sonochemical method is an additional important approach for synthesizing nanomaterials. This method has significant advantages such as fast reaction, harmless, easy to control the reaction conditions, and it produces monosize and uniform NPs with high purity [23].

In the following, we will explain the sonochemical method by an example of preparing highly water-soluble silver nanoparticles (AgNCs) by Liu et al. [23].

In this work, the BSA-stabilized AgNCs were synthesized by a facile, fast, green sonochemical synthesis method. In this method, briefly, 250 mg of BSA was dissolved in 9 mL water and mixed with aqueous $AgNO_3$ solution (1 mL, 100 mmol/L), reacting at ambient temperature for 5 min with vigorous stirring, then 0.50 mL, mol/L NaOH was added to the solution to adjust the pH to 12, where BSA acted as a stabilizing agent and reducing agent in this condition; finally, the mixture was exposed to ultrasonic irradiation (50 W/cm$^2$) under low temperature (15°C) for 4 h. During this period, the color of the colloid solution changed from colorless to yellow, providing clear evidence for the formation of AgNCs [24]. The AgNCs solution is purified via dialysis using 7000 Da molecular-weight cutoffs (MWCO) dialysis bag.

Sonochemical reduction of $Ag^+$ needs the use of a template or capping agent to avoid the aggregation of Ag nanoclusters to form large Ag nanoparticles. Xu et al. [25] used a simple polyelectrolyte, polymethylacrylic acid (PMAA), as a stabilizer agent. It should be noted that PMAA contains carboxylic acid groups, which have a great tendency to silver ions and silver surfaces; therefore, PMAA is a preferable capping agent for synthesis of Ag nanoclusters [26,27]. The charged carboxylate groups afford stability for Ag nanoclusters and avoid extra growth of nanoclusters to large nanoparticles.

To synthesize Ag nanoclusters, a fresh solution of $AgNO_3$ as $Ag^+$ precursor was mixed with an aqueous PMAA solution. In this study, the molar ratio of carboxylate groups (from the methacrylic acid units) to $Ag^+$ was 1:1. Then the pH was adjusted to 4.5 due to formation of compacted coil PMAA, which has been reported as the best reagent to produce Ag nanoclusters [25]. The solution was purged with Ar for 2 h and then sonicated for a different period of time. As shown in Figure 3.3a, under sonication, the first colorless solution slowly changes to pink (90 min) and then dark red (180 min). The producing Ag nanoclusters have a high fluorescence (Figure 3.3b). As it is illustrated in Figure 3.3c and e, the resulting Ag nanoclusters are less than 2 nm in diameter [25].

In addition to the methods of synthesis mentioned above, other techniques may also be used for preparing AuNCs, for example:

- *Photoreductive synthesis*: In this technique, ultraviolet light is employed as a reducing agent for preparation of AuNCs from Au(III) precursors.
- *Etching-based technique*: In this method, the AuNPs are etched with extra amounts of molecules, such as dendrimers, thiols, and $Au^{3+}$ ions to form the AuNCs [28].
- *Microemulsion method*: In this method, AuNCs is obtained by adding aqueous $NaBH_4$ solution to a microemulsion containing methanol, thiolates, and $HAuCl_4$ under vigorous stirring [29–32]. The microemulsion method will be explained in detail in the next section.

Figure 3.4 illustrates a schematic of different methods for preparation of AuNCs. As it is shown in this figure, $Au^{3+}$ precursors and appropriate stabilizer agents are essential for synthesis of fluorescent AuNCs [14].

### 3.2.2  SYNTHESIS OF QUANTUM DOTS

Semiconductor quantum dots (QDs) are nanoparticles with very high luminescence that have received great attention in bioanalysis, bioimaging, and optoelectronics. These colloidal nanocrystalline semiconductors are small and spherical particles or nanocrystals of a semiconducting material with diameters in the range of 1–12 nanometers (10–50 atoms). At such small sizes (smaller than

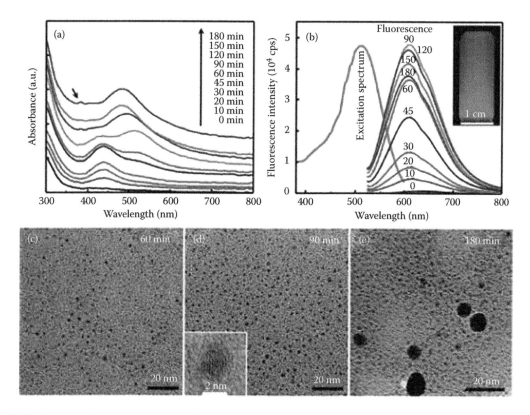

**FIGURE 3.3** (**See color insert.**) (a) UV–vis spectra and (b) fluorescence emission spectra of the solution containing PMAA and AgNO₃ in different length of sonication time; TEM images of as prepared Ag nanoclusters from different lengths of sonication: (c) 60 min, (d) 90 min, and (e) 180 min. (Reprinted with permission from Xu Hangxun, S. Kenneth Suslick, Sonochemical synthesis of highly fluorescent nanoclusters, *American Chemical Society*, 4, 3209–3214. Copyright 2010 American Chemical Society.)

the dimensions of the exciton Bohr radius), these nanocrystals act differently from bulk solids due to quantum confinement effects [33,34].

QDs have been prepared from a wide range of semiconductor materials. CdSe, CdTe, and their core/shell analogs, CdSe/ZnS and CdTe/ZnS, have received the most attention among QDs due to their well-established synthetic approaches [35].

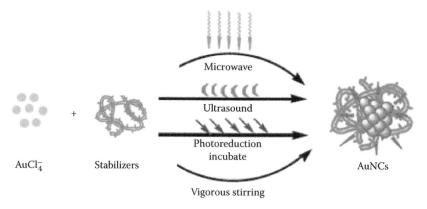

**FIGURE 3.4** (**See color insert.**) Synthesis of gold nanoclusters (AuNCs). (Adapted from M. Cui, Y. Zhao, Q. Song, *Trends in Analytical Chemistry*, 57, 2014, 73–82.)

However, there are frequent efforts and reports on preparation of QDs in aqueous media using suitable air-stable precursors, the actual breakthrough in preparation of highly fluorescent colloidal QDs happened when the Bawendi group reported [36] preparation of CdSe QDs with highly crystalline cores and size distributions of 8%–11% using a mixture of trioctyl phosphine/trioctyl phosphine oxide, TOP/TOPO with pyrolysis of organometallic precursors.

In the following, we will use this report as an example to exemplify the general method.

Dimethylcadmium (Me$_2$Cd) was utilized as the Cd precursor and bis(trimethylsilyl)sulfide ((TMS)$_2$S), trioctylphosphine selenide (TOPSe), and trioctylphosphine telluride (TOPTe) were used as S, Se, and Te sources, respectively. Mixed tri-n-octylphosphine (TOP) and tri-noctylphosphineoxide (TOPO) solutions were used as solvents and stabilizer agents known as coordinating solvents.

The process for synthesizing of TOP/TOPO capped CdSe nanocrystallites is briefly outlined below [36]. Fifty grams of TOPO is dried and purged in the reaction vessel via heating to 200°C at 1 torr for 20 min, and then it was purged with argon intermittently. The temperature of the reaction flask is then adjusted at 300°C under 1 atm of argon. One milliliter of Me$_2$Cd is added to 25.0 mL of TOP in the dry box, and 10.0 mL of 1.0 M TOPSe stock solution is added to 15.0 mL of TOP. Two solutions are then mixed and added to a syringe in the dry box. The heat is removed from the reaction vessel. The syringe holding the reagent mixture is rapidly removed from the dry box and its content is strongly stirred. The rapid introduction of these reagents results in an orange solution with an absorption feature at 440–460 nm. This is also accompanied by a sudden decrease in temperature to 180°C. Heating is restored to the reaction flask and the temperature is gradually raised to and aged at 230–260°C. Depending on the aging time, CdSe nanoparticles with a series of sizes ranging from 1.5 nm to 11.5 nm in diameter are prepared.

Produced colloidal dispersion is purified by cooling to 60°C, above the melting point of TOPO, and adding 20 mL of anhydrous methanol, which results in the reversible flocculation of the nanocrystallites. The flocculate and supernatant were then separated using centrifugation. Then flocculation was dispersed in 25 mL of anhydrous 1-butanol and a clear solution of nanocrystallites and a gray precipitate was obtained by additional centrifugation. In the next step, 25 mL of anhydrous methanol was added to the supernatant to flocculate the crystallites and remove excess TOP and TOPO. A final washing of the product with methanol followed by vacuum drying produces 300 mg of free flowing TOP/TOPO capped CdSe nanoparticles.

These CdSe nanocrystals are subsequently dispersed in anhydrous 1-butanol to form a transparent solution. Subsequently, methanol was added to the dispersion until opalescence persists. Finally, supernatant and flocculate were separated by centrifugation to produce a precipitate enriched of CdSe nanocrystallites. This dispersion and size-selective precipitation with methanol is repeated until no further narrowing of the size distribution as indicated by sharpening of the optical absorption spectrum.

A similar method joined with suitable organometallic reagents was used to coat the native CdSe core with different semiconducting, which has wider band gaps (e.g., ZnS and CdS) [37,38]. A summary of TOP/TOPO method is illustrated in Figure 3.5 [39].

It should be noted that QDs that are synthesized using this procedure are not aqueous soluble; therefore, phase transfer to aqueous solutions with surface modification using hydrophilic ligands or by encapsulating these nanocrystals in a thick hetero functional organic coating are required. These ligands facilitate both the colloid's solubility and act as a point of chemical attachment for biomolecules [40].

The microemulsion method is another prevalent approach for preparation of QDs at room temperature. In this technique, reactions occur among the reagents inside the micelles. The formation of micelles is explained in the following.

When surfactants or block polymers, which generally contain two parts of hydrophilic head and hydrophobic chains, are dissolved into a solvent, they typically self-assemble at air/aqueous solution or organic/aqueous solution interfaces. The hydrophilic chains are turned to the aqueous

(a)

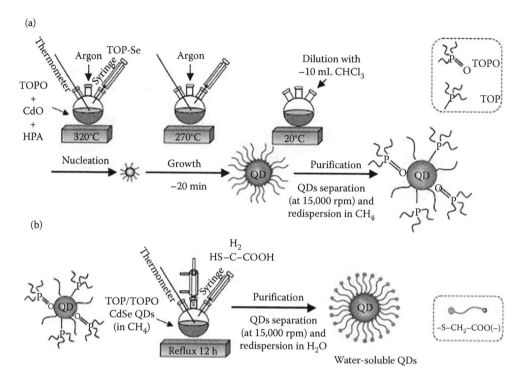

(b)

**FIGURE 3.5** (See color insert.) Schematic illustration of a typical synthesis process and surface-modification of a luminescent QD based on the use of CdO as precursor. (a) Nanoparticle synthesis and (b) surface modification. (Adapted from M. Costa-Fernandez, R. Pereiro, A. Sanz-Medel, *Trends in Analytical Chemistry*, 25, 2006, 207–218.)

solution. When the concentration of the surfactants or block polymers is higher than a critical amount, they self-assemble to create micelles. Surfactants or block polymers remain at the interface of organic and aqueous solutions. Finally, a microemulsion is a distribution of fine liquid droplets of an organic solution in an aqueous solution which could be considered as normal microemulsions (oil-in-water emulsion) or inverse microemulsions (water-in-oil emulsion). In some conditions, other polar solvents such as alcohol could be applied instead of water. The microemulsion method could be employed for the preparation of nanoparticles. The reverse micelle process received the most attention for preparation of QDs, which two immiscible liquids such as water and long-chain alkane are mixed to create the emulsion. Nanoscale water droplets dispersed in different organic solutions might be achieved using added surfactant such as aerosol OT (AOT), cetyl trimethyl-ammonium bromide (CTAB), or sodium dodecyl sulfate (SDS) ortriton-X.

These micelles are thermodynamically stable and could behave as "nanoreactors" [41]. The microemulsion technique needs the reaction of suitable initial resources, as an example, the reduction of metal ions or the decomposition of a single organometallic reagent could occur in the presence of a surfactant or polymer that avoids the particles growing and aggregating into bigger sizes. On the other hand, the nucleation and growth of the nanoparticles could happen in the presence of the surfactant. In this method, we could control size, shape, and feature of nanoparticles by adjusting the surfactant concentration and other factors such as reaction time, temperature, and concentrations of reagents [42].

### 3.2.3 SYNTHESIS OF NANOPARTICLES

The preparation procedures for nanoparticles (NPs) could be classified into three major collections. The first group includes the liquid-phase approaches, which employ chemical processes in solvents.

This results in colloids, in which the produced nanoparticles could be stabilized against agglomeration by surfactants or ligands.

The second group contains techniques based on surface growth under vacuum environments. In this method, diffusion of atoms or small clusters on appropriate substrates result in island creation, which can be consider nanoparticles. An important example of this method is Stranski-Krastanow, which is applied for growing III–V QDs. Finally, the last group is gas-phase preparation [43]. In the following, we will explain all these approaches by some example of metal, ceramic, and polymeric nanoparticles separately. First, we will emphasize the preparation of different sorts of nanoparticles using solution procedures. Synthesis of nanoparticles dispersed in a solvent is the most popular method, which has numerous advantages.

### 3.2.3.1  Synthesis of Metal Nanoparticle

Noble metal nanoparticles like gold, silver, and platinum have received great interest because of their size and shape and unique optoelectronic properties. These noble metal nanoparticles, especially gold nanoparticles, have obtained great attention for significant biomedical requests due to facile preparation, characterization, and surface modification. Preparation of noble metal nanoparticles such as Pt, Au, and Ag has attracted the most attention in recent decades. The most prevalent methods are chemical reduction, physical processes, and biological approaches.

#### 3.2.3.1.1  Chemical Reduction

The general procedure in the preparation of metal colloidal dispersions is reduction of metal complexes in dilute solutions, and different approaches have been advanced to start and adjust the reduction reactions [44]. Since this method is the most common method of synthesis for AuNP, we will exemplify it by the synthesis of AuNP in the following.

The formation of monosized AuNP is achieved by the reduction of gold salts in the presence of a reducing agent such as sodium citrate and a stabilizer [45]. Sodium citrate reduction of chlorauric acid ($HAuCl_4$) at 100°C was established more than 50 years ago [46] and remains the most frequently used technique. In this technique, $HAuCl_4$ dissolves into water to form a sufficiently dilute solution. Then sodium citrate is added into the boiling solution. The solution is kept at 100°C until color changes, while maintaining the overall volume of the solution by adding water. This synthesis of citrate stabilized AuNPs was based on a single-phase reduction of $HAuCl_4$ by sodium citrate in an aqueous medium and produced particles about 20 nm in size.

One of the most important reports on synthesis of AuNP was published in 1994 and is known currently as the Brust-Schiffrin method [47]. This method employed a two-phase synthesis that applied thiol ligands which could attach to gold intensely because of the soft properties of both S and Au. First, gold precursor is added to an organic solvent such as toluene with a phase transfer agent like tetraoctylammonium bromide, and then an organic thiol is added. Finally, an extra amount of an intense reducing precursor, such as sodium borohydride, is added to create AuNPs capped with thiol ligands [47]. The main benefits of this technique are the facile synthesis, uniform size distribution, thermally stable NPs, and ease of size adjustment [48].

Silver nanoparticles are an additional distinguished sample of chemical reduction technique. The easiest approach to produce silver nanoparticles is using reduction of silver nitrate ($AgNO_3$) as $Ag^+$ precursor in ethanol in the presence of a surfactant [49]. Aggregation of Ag NPs could be prohibited by using some common stabilizers agents such as poly(vinylpyrrolidone) (PVP), bovine serum albumin (BSA), polyvinyl alcohol (PVA), and citrate and cellulose.

Dong's group [50] reported work on controlling the size and morphology of silver nanoparticles via sodium citrate reagent in the pH range of 5.7–11.1. Reduction of the silver salt ($Ag^+$) was increased by increasing the pH, due to the higher activity of the citrate reagent in high pH value. It is interesting that at higher pH, the morphology of nanoparticles was a mix of spherical and rod-like due to the rapid reduction rate of the precursor. However, in lower pH value, morphology of

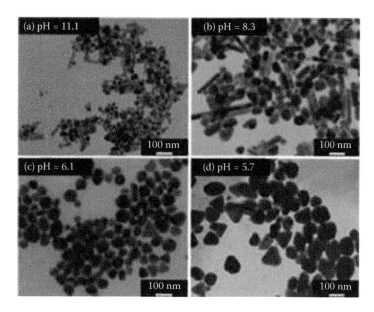

**FIGURE 3.6** Transmission electron images for the silver nanoparticles synthesized under pH values of 11.1, 8.3, 6.1, and 5.7. (Adapted from X. Dong et al., *The Journal of Physical Chemistry C*, 113, 2009, 6573–6576.)

nanoparticles was mostly triangle or polygon because of the slow reduction rate of the precursor. Figure 3.6 illustrates the morphology of produced nanoparticles in different pH values.

### 3.2.3.1.2 Direct Laser Irradiation

Some other significant methods to prepare noble metal nanoparticles are UV irradiation, laser irradiation, and microwaves, which may not use reducing agents. For example, a new technique was reported [51] to prepare silver nanoparticles with a well-defined size and shape distribution using direct laser irradiation of an aqueous solution including a silver precursor and a surfactant without any reducing agents. The main purpose of this report was suggestion of a new technique for the synthesis of metallic nanoparticles using the laser irradiation of a metallic precursor solution without any reducing agent. In this work, about 3 mL of an aqueous solution of silver nitrate with concentrations of 0.833–4.166 mM was placed in a closed spectrophotometric cuvette and irradiated by laser beam in the presence of the surfactant sodium dodecyl sulfate (SDS) [51].

The mechanism of this synthesis method is explained by formation of radicals in the aqueous solution due to laser irradiation of the metal salt precursor in the absence of any reducing agent [52].

Photochemical reduction of gold precursor has also been applied to preparation of AuNPs [53]. This technique uses a continuous wave UV irradiation (250–400 nm), PVP as the stabilizer agent, and ethylene glycol as the reducing agent. The preparation rate of AuNPs with this technique is reliant on the glycol concentration as well as the viscosity of the solvent mixture [53].

### 3.2.3.1.3 Biological Methods

Another interesting method for nanoparticle fabrication is using biomolecules extracted from plants as a reducing agent of metal ions in a single-step green synthesis process. This method has a lot of significant advantages such as fast reaction, easy to perform in room temperature and pressure, and easily scaled up. It should be noted that preparation with plant extracts is an environmentally friendly approach. The extracted reducing agents comprise several water-soluble plants like alkaloids, phenolic compounds, terpenoids, and coenzymes. Silver (Ag) and gold (Au) nanoparticles have received the most important attention in the field of plant-based syntheses methods. Extracts of a different plant species have been effectively applied in producing nanoparticles. In addition to plant extracts, live plants can be employed for the preparation of nanoparticles [54]. Up to now, different biomolecules such as microorganisms, plant tissue, their fruits, plant extracts, and marine algae [55] have been applied to preparation of nanoparticles. Biological synthesis methods are valuable not only due to their decreased environmental effect [56] contrary to many physicochemical synthesis approaches, but also because they could be applied to yield a lot of nanoparticles without any contamination with a suitable size and morphology [57].

It is claimed that biosynthetic approaches can really produce nanoparticles with even better size and morphology than some of the physicochemical techniques [58].

In preparation of nanoparticles by plant extracts, the extracted agent is easily mixed with a solution of the metal precursor at room temperature. The reaction is complete within minutes. Different nanoparticles of silver, gold, and many other metals have been prepared with this method [59].

Several plants such as tea (*Camellia sinensis*), aloe vera, neem (*Azadirachta indica*), and *Catharanthus roseus* have been applied for the biosynthesis of nanoparticles.

In this method, the production rate and properties of nanoparticles depend on some factors such as the nature of the plant extract, concentration of extract and metal precursor, the pH value, temperature, and reaction time [59].

### 3.2.3.2  Synthesis of Ceramic Nanoparticles

Recently, a new group of nanomaterials, ceramic materials, for biomedical application are developing rapidly. Nanoscale ceramics such as hydroxyapatite (HA), zirconia ($ZrO_2$), silica ($SiO_2$), titania ($TiO_2$), and alumina ($Al_2O_3$) have received great attention to produce new synthetic methods to increase their physical–chemical features toward decreasing their cytotoxicity in biological applications [60].

#### 3.2.3.2.1  *Sol–Gel Method*

The sol–gel process is the most widely used technique to prepare pure silica particles because of some significant advantages such as easy to control the particle size, size distribution, and morphology nanoparticles by systematic adjustment of reaction condition [61]. For a long time, the sol–gel approach was broadly used to prepare silica, glass, and ceramic materials because of its ability to form pure and homogenous products in moderate situations. The sol–gel method is a wet chemical procedure for the preparation of colloidal dispersions of inorganic and organic–inorganic hybrid materials, especially oxides and oxide-based hybrids.

This procedure includes hydrolysis and condensation of metal alkoxides ($Si(OR)_4$) like tetraethylorthosilicate (TEOS, $Si(OC_2H_5)_4$) or inorganic precursors such as sodium silicate ($Na_2SiO_3$) in the presence of mineral acid (e.g., HCl) or base (e.g., $NH_3$) as catalyst [62–64]. A general flowchart for the sol–gel procedure, which leads to the fabrication of silica using silicon alkoxides ($Si(OR)_4$), is illustrated in Figure 3.7.

The general reactions of TEOS make the formation of silica particles in the sol–gel procedure as the following reaction [61].

$$Si(OC_2H_5)_4 + H_2O \xrightarrow{\text{Hydrolysis}} Si(OC_2H_5)_3OH + C_2H_5OH \tag{3.1}$$

$$\equiv Si-O-H + H-O-Si \equiv$$

$$\xrightarrow{\text{Water condensation}} \equiv Si-O-Si \equiv + H_2O \tag{3.2}$$

$$\xrightarrow{\text{Alcohol condensation}} \equiv Si-O-Si \equiv + C_2H_5OH \tag{3.3}$$

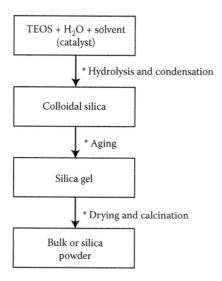

**FIGURE 3.7** Flowchart of a typical sol–gel process.

Alumina is one of the most important materials as inert biomaterials applied in implants that has been produced by the sol–gel technique.

It is therefore a biodegradable material, well endured by the biological environment. Different work on producing $Al_2O_3$ by the sol–gel technique using various precursors such as aluminum tri-isopropylate in a hydrolysis system consisting of octanol and acetonitrile [62], aluminum nitrate—in aqueous medium [63,64], aluminum secondary butoxide—in an alcoholic medium [65] have been reported thus far. To a general group of precursors include inorganics such as aluminum chloride ($AlCl_3$) and organics such as aluminum triisopropylate ($(C_3H_7O)_3Al$ [66] have been used for synthesis of $Al_2O_3$ by the sol–gel method.

We explain the synthesis of $Al_2O_3$ by the sol–gel technique with an example of using an inorganic precursor such as $AlCl_3$ in the following.

In this method, at first a solution of 0.1 M $AlCl_3$ in ethanol should be prepared. Then a 28% $NH_3$ solution was added to form a gel. This gel was kept at room temperature for 30 h and then dried at 100°C for 24 h.

Another example is using $C_9H_{21}AlO_3$ as a precursor, again first 0.1 M $(C_3H_7O)_3Al$ solution in ethanol was prepared, then $NH_3$ solution (28%) was added in order to form a gel. It was put under slow stirring at 90°C for 10 h. This gel was kept at room temperature for 24 h and then dried at 100°C for 24 h. The produced gels were calcined in a furnace for 2 h (heating rate 20°C/min), at temperature values of 1000°C and 1200°C.

### 3.2.3.2.2 Coprecipitation Method

Coprecipitation (CPT) method is applied by a precipitate of materials, which is generally soluble under the conditions employed. Magnetite nanoparticles were prepared by the chemical coprecipitation technique by using ammonium hydroxide as the precipitating reagent.

We could adjust the particle size by changing the temperature of reaction and also by surface functionalization.

Nano- and micro-sized magnetic particles have received great attention in the field of biomedical applications and magnetic recording [67].

In the following, we will exemplify the coprecipitation method by synthesis of functionalized magnetic core–shell $Fe_3O_4@SiO_2$ nanoparticles [68]. $Fe_3O_4$ nanoparticles were produced by the coprecipitation of $Fe^{2+}$ and $Fe^{3+}$ precursor (molar ratio 1:2) in an alkali solution. A black precipitation of $Fe_3O_4$ was obtained and then was constantly stirred for 1 h at room temperature and then heated to 80°C for 2 h. The produced $Fe_3O_4$ was collected by a permanent magnet after washing with deionized water and EtOH. Subsequently, it was dried at 100°C in a vacuum for 24 h. in the next step, silica coated magnetic nanoparticles was synthesized, in this way, 1 g of produced $Fe_3O_4$ under ultrasonic bath for 1 h at 40°C was dispersed in 80 mL of methanol. Then, concentrated ammonia solution was added to this dispersion and stirred at 40°C for 30 min. Then, tetraethylorthosilicate (TEOS, 1.0 mL) was added to the obtained mixture, and continuously stirred at 40°C for 24 h. Finally, the silica coated magnetic nanoparticles were collected by a permanent magnet, then washed with and dried at 60°C in vacuum for 24 h.

### 3.2.3.2.3  Aerosol Method

The aerosol-based method is one of the most attractive and simple methods for synthesis of nanoparticles. In contrast to the conventional solution methods, the aerosol process has many advantages, such as fast particle production, low fabrication cost, green method, simple and cheap particle collection, and produce material with high purity. Generally, this technique is based on homogeneous nucleation, condensation, and consequent coagulation in the continuous gas phase. Nanoparticles can also be produced by the ablation of a solid source with a pulsed laser [69].

Spray drying is another aerosol-based method that produces nanoparticles with a few process steps and high purity. In this method, liquid precursors were atomized directly into the furnace with high temperature to evaporate the solvent and resulting nanoparticles were collected on the filter. It should be noted that this gas phase method has the ability to produce dry powder with mesoporous structure and spherical shape [70].

In addition to the aforementioned methods, aerosol-polymerization is a simple and novel method to produce core–shell nanoparticles, consisting of an inorganic scaffold and polymeric shell, in a continuous gas-phase process. In this method, the polymerization was initiated in "flight" and avoids the need of surfactants and solvent. Furthermore, the resulting core–shell nanoparticles, leaving the aerosol set-up, were collected directly on the filter. Recently, Poostforooshan et al. have published an article about the in situ coating of inorganic nanoparticles with a polymer shell in a continuous aerosol-based synthesis [71].

In this method, silver and silica nanoparticles were initially produced in the gas-phase as inorganic cores by spark discharge and nebulization, respectively, and the particle-laden nitrogen gas flow was then passed through the saturator containing the glycidyl methacrylate (GMA) as organic monomer at 80°C. Behind the saturator, when the gas temperature dropped to room temperature, a super saturation was achieved resulting in the heterogeneous condensation of GMA vapor on the inorganic nanoparticle surface. Subsequently, the monomer coating was chemically polymerized to form a solid polymer shell by addition of ammonia vapor as initiator. The average aerosol residence time in the reactor is about 2 min. The continuous experimental setup consists of four main components: a core particle generator, a GMA monomer saturator, an ammonia addition, and a polymerization reactor, as illustrated in Figure 3.8.

### 3.2.3.3  Synthesis of Polymeric Nanoparticles

Recently, polymer nanoparticles have been considered in numerous applications. Two main procedures are used for their synthesis: (1) polymerization of monomers and (2) dispersion of synthesized polymers. There are various techniques to produce polymer nanoparticles including solvent evaporation, salting out, dialysis, supercritical fluid technology, microemulsion, miniemulsion, surfactant-free emulsion, and interfacial polymerization. A number of factors such as particle size, particle size distribution, and area of application as final product properties should be considered to affect choice of synthesis methods. In the next sections, methods for preparation of such particles will be discussed.

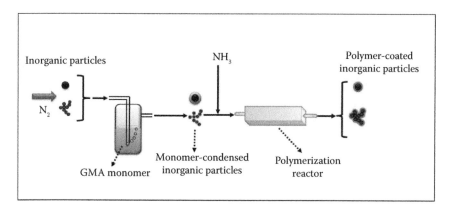

**FIGURE 3.8** Scheme of the experimental setup for encapsulation of inorganic nanoparticles with a polymer shell in a continuous aerosol-based process.

### 3.2.3.3.1 Emulsion Polymerization Method

The inverse emulsion polymerization of polymer nanoparticles is carried out in a 250 mL five-neck reactor equipped with a mechanical stirrer, heating system, reflux condenser, and submicron dropping injector for reaction ingredients and nitrogen gas as shown in Figure 3.9. A typical recipe for the inverse emulsion polymerization is presented in Table 3.1.

Tamsilian and Ramazani S.A. have synthesized polymeric core–shell nanostructure by in situ inverse miniemulsion to use in the enhanced oil recovery (EOR) process [72–78]. Their works are divided into four categories: (1) synthesis of polyacrylamide (PAM) nanoparticles [72,73], (2) synthesis of PAM-polystyrene (PS) core–shell nanostructure [74,75], (3) synthesis of PAM/PAM-c-PS [76,77], and (4) synthesis of thermoviscosifying polymer (TVP)/PAM-c-PS [78].

To produce PAM NPs based on inverse emulsion polymerization procedure, first, organic solvent and dispersion stabilizer with a modified hydrophilic–lipophilic balance (HLB range: 4–6) are introduced into the five-neck flask. The flask is then placed in a water bath at a constant temperature of 30°C with a stirring rate of 2000 rpm. After complete dissolution of the dispersion stabilizer in

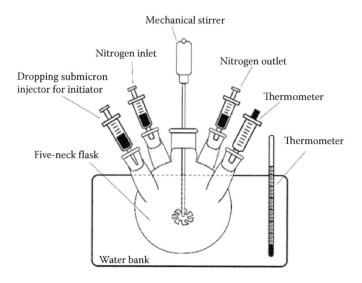

**FIGURE 3.9** The schematic diagram for inverse emulsion polymerization.

**TABLE 3.1**

**Typical Recipe for the Inverse Emulsion Polymerization of Polyacrylamide**

| Ingredients | Weight% |
|---|---|
| Acrylamide monomer | 15 |
| Water | 35 |
| Hexane solvent | 45 |
| Span80 surfactant | 5 |
| AIBN initiator | 0.0068 |

the solvent for 1 h under nitrogen gas bubbling, hydrophilic monomer such as polyacrylamide dissolved in deionized water is dripped into the reactor by submicron dropping injector within a specified time to form inverse water in oil (W/O) emulsion. Then, temperature and stirrer speed are set at 60°C and 400 rpm, respectively, and azobisisobutyronitrile (AIBN) initiator solution in hexane is then charged into the reaction mixture. The polymerization reaction is performed for 150 min at 60°C and stirrer speed of 400 rpm [72,73].

To synthesize PAM-PS core–shell nanostructure, 60 mL of hexane solvent and 0.0035 mL of span80 surfactant are mixed in a reactor with three necks, and after mixing those by using a mechanical mixer with speed around 2000 rpm, the water phase, including 5 g hydrophilic monomer of acrylamide and 20 mL of deionized water, is dispersed to the previous solution. Then, the water phase is injected to the mixed organic phase in the reactor via a microinjection. After the mixing water and organic phases, the initiator system (redox), including ferrous sulfate and potassium persulfate, is injected (5:2 ratio) in temperature condition –15°C by entering this material into the reactor. The first time period of polymerization is selected that the reactor remains under mentioned conditions for 30 min and after that immediately is moved to the very low temperature condition for 3 or 4 days without mechanical mixing. After this time, nanoparticles of polyacrylamide with high molecular weight are produced and are the time to inject the second initiator and monomer of nanolayer with hydrophobicity properties. Therefore, in the second steps of the process, the redox initiator and styrene monomer are injected for making a nanolayer in low temperature conditions. It is interesting to point out that the initiator of the second steps and the organic monomer must be injected simultaneously, and the initiator on the surface of the polymer nanoparticle causes a chain of polystyrene to fprm, and this chain is propagated continuously. The transfer of the chain must be done in a special time and extra propagation of a polymer nanolayer chain is prevented. In this way, by considering the short time for a second polymerization process (about 30 min), the thin layer of polystyrene is made on nanoparticles of polyacrylamide and after this time, the reaction process is terminated. Further purification must be considered to prevent core–shell nanostructure agglomeration. Finally, the synthesized powders have core–shell nanostructure that its nanocore of polyacrylamide with 10 million Dalton (molecular weight) and its size is 80 nanometers, and its shell is a nanolayer of polystyrene with 40,000 Dalton (molecular weight) and its size is 10 nanometers.

As mentioned before, a third category is due to synthesize core–shell nanostructure with binary properties. This type of core–shell nanoparticle is the future work of Ramazani S.A. and coworkers. The object of this section is design and preparation of a smart system, including a core made up of a hydrophilic polymer nanoparticle such as PAM, HPAM, TVP, and a hydrophilic–hydrophobic coating nanolayer (surfactant properties) such as PAM-b-PS copolymer made up of long hydrophilic and short hydrophobic blocks that are grafted chemically or inverse polymerized surfactant (surfmer). Two procedures to prepare the smart system are described. The first one involves inverse emulsion polymerization of polymeric cores using the surfmer as the emulsifier, whereas the shell is obtained via polymerization using the surfmer as the hydrophobic monomer of the shell. Consequently, the

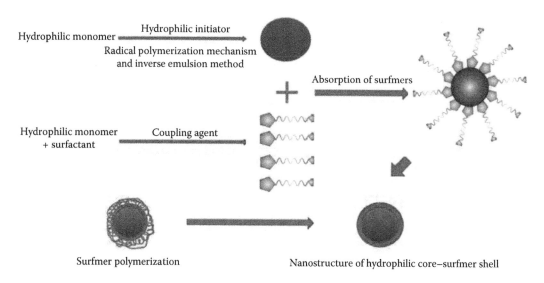

**FIGURE 3.10** (**See color insert.**) The schematic diagram for inverse emulsion polymerization of hydrophilic polymer–surfmer core–shell nanostructure. (Adapted from Y. Tamsilian, A. Ramazani S.A., *The International Conference on Nanotechnology: Fundamentals and Applications (ICNFA 2013)*, Canada, 2013.)

term nanostructure of hydrophilic core–surfmer shell is proposed to demonstrate the application of the double function surfactant-monomer (Figure 3.10).

The second one relates to a nanostructure of the hydrophilic core-block copolymer shell. The polymer cores are obtained via inverse emulsion polymerization using some hydrophilic monomers that are polymerized and terminated, whereas the shell is obtained via polymerization using other live radical chains which are propagated without any termination agents. Hydrophobic monomers are grafted with the hydrophilic open end chains to produce the monolayer of block copolymers (Figure 3.11).

This core–shell system consists of hydrophilic polymer-surfmer or hydrophilic polymer-block copolymer nanocomposites and not only acts as a protective layer for water soluble polymer from

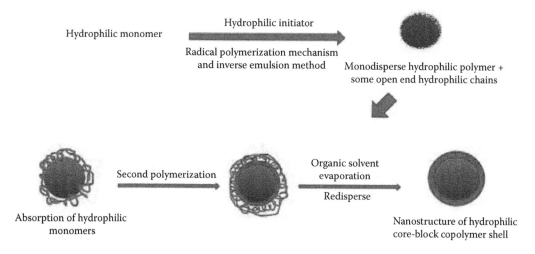

**FIGURE 3.11** (**See color insert.**) The schematic diagram for inverse emulsion polymerization of hydrophilic polymer-block copolymer core–shell nanostructure. (Adapted from Y. Tamsilian, A. Ramazani S.A., *The International Conference on Nanotechnology: Fundamentals and Applications (ICNFA 2013)*, Canada, 2013.)

degradations but also targeting delivery is another approach in some applications such as the enhanced oil recovery process. The investigation and characterization of the core–shell nanoparticle properties are done by DSC, IR, SEM, EDX, NMR, UV, XPS, and AES. It can be seen that the nanoscale coating of hydrophilic polymer nanoparticles is successfully done and there is no effect of virgin monomers in the reaction environment. In addition, the behavior of polyacrylamide release from its nanolayer coating and effects on rheological properties of the water phase in underground reservoirs with wettability variable is investigated via dilute viscometer. The obtained results show that the dissolution of pure polyacrylamide with 6 million Dalton (molecular weight) under the temperature 90–100°C takes 6 days (Figure 3.12a), but the total release of nanoparticles of polyacrylamide from a nanolayer of polystyrene to dissolve in the water phase under similar condition takes 21 days (Figure 3.12b).

Due to the initial experiments by Tamsilian and Ramazani S.A., the obtained results show that the core–shell nanostructure idea can considerably optimize the EOR process and remove the existing pitfalls. However, viscosity of aqueous polymer solution including PAM nanoparticles as core nanomaterial in the previous works decrease by temperature elevation due to PAM chain degradation. So, the best selection for the core material could be thermoviscosifying polymers (TVPs) whose viscosity increases on increasing temperature and salinity to optimize mobility ratio in the oil reservoir, may overcome the deficiencies of most water soluble polymers during EOR. In last study by the mentioned group, a novel core–shell nanostructure of TVP as an active thermosensitive polymer and an organic material as a nanolayer for the core material protection is designed and prepared to intelligently control mobility ratio in oil reservoirs and overcome weakness and limitations of classical polymer flooding such as mechanical, bacterial, or thermal degradations. This intelligent nanostructure releases TVP in oil–water interface after dissolution of organic nanolayer in oil phase, which could result in dramatically increasing water viscosity in the interface layer. This study is divided into three phases, including (1) synthesis of TVP nanocore-organic nanoshell, (2) characterization and rheology studies to investigate the structure of the synthesized nanomaterials and thermo-thickening properties, respectively, and finally (3) smart EOR study by the prepared materials [78].

Another example of polymer NP synthesis by emulsion mechanism is starch nanoparticles. A starch solution is prepared by dissolution of 0.5 g native sago starch powder in 50 mL of NaOH solution with concentration of 0.5 M. The mixture is heated to 80°C for 1 h under magnetic stirring

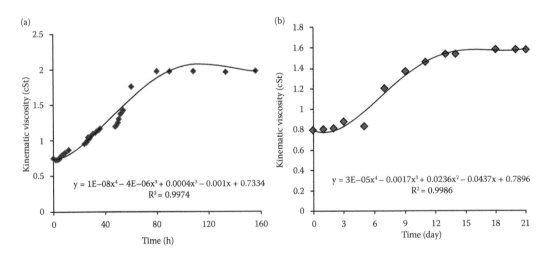

**FIGURE 3.12** Viscosity of (a) PAM-PS and (b) pure polyacrylamide in water phase versus time. (Adapted from Y. Tamsilian, A. Ramazani S.A., Smart polymer flooding process, U.S. Patent, US20150148269, United States, 2015.)

(a)  (b)  (c)

**FIGURE 3.13** SEM micrographs of (a) native sago starch; starch nanoparticles prepared (b) without stirring; and (c) with stirring rate at 900 rpm. (Adapted from Y. Tamsilian, A. Ramazani S.A., *Scientia Iranica F*, 21, 2014, 1174–1178.)

to obtain a homogeneous starch solution. One milliliter of starch solution after cooled to room temperature is added drop-wise to an organic phase (e.g., 15 mL of cyclohexane, 5 mL of ethanol, and a certain amount of surfactant) under mixing at 900 rpm for 1 h. The same procedure is repeated by varying the surfactant's concentrations, oil-co-surfactant ratios, oil phases (hexane, olein palm oil, sunflower oil, and oleic acid), cosurfactants (methanol, propanol, butanol, and acetone), and water/oil ratios [79].

Based on Figure 3.13a, native sago starch particles are mostly of large, oval granular shape with smooth surface in the range of a diameter size around 20–40 μm. The microsizes of starch granules have converted into nanoparticles by dissolution of the prepared native starch into aqueous solution and also reprecipitation of the starch solution into ethanol. Thus, the nanoprecipitation technique is suitable for nanoparticle formation. The mixing rate can affect the morphology of starch nanoparticles to aggregate and have wider particle size distribution (Figure 3.13b) as compared to using mixing during the synthesis (Figure 3.13c). At low mixing rates, the nucleation species are not dispersed uniformly through the solution; leading to agglomerate particles and higher stirring rates, enhanced mobility of nucleated species causes to be uniform, homogeneous dispersion and smaller nanoparticles [80].

### 3.2.3.3.2 Microfluidic Method

This section shows a fantastic tubing technique to irreversibly interconnect polydimethylsiloxane (PDMS) microfluidic devices with external equipment. This structure can keep a pressure of up to 4.5 MPa by experimental and theoretical investigations [81].

During leakage tests, the microfluidic chip has a straight channel with dimensions $50 \times 50 \times 50 \ \mu m^3$ (Figure 3.14a). To prepare the nanoparticles, the semicircular channel is 300 and 50 μm for wide and deep dimensions, respectively, and also 5 cm for the total length. Degassed PDMS is injected into the mold and heated to 80°C for 2 h. The PDMS slab is removed from the silicon substrate by a blade, and then a flat-tipped needle is used to make a hole with diameter 0.5 mm in the channel inlet. The PDMS chip is pressed down in a Petri dish and contact between the embedded microchannel and the dish surface is made. An adhesive sealant is smearing the end of the tube before inserting a plastic tube into the hole (Figure 3.14b). Another PDMS layer on top of the PDMS chip is placed to protect this tubing interconnection and then the device placed in the Petri dish is baked at 80°C for 2 h.

Poly(lactic-co-glycolic acid) (PLGA) nanoparticles is synthesized with varying flow rate rations (FRs) based on the fabricated microfluidic device. Small size of PLGA nanoparticles, approximately 55 nm, is synthesized at a high flow rate. The sizes of the nanoparticles show a variation by adjusting the FR (Figure 3.15) and a good dispersion of PLGA nanoparticles are presented by TEM and DLS results. Table 3.2 shows FR effects on particle size and polydispersity (PDI). This size difference is because of diffusion mixing against convective mixing. The convective mixing

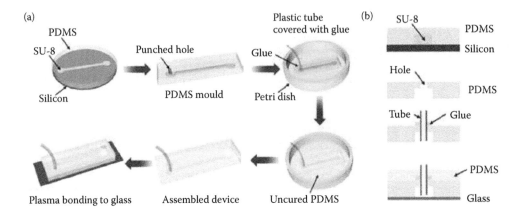

**FIGURE 3.14** (**See color insert.**) (a) Schematics of fabricating a tubing interconnection for PDMS micro-fluidic devices. (b) A cross-sectional view of the tubing interconnection fabrication. (Adapted from J. Wang et al., *Lab on Chip*, 14, 2014, 1673–1677.)

can provide a rapid interfacial deposition of small size PLGA nanoparticles. Moreover, choice of the flow rate can directly influence the particle size distribution to prepare a narrow size distribution by high flow rate and vice versa.

### 3.2.3.3.3  Grafting Method

Grafting of N-isopropylacrylamide and acrylic acid to produce poly(NIPAM-co-AA) nanoparticles as thermosensitive polymer is synthesized by grafting onto route [82] (Figure 3.16). In this way, NIPAM monomer is grafted onto AA and is subsequently conjugated with potassium persulfate (KPS) initiator at 80°C to synthesize poly(NIPAM-co-AA) nanoparticles.

The solution of poly(NIPAM-co-AA) nanoparticles is significantly affected by temperatures. The obtained results represent that the solution of poly(NIPAM-co-AA) nanoparticles at 37.2°C is opaque, whereas it changes to become transparent at room temperature. Ultraviolet (UV)-visible results in Figure 3.17 demonstrate that absorbance of poly(NIPAM-co-AA) nanoparticles is abruptly increased over 37.2°C, whereas the absorbance below 37.2°C is steadily constant, showing low critical solution temperature (LCST) in 37.2°C.

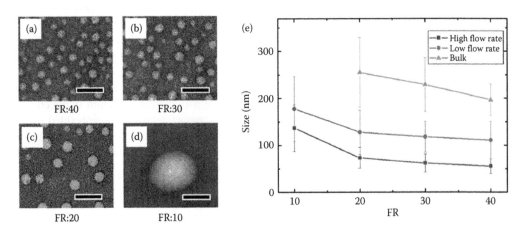

**FIGURE 3.15** (**See color insert.**) TEM images of PLGA nanoparticles with (a) FR: 40, (b) FR: 30, (c) FR: 20, (d) FR: 10, and (e) size distribution of PLGA nanoparticles versus FR. (Adapted from J. Wang et al., *Lab on Chip*, 14, 2014, 1673–1677.)

**TABLE 3.2**

**Comparison of Precipitated PLGA Nanoparticles Using Microfluidic (High and Low Flow Rates)**

|  | Total Flow Rate (mL/h) | Minimum Size (nm) | PDI |
|---|---|---|---|
| Microfluidics | 41–44 | 110 | 0.2 |
| Microfluidics | 410 | 55 | 0.1 |

*Source:* Adapted from J. Wang et al., *Lab on Chip*, 14, 2014, 1673–1677.

### 3.2.3.3.4 In-Situ Polymerization Method

Most researchers have good attention to polymer/clay nanocomposites in research and industrial applications due to the effective optimization of different properties of polymer nanoparticles. These nanocomposites show improvements to mechanical and barrier properties, flammability and solvent resistance, and environmental stability [83–86]. Ramazani S.A. and coworkers have synthesized polyethylene/clay nanocomposites from mineral clay by Ziegler-Natta catalyst system with exfoliated structure of clay. To achieve this goal, they have used an acid treatment method to improve clay structure for production of highly active catalyst to produce polyethylene clay nanocomposites [87–89].

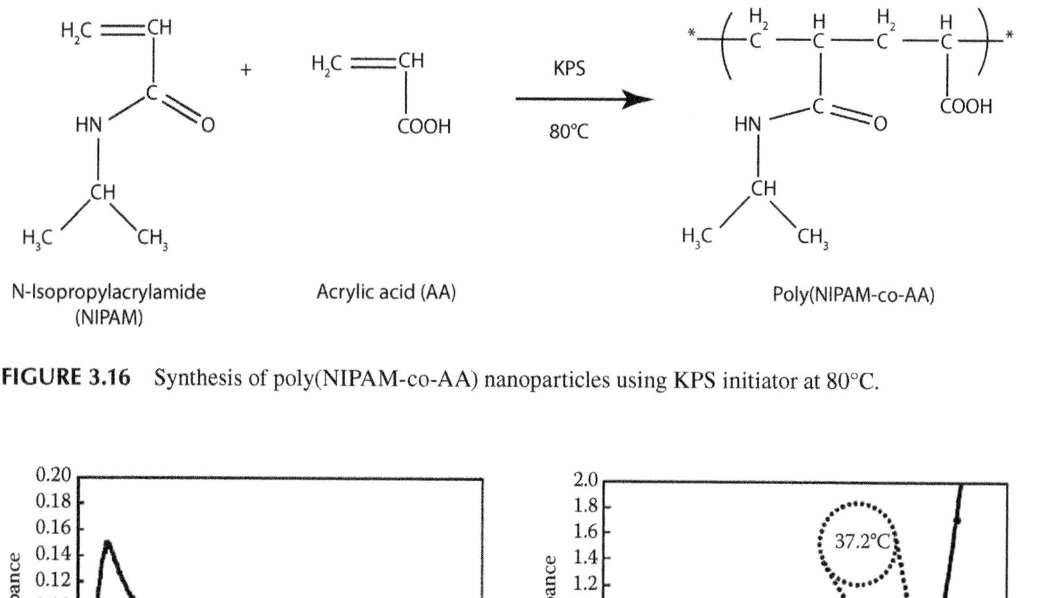

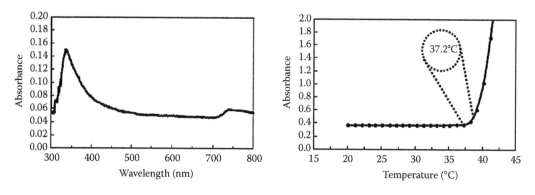

**FIGURE 3.16** Synthesis of poly(NIPAM-co-AA) nanoparticles using KPS initiator at 80°C.

**FIGURE 3.17** UV–visible analysis of poly(NIPAM-co-AA) nanoparticles. (Adapted from B. Ku, H. Seo, B.G. Chung, *BioChip Journal*, 8, 2014, 8–14.)

To remove the OH groups on the silicate layers, clay is dehumidified for 6 h at 400°C. Then $Mg(OEt)_2$ and toluene are added to the clay at continuous argon flow. The slurry is heated to 80°C under mixing conditions. In the next step, $TiCl_4$ and electron donor are added to the slurry and heated to 115°C. The product is washed several times with fresh toluene and treated with $TiCl_4$ for 4 h at 115°C.

To prepare polyethylene/clay nanocomposites by *in situ* polymerization method, propylene in hexane is synthesized by slurry polymerization procedure in the Buchi (1 L) type reactor at a pressure of 4 bars and boiling point of hexane. To inject into the reactor, triisobutylaluminum (TIBA), DMCHS, and hydrogen are used as cocatalyst, external donor, and chain transfer agent, respectively. In the final step, the polymer is washed with ethanol several times, filtered, and dried in a vacuum oven at 70°C for 24 h.

The final properties of produced nanocomposite such as mechanics, morphology, and dispersion are investigated by a wide range of characteristic techniques. The obtained results show that the thermomechanical properties of the nanocomposites are considerably improved by introducing a small amount of clay. However, the clay must be significantly intercalated and/or exfoliated during the preparation process of the nanocomposites.

### 3.2.3.3.5 Synthesis of Conductive Polymer Nanoparticles

To synthesize conducting polymers, there are a number of methods including electrochemical oxidation of the monomers, chemical synthesis, and some less common ones such as enzyme-catalyzed and photochemically initiated polymerization [90]. Chemical or electrochemical polymerization is widely used to synthesize polyaniline (PANI) in the aqueous acid media. The synthesized polymer is called an emeraldine salt. For bulk production, the chemical oxidation of aniline is the more feasible method. The limitation of this method is the poor processibility of the obtained polymer due to its insolubility in common solvents, although it can be improved by using different dopants [91,92].

In electrochemical oxidation synthesis, PANI is synthesized in acidic media by constant potential and current and on an inert metallic electrode, for example, Pt or conducting indium tin oxide (ITO) glass.

In chemical synthesis, monomer (aniline) is synthesized in aqueous solution containing oxidant, for example, ammonium peroxydisulfate and acid, for example, hydrochloric. In this type of synthesis, the monomer is converted directly to conjugated polymer by a condensation process. However, an excess of the oxidant leads to materials that are essentially intractable, which is one of its disadvantages. By progressing the oxidative condensation of aniline, the color of solution turns to black which probably is due to the soluble oligomers. The type of the medium and the concentration of the oxidant are the effective parameters on the intensity of coloration. The major effective parameters on the course of the reaction and on the nature of the final product are as follows: type and temperature of medium, concentration of the oxidant, and time. To obtain desirable results, some factors such as low ionic strength, volatility, and noncorrosive properties should be controlled although no medium satisfies all of these requirements [93].

In the polymerization mechanism, there is a close similarity in the electrochemical or chemical polymerization mechanism of aniline. The following mechanism proceeds in both cases: the first step is the formation of the aniline radical cation, which has several resonant forms and is formed by transferring the electron from the 2 s energy level of the nitrogen atom.

Among the different resonance forms shown in Figure 3.18, form (c) is the more reactive one due to its important substituent inductive effect and its absence of steric hindrance. The next step is the dimer formation between the radical cation and its resonant form.

Then, as shown in Figure 3.19, a new radical cation dimer is formed by oxidizing the dimer with having two possible reactions including the reaction with the radical cation monomer and with the radical cation dimer to form a trimer or a tetramer, respectively. PANI polymer is formed with following steps (Figure 3.20).

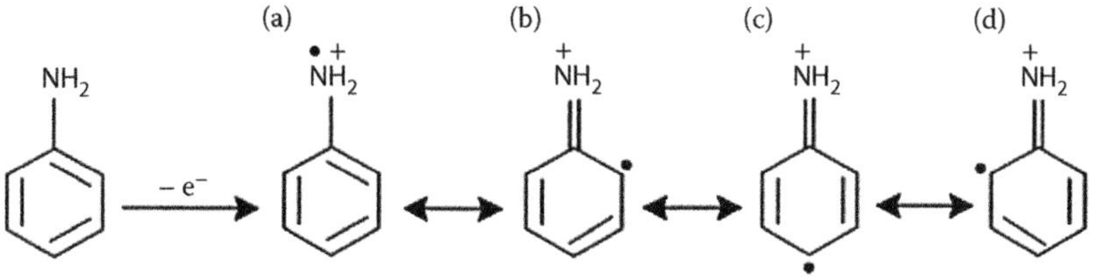

**FIGURE 3.18** The formation of the aniline radical cation and its different resonant structures. (Adapted from I. Harada, Y. Furukawa, F. Ueda, *Synthetic Metals*, 29, 1989, 303–312.)

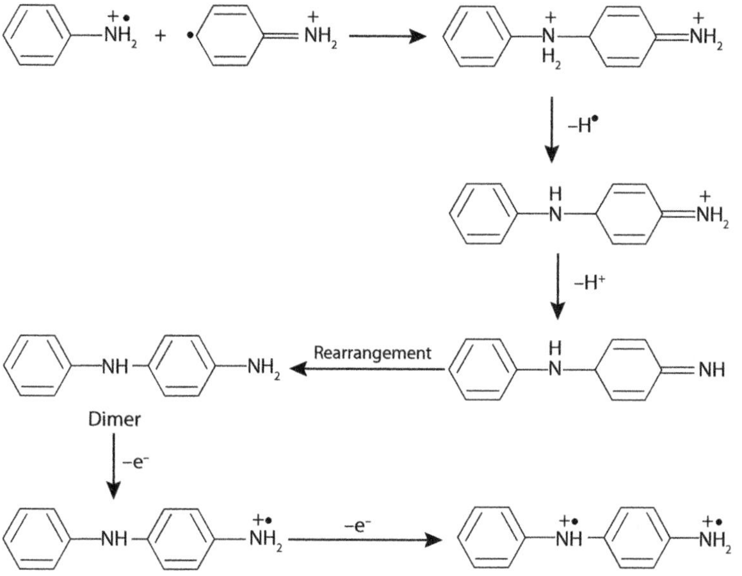

**FIGURE 3.19** Formation of the dimer and its corresponding radical cation. (Adapted from I. Harada, Y. Furukawa, F. Ueda, *Synthetic Metals*, 29, 1989, 303–312.)

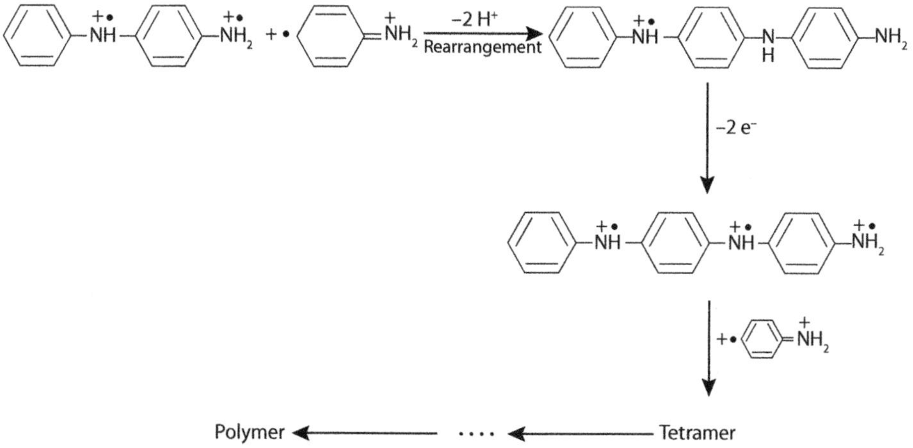

**FIGURE 3.20** One possible way of PANI polymer formation. (Adapted from I. Harada, Y. Furukawa, F. Ueda, *Synthetic Metals*, 29, 1989, 303–312.)

It is known that only the emeraldine state of PANI can be used for nonredox doping process to produce conductive polymer. Charge transfer is a kind of doping process with the number of electrons of the polymer unchanged. Angelopoulos et al. for the first time converted emeraldine base form of PANI to highly conducting metallic regime by this doping method [93]. They did this doping process by treating emeraldine base with aqueous protonic acids as shown in Figure 3.21. It is known that the conductivity of PANI when doped with this method is about 9–10 times greater than that of nondoped ones.

Earlier studies show the formation of a stable polysemiquinone radical cation as shown in Figure 3.22 [85].

Here PANI (emeraldine salt, ES) is synthesized by two chemical oxidation methods: (1) conventional emulsion polymerization to produce binary doped PANI and (2) homogeneous solution polymerization to produce single-doped PANI.

In a conventional emulsion system 5.768 g of SDS was dispersed in 40 mL HCl (1 M) in a two-necked round bottom flask, then 0.745 g aniline in 10 mL HCl (1 M) was introduced to the mixture with vigorous stirring at room temperature under nitrogen atmosphere for 30 min. Then, 10 mL HCl (1 M) aqueous solution with 0.923 mL ammonium persulfate (APS) as an oxidant were added drop-wise into 100 mL of reaction mixtures during 20–30 min. After the purging period of about 20–40 min, the homogeneous recipes were turned into a bluish tint and the coloration was pronounced as polymerization proceeded without agitation for 24 h at room temperature. Finally, dark green colored PANI dispersions were obtained without any precipitation. In the experiments, the molar ratios of APS to aniline and SDS to aniline were kept 0.5 and 2.5, respectively. Excess amounts of methanol were added into the SDS–HCl binary doped PANI dispersion to precipitate PANI powder by disrupting the hydrophilic–lipophilic balance of the system and stopping the reaction. After that, the solution was centrifuged for 20 min at 8000 rpm. The precipitation was washed with methanol, acetone, and water to remove unreacted materials, aniline oligomers, and initiators. The obtained binary-doped emeraldine salt PANI cakes were dried in a vacuum oven at 50°C for

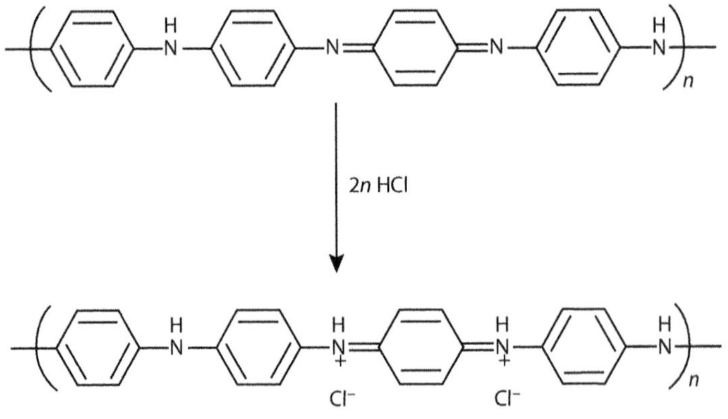

**FIGURE 3.21** Protonic acid doping of PANIs. (Adapted from I. Harada, Y. Furukawa, F. Ueda, *Synthetic Metals*, 29, 1989, 303–312.)

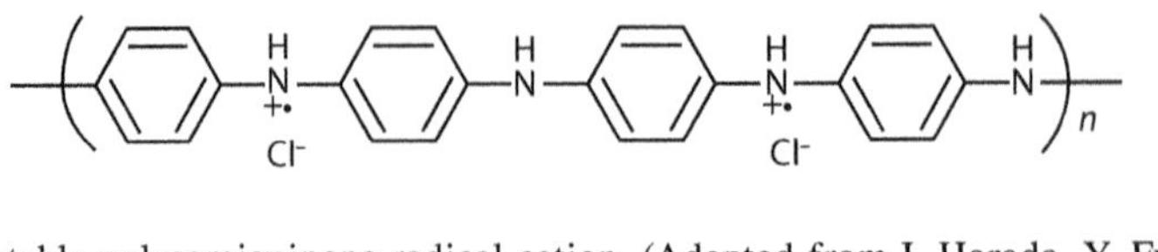

**FIGURE 3.22** A stable polysemiquinone radical cation. (Adapted from I. Harada, Y. Furukawa, F. Ueda, *Synthetic Metals*, 29, 1989, 303–312.)

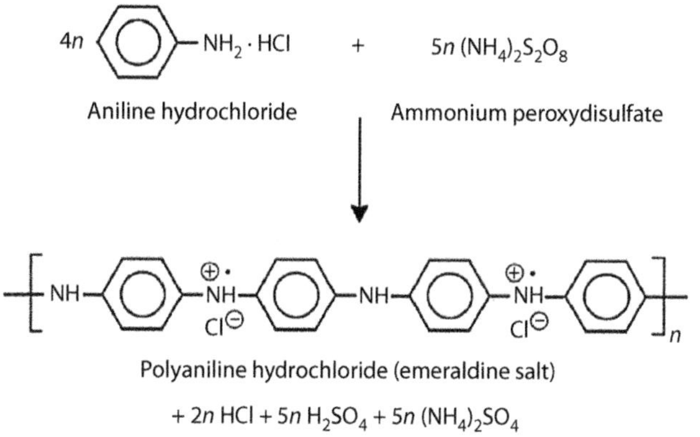

**FIGURE 3.23** Oxidation of aniline hydrochloride with APS. (Adapted from I. Harada, Y. Furukawa, F. Ueda, *Synthetic Metals*, 29, 1989, 303–312.)

48 h. Figure 3.23 schematically shows what happens during oxidation of aniline monomer with APS in acidic media.

The solution polymerization prepared with the same molar ratio of oxidant to monomer was stirred for 24 h at room temperature under N2 purging, to obtain PANI by the same procedure. The obtained product (single-doped emeraldine salt PANI) was dried in a vacuum oven at 50°C for 48 h. Emeraldine base (EB) PANI also was prepared as a control by suspending prepared PANI-ES with 100 mL of $NH_4OH$ (24%) solution to convert the PANI hydrochloride (emeraldine salt) to PANI (emeraldine base), shown in Figure 3.24.

To increase electrical conductivity of polyaniline nanoparticles, one can add highly conductive nanoparticles such as graphene. The PANI/graphene nanoparticles were prepared with a similar method described for aniline with the difference that prescribed amounts of graphene were dispersed in 1 M HCl solution of aniline monomer with sonication and the monomer-graphene dispersion was used for both emulsion and solution polymerization. Figure 3.25 shows schematically the preparation of PANI/graphene nanocomposite. The graphene amounts were 0, 0.1, 0.2, 0.3, 0.4, 0.5, 0.7, and 1 wt% according to the monomer net weight [95].

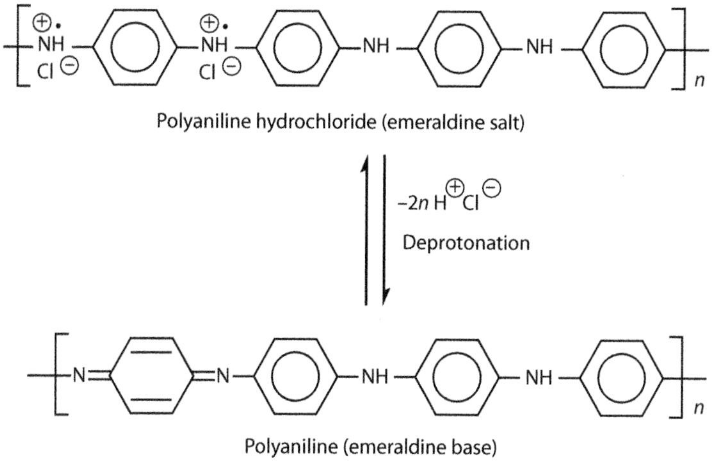

**FIGURE 3.24** PANI emeraldine salt is deprotonated in the alkaline medium to PANI emeraldine base. (Adapted from I. Harada, Y. Furukawa, F. Ueda, *Synthetic Metals*, 29, 1989, 303–312.)

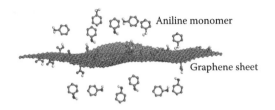

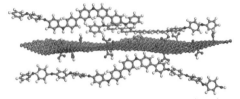

Polyaniline/graphene nanocomposite (PAG)

**FIGURE 3.25**  (**See color insert.**) Schematic process of preparing PANI/graphene nanocomposites. (Adapted from I. Harada, Y. Furukawa, F. Ueda, *Synthetic Metals*, 29, 1989, 303–312.)

## 3.3  ONE-DIMENSIONAL NANOMATERIALS

### 3.3.1  SYNTHESIS OF NANOTUBES

CNTs have the simplest chemical composition and atomic bonding configuration, representing the extreme versatility and enrichment in structures and structure–property relations [96]. There are several main methods to produce CNTs including CVD, electric field nanotube growth, arc-discharge, and laser ablation. Tangled nanotubes could be produced by the last two methods and using CVD and electric field methods on catalytic patterned substrates grow nanotube arrays at controllable locations that are site selective.

#### 3.3.1.1  Chemical Vapor Deposition

CNT synthesis by CVD involves heating a catalyst material (metal nanoparticles typically supported on high surface area materials such as alumina materials) and flowing a hydrocarbon gas through the tube reactor for a period [97]. The catalyst particles are considered as seeds to nucleate the growth of nanotubes. Today, patterned growth approaches have developed to obtain organized nanotube structures, positioning catalyst in arrayed locations for the growth of nanotubes from specific catalytic sites [98].

The earlier work has presented the ordered arrays of CNT consisting of multiwalled nanotubes (MWNTs) by CVD growth on porous silicon and silicon substrates patterned with iron particles in square regions (700°C; carbon source, $C_2H_4$; alumina-supported iron catalyst) [99]. The nanotubes are well aligned along the direction perpendicular to the substrate surface, resulting from nanotubes grown by closely spaced catalyst particles due to strong intratube van der Waals binding interactions. The arrayed nanotubes exhibit excellent characteristics in electron field emission to create the spatially defined massive field emitter arrays derived by simple chemical routes to apply in flat panel displays.

Dai at Stanford University [99] with his colleagues have carried out arrayed patterned growth for both multiwalled and single-walled nanotubes, including self-assembly and active electric field control to manipulate the orientation of nanotubes. The ordered nanotube arrays and networks were formed at the synthesis stage of nanotubes as results of these works.

Contact printing technique is used to transfer catalyst materials onto the tops of pillars, and SWNTs with ordered networks with the desired nanotube orientations are formed by the CVD method on the substrates (900°C; carbon source, $CH_4$; supported iron catalyst) [100,101]. In this case, the arrayed nanotubes are self-orientated by van der Waals forces between nanotubes and the silicon posts.

#### 3.3.1.2  Electric Field-Directed Nanotube Growth

Zhang's group has controlled the growth directions of SWNTs by electric fields [102], produced a large alignment torque to direct the nanotube parallel to the electric field. One of the advantages of using electric field against thermal and gas flow fluctuations is stability of the alignment during the growth procedure. However, further studies of both suspended molecular wires and complex nanotube fabric structures on flat substrates can be interested by applied fields varying during nanotube synthesis.

### 3.3.1.3 Molecular Seeds Self Assembly

The CNTs self-assembled by organic precursor molecules on a platinum surface is represented by the researchers in the *Nature* journal [103]. In this research, starting molecule into a three-dimensional object is transferred, named as germling, on a hot platinum surface using a catalytic reaction to split off hydrogen atoms and form new carbon–carbon bonds. The germ, the defined parameter for the nanotube's atomic structure, is folded out of the flat molecule. The lid of the growing SWCNT is formed by this end cap. More carbon atoms attach and originate from the catalytic decomposition of ethylene on the platinum surface (Figure 3.26). Scanning helium ion microscope (SHIM) shows that the produced SWCNTs have lengths greater than 300 nm.

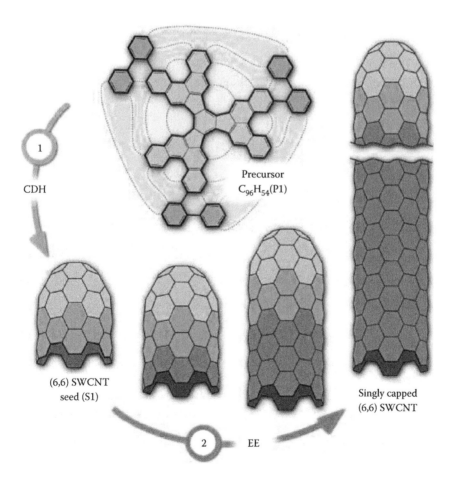

**FIGURE 3.26** **(See color insert.)** Two step bottom-up synthesis of SWCNTs: (1) singly capped ultrashort (6,6) SWCNT seed formation via cyclodehydrogenation (CDH) as polycyclic hydrocarbon precursor $C_{96}H_{54}$ (P1) and (2) nanotube growth via epitaxial elongation (EE). Orange and blue: short CNT segment of the seed S1; red dashed lines: new C–C bonds; green: epitaxial elongation. (Adapted from J.R. Sanchez-Valencia et al., *Nature*, 512, 2014, 61–64.)

### 3.3.1.4 Microwave Heating Method

MWCNTs are synthesized using a microwave oven to calcinate polystyrene with nickel nanoparticles under $N_2$ purging at 700 or 800°C for 15 or 10 min [104]. The synthesized MWCNTs are characterized by TEM (Figure 3.27), a Raman spectrophotometer, and a wide-angle x-ray diffractometer. The obtained results show that a relationship between the outer diameter of the carbon nanotubes and the diameter of catalytic nickel nanoparticles is correlated by a linear function, $D_{CNT} = 1.01D_{Ni} + 14.79$ nm and $D_{CNT} = 1.12D_{Ni} + 7.80$ nm for the calcination condition of 800°C, 10 min and 700°C, 15 min, respectively.

### 3.3.1.5 Laser Ablation Method

In the laser ablation method, a high power laser is used to vaporize carbon from a graphite target at high temperature to produce both MWNTs and SWNTs using metal particles as catalysts to generate SWNTs process [105,106]. The laser is focused onto carbon targets, placed in a 1200°C quartz tube furnace under the argon atmosphere (~500 torr) [107]. Argon gas carries the vapors from the high temperature chamber into a cooled downstream. The nanotubes will self-assemble from carbon vapors and condense on the walls of the flow tube to produce SWNTs with a varied diameter distribution between 1.0 and 1.6 nm. The quantity and quality of these carbon nanotubes depend on several synthesis parameters such as amount and type of catalysts, laser power and wavelength, inert gas type, temperature, pressure, and the fluid dynamics near the carbon target.

### 3.3.1.6 Arc Discharge Method

The carbon arc discharge is the most common and easiest method to produce CNTs or complex mixtures of components through arc-vaporization of two carbon rods placed end to end with

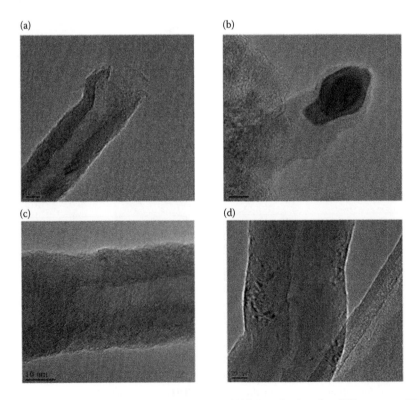

**FIGURE 3.27** TEM images of CNTs for calcination at 700°C for 15 min using different nano Ni diameters (DNi), (a) 10, (b) 20, (c) 50, and (d) 90 nm. (Adapted from K. Ohta et al., *Journal of Materials Chemistry A*, 2, 2014, 2, 2773–2780.)

approximately a 1 mm gap, in a chamber filled with inert gas at low pressure. It requires further purification to remove catalytic metals from the produced CNTs. A potential difference of 20 V is applied to drive a direct current of 50–100 A, following a high temperature discharge between the two electrodes to vaporize the surface of one of the carbon electrodes, and forms a small rod-shaped deposit on the other electrode. To obtain high yield CNTs, the uniformity of the plasma arc and the temperature of the deposit are two of the most important parameters [107,108].

### 3.3.1.7 Bottom-Up Chemical Synthesis

Two basic areas of research are considered for the bottom-up synthesis of CNTs: (1) using aromatic macrocyclic templates and (2) development of these templates to obtain longer CNTs by polymerization reactions. This template approach is an attractive method to produce both zigzag and armchair CNTs with different diameters, as well as to chiral CNTs with different helical pitches. There are several advantages for the organic synthesis approach to control CNT chirality compared with the current methods as follows:

- First, the process is easier to troubleshoot and optimize because of its mechanistically well-understood reactions.
- Second, the synthesis condition for organics is typically at temperatures below 200°C compared with temperatures closer to 1000°C for the current techniques.
- Last, it is possible to produce CNTs with incorporation of nitrogen, boron, or sulfur by the organic synthesis approach.

Omachi et al. initiated well-defined carbon nanorings to grow carbon nanotubes [109]. In this study, a solution of cycloparaphenylenes (0.5 mM in toluene) was spin-coated at 4000 rpm on a C-plane sapphire substrate plate (5 mm × 5 mm). The reaction plate was placed in the chamber, followed by heating at 500°C for 15 min under ethanol gas purging. The CNTs formed were analyzed by TEM and Raman spectroscopy.

After extensive investigations, it was found that CNTs could be grown from CPPs by simply heating these seed molecules with ethanol (Figure 3.28a). Figure 3.28b and c are TEM images of CNTs synthesized under the previously mentioned conditions. Regarding the Raman spectroscopy result in Figure 3.28d and e, residual cycloparaphenylenes was not found on the reaction plate. Figure 3.28d shows the radial breathing mode regions and a relationship between diameter (d) and frequency of radial breathing mode. Raman spectroscopy results indicate that the CNTs were distributed in the diameter range 1.3–1.7 nm, very close to that of CPP (1.7 nm). Also, the diameter distribution histogram in Figure 3.28f was derived from the number of nanotubes with different diameters observed by TEM. It is also consistent with this size regime to conclude that the high percentage of the CNTs was between 1.7 and 1.3 nm in diameter.

In addition, Mogilevsky and his coworkers succeeded to prepare a hybrid nanostructure including graphene and $TiO_2$ by a bottom-up synthetic approach of alizarin and titanium isopropoxide [110]. It would be amenable to synthesize photocatalytically active materials with maximized graphene–$TiO_2$ interface (see Figure 3.29).

Special advantage of a photoactive material (e.g., $TiO_2$) is the charge transfer degree after photo-excitation at the interface with graphene. Figure 3.30 shows the movement of electrons away from, or toward the graphene at the interface. It was found that contacting the graphene patch with the $TiO_2$ surface causes to transfer electrons from graphene to the oxygen molecules in the top layer of $TiO_2$ which were also confirmed for the $TiO_2$-sandwiched graphene and graphene ribbons. Positive and negative of the charge difference density (CDD), charge density in the interaction of the two substrates, denote the addition of electrons and the removal of electrons, respectively. The CDD measurement in this case shows clearly charge transfer from graphene to the top oxygen layer of $TiO_2$, to produce positive graphene and negative $TiO_2$, which the total charge transfer is approximately 0.017 $e^-$ per carbon atom.

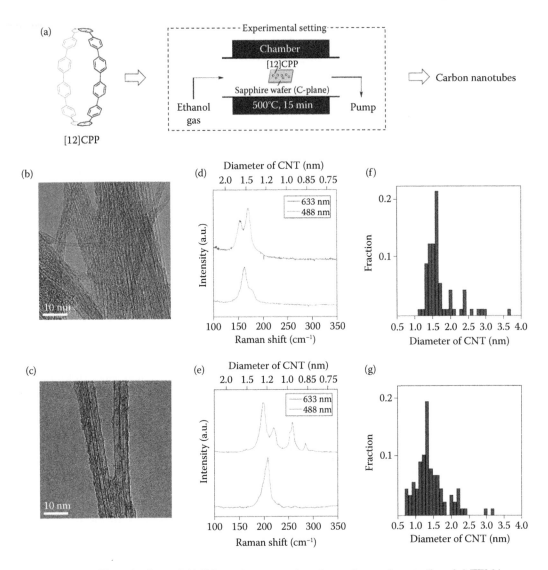

**FIGURE 3.28** (**See color insert.**) (a) Schematic presentation of growth experiments, (b and c) TEM images of CNTs synthesized from CPP, (d and e) Raman spectra of CNTs, and (f and g) diameter distribution histograms. Data for CNTs grown from CPP are given in (d and f) and data for CNTs grown from CPP are given in (e and g). (Adapted from H. Omachi et al., *Nature Chemistry*, 5, 2013, 572–576.)

### 3.3.2 SYNTHESIS OF NANORODS

#### 3.3.2.1 Simple Hydrothermal Method

Using hydro/solvothermal treatment of salen ligand with nickel nitrate is of particular interest to synthesis solid Ni (salen) complexes at different temperatures (120–180°C), times (6–24 h), and solvents (water, ethanol, and $H_2O$, $C_2H_5OH$, and acetonitrile) [111]. In a typical synthesis, 0.75 mmol of $Ni(NO_3)_2 \cdot 6H_2O$ dissolved in 30 mL of $H_2O$ is added to 0.75 mmol of N,N′-bis(salicylidene)eth-ylenediamine-70 mL of $H_2O$ as aqueous solution, followed by mixing for 15 min and transferred into a stainless steel autoclave. The autoclave is cooled down to room temperature after hydrothermal treatment. The red-orange color precipitated material is collected and washed several times with distilled water and ethanol. The product obtained after 6 h consisted of separated nanorods (Figure 3.31a and b). It is interesting to point out that synthesis after reaction time to 13, 18, and

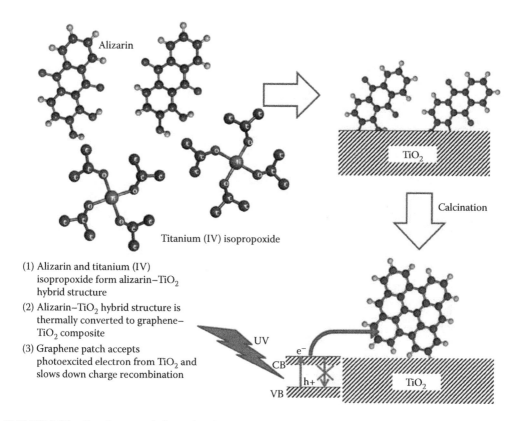

**FIGURE 3.29** Graphene patch formation from alizarin. (Adapted from G. Mogilevsky et al., *ACS Applied Materials & Interfaces*, 6, 2014, 10638–10648.)

24 h produced some agglomeration with nanoparticles (Figure 3.31c and d), a mixture of nanorods and nanoparticles (Figure 3.31e and f), and the more agglomerated nanorods with shorter length (Figure 3.31g and h), respectively.

Figure 3.32 shows the formation mechanism of a nanorod Ni(salen) complex as a sequential two-step growth mechanism, heterogeneous complex nucleation, and directional growth route due to the morphology observation.

Under hydrothermal conditions, the soluble $Ni^{2+}$ cation reacted with salen ligand to form an insoluble Ni(salen) nucleus wherein it can be led to the precipitation transformation process in $H_2O$ to form more insoluble Ni(salen). In the first stage, Ni(salen) nucleates heterogeneously due to a lower energy barrier than that of the nucleation in solution. Based on all reported crystal structures, the <100> and/or <010> directions are the favored directions for crystal growth. Under hydrothermal conditions, nanorod formation starts by the nucleation and spontaneous aggregation on the formed nuclei to decrease their surface area during the oriented attachment process. By domination of Ostwald ripening process during the reaction time, nanorods with smooth surfaces were produced.

### 3.3.2.2 Seed-Mediated and Seedless Methods

In the seed mediated route [112], 9.91 mL of 0.2 M CTAB is mixed with $HAuCl_4$. Then, 0.006 M $NaBH_4$ is added under magnetic mixing for 2 min. The growth solution of the gold nanorods is prepared by mixing 0.1 M of CTAB, 25.4 mM of $HAuCl_4$, 0.5 M of $H_2SO_4$, 10 mM of $AgNO_3$, and 0.1 M of ascorbic acid. Then, the mixture of 4 mL raw seed solution and 6 mL deionized water is centrifuged with 10,000 rpm for 7 min to remove aggregated seeds. Finally, the growth of the gold nanorods starts by addition of 30 μL seed solution to the recipe. The resulting mixture is stirred for

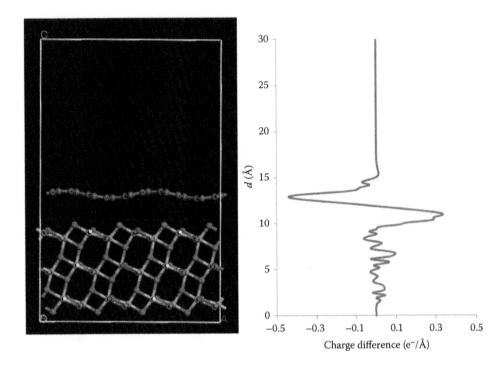

**FIGURE 3.30** Model cell alongside with the planar averaged charge difference. $d$ (Å) is the position on the z-axis of the unit cell and the y-axis of the graph on the right side of the figure. (Adapted from G. Mogilevsky et al., *ACS Applied Materials & Interfaces*, 6, 2014, 10638–10648.)

30 s. After 12 h, the reaction is stopped by centrifugation with 10,000 rpm for 10 min. The precipitate is redispersed in water.

In the seedless route [113], gold nanorods are prepared by mixing 0.15 M of CTAB, 25.4 mM of $HAuCl_4$, 100 mM of $AgNO_3$, 0.05 M of paradioxybenzene, and 1.19 M of HCl. The reaction is started by adding 0.01 M of $NaBH_4$ solution. Finally, the solution is stirred for 30 s to make it homogeneous. After 36 h, the reaction is stopped by centrifugation with 10,000 rpm for 10 min and the precipitate is redispersed in water.

The gold ions concentration also was tried to downregulate the width (Figure 3.33). The product yield began to decrease by reduction of the gold ions concentration to 0.3 mM (Figure 3.33e) whereas the width of rods still was 13.4 nm. High concentrations of $Ag^+$ (0.12 –0.5 mM) and CTAB (0.1–0.15 M) would be favored to form rods with higher aspect ratio using ascorbic acid, tartaric acid, paradioxybenzene, gallic acid, and a mixture of ascorbic acid and paradioxybenzene as the reducing agents.

The seeded or secondary growth method is manageable to scale up larger amounts of gold nanorods by the primary uniform gold nanorods solution [114]. This is possible to increase the reaction volume from 100 mL to 1 L. It should be noted that the primary uniform gold nanorods solution for all of the experiments was prepared by an identical procedure, except for the KBr removing in the primary growth solution through utilizing benzyldimethylhexadecylammonium chloride hydrate (BDAC). BDAC helps direct the shape into nanorods with higher aspect ratios to complete the primary growth phase (Figure 3.34).

### 3.3.2.3 Metal–Organic Chemical Vapor Deposition Method

Metal–organic chemical vapor deposition (MOCVD) has also been used for ZnO nanorods, thin film, and QD growth without any catalyst. High purity product and easy fabrications of nanorod

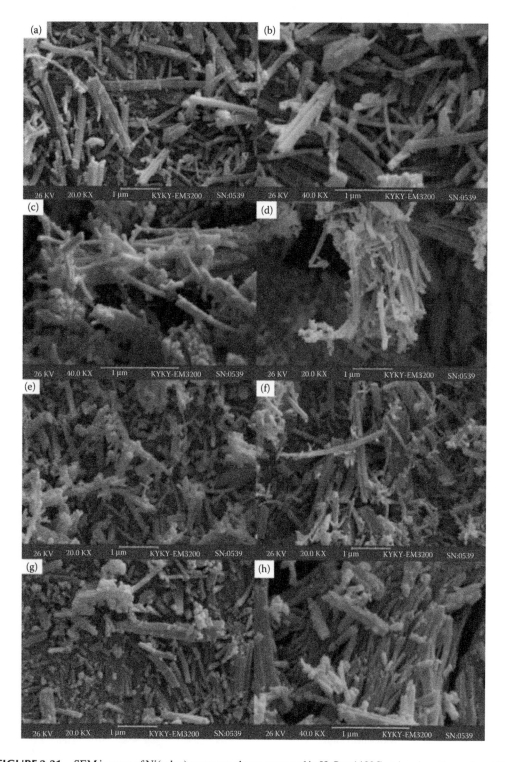

**FIGURE 3.31** SEM images of Ni(salen) nanocomplexes prepared in $H_2O$ at 140°C and various times: (a and b) 6 h, (c and d) 13 h, (e and f) 18 h, and (g and h) 24 h. (Adapted from M. Mohammadikish, *CrystEngComm*, 16, 2014, 8020–8026.)

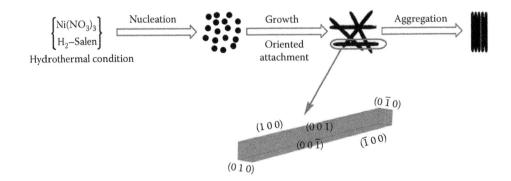

**FIGURE 3.32** Schematic illustration of a proposed mechanism for the formation of the Ni(salen) nanorods.

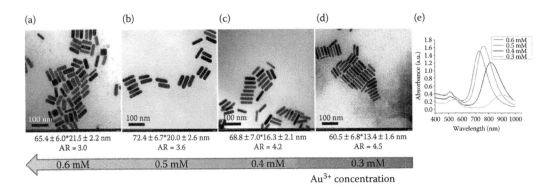

**FIGURE 3.33** TEM images of gold nanorods synthesized with different $Au^{3+}$ concentration (a) 0.6 mM, (b) 0.5 mM, (c) 0.4 mM, and (d) 0.3 mM. (e) Corresponding UV–vis–NIR spectra. (Adapted from X. Xu et al., *Journal of Materials Chemistry A*, 2, 2013, 3528–3535.)

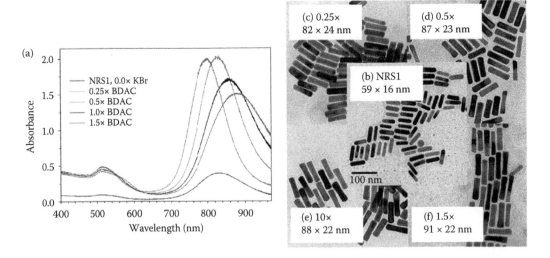

**FIGURE 3.34** Secondary growth of the primary and final gold nanorods with BDAC in addition to the standard amount of CTAB: (a) optical absorbance spectra and TEM images of (b) the primary and final gold nanorods using (c) 0.25×, (d) 0.5×, (e) 1.0×, and (f) 1.5× BDAC. (Adapted from K.A. Kozek et al., *Chemistry of Materials*, 25, 2013, 4537–4544.)

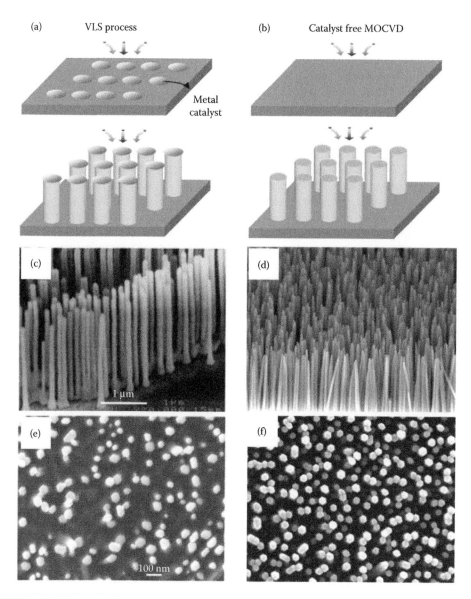

**FIGURE 3.35** Schematic diagrams illustrating the growth of ZnO nanorods (nanowires) from (a) the VLS process and (b) catalyst-free MOCVD, FE-SEM images of (c) and (e) VLS-grown and (d) and (f) catalyst-free MOCVD grown ZnO nanorods. (Adapted from W.I. Park et al., *Applied Physics Letters*, 80, 2002, 4232–4234.)

quantum structures and heterostructures are the major advantages for this case (see Figure 3.35). Although the catalyst-free growth mechanism of ZnO nanorods has not been thoroughly investigated, it can be pointed out that anisotropic surface energy of wurtzite ZnO is the main reason for anisotropic growth. In addition, adsorption of fresh reactant gases just on nanorod tips is obtained by high-speed laminar gas flow, inducing turbulent flow between the nanostructures.

### 3.3.2.4 Vapor Phase Synthesis

The most extensively explored approach to the formation of 1D nanostructure including whiskers, nanorods, and nanowires is probably vapor phase synthesis where vapor species are generated by several methods (including evaporation, chemical reduction, and gaseous reaction), subsequently transported and condensed onto a solid substrate surface with a temperature lower than that of the source material. With proper control over the super-saturation factor, 1D nanostructure can be obtained in large quantities. For example, Yi et al. [116] have written a review paper on ZnO nanorods works, focusing on current research activities, and investigated the physical and chemical methods of ZnO nanorods and nanowires.

### 3.3.3  Synthesis of Nanowires

Generally, semiconductor NWs can be synthesized by metal nanoclusters as catalysts via vapor–solid–solid (VSS) and vapor–liquid–solid (VLS) growth mechanisms. In a VLS growth, vapor-phase precursors are introduced at temperatures that should be higher than the eutectic point of the metal–semiconductor system, resulting in liquid droplets of the metal–semiconductor alloy. Continuous feeding of the precursors leads to the super saturation of catalytic droplets, on which the semiconductor material starts to congregate and grow into crystalline NWs. Unlike the VLS mechanism, in VSS, the metal nanoclusters exist as solid particles. The particles can provide low-energy interface for trapping the precursor materials and also yield high epitaxial growth rates. Specifically, various techniques have been applied to produce vapor-phase precursors for the NWs growth, including chemical vapor deposition (CVD), pulsed laser ablation, and molecular beam epitaxy (MBE). Among all, a CVD growth method utilizing the conventional tube furnace and solid powder source (solid-source CVD) has been widely explored in recent years for the growth of various III–V NWs because of its relatively low cost, simple growth procedures, and, importantly, no entanglement of toxic gas precursors, compared to other feigned growth systems such as MBE and metal–organic CVD. Source powder is put at the upstream of a two-zone tube furnace, while hydrogen is used as a carrier gas to forwarding the evaporated source material into the downstream, and a substrate precoated with catalysts is positioned at the downstream for the NW growth.

#### 3.3.3.1  Subdiffraction Laser Synthesis

Using the interposition of incident laser radiation and surface scattered radiation is an alternative method to produce nanowires with diameters of 60 nm. This condition causes spatially confined nanowires, which periodical heating is needed for the high resolution chemical vapor deposition [117]. Also, multiple parallel nanowires are produced on a dielectric substrate with controlled properties (i.e., diameter, length, and orientation), which can be controlled by the intensity and polarization direction of the incident radiation for the device fabrication.

Figure 3.32 illustrates the radiation interference for silicon nanowire diameter control, which the high numerical aperture of the zone plates is a critical point to forming a single nanowire. This is because the intensity peak of the focal spot must be narrow enough to produce sufficient intensity for heating the substrate.

SEM images of nanowires grown from horizontally polarized laser radiation in Figure 3.36 show single (a), double (b), and triple nanowires (c) with width of 60 nm. Also, these images show the polarization dependence of nanowire patterns for linear polarization 45° from the scanning direction (d), at linear polarization 90° from the scanning direction (e), and for circular polarization (f).

#### 3.3.3.2  Electrophoresis-Assisted Electroless Deposition

Another attractive structure is synthesis of core–shell nanowire (NW) arrays of ZnO/CuO with high aspect ratios by electrophoresis-assisted electroless deposition of CuO onto ZnO NWs [118]. During this procedure, CuO seeds are successfully manufactured on long ZnO NWs by the electrostatic attachment of colloidal $Cu_2O$ NPs under thermal oxidation. The core–shell ZnO/CuO nanowires are fabricated by kinetic-limited deposition in the solution to prevail the inherent diffusion limitations in conventional electroless deposition method. The experimental observations and thermodynamic modeling results are presented to validate the supposed dielectrophoresis-assisted electroless deposition mechanism.

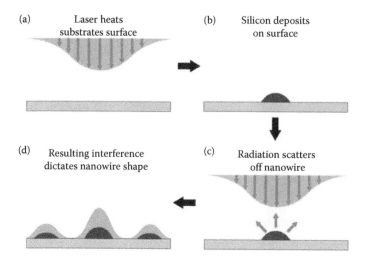

(a) Laser heats substrates surface

(b) Silicon deposits on surface

(d) Resulting interference dictates nanowire shape

(c) Radiation scatters off nanowire

**FIGURE 3.36** Schematic of radiation interference for silicon nanowire diameter control. (Adapted from J.I. Mitchell et al., Scientific Reports, *Nature*, 4, 3908–3912.)

### 3.3.3.3 Microwave Heating Growth Method

ZnO nanowires for electrical and optical devices are extensively synthesized by the hydrothermal method. The main challenge in this method is to obtain a rapid procedure for the synthesis of long vertically aligned ZnO nanowire arrays on a transparent conductive oxide substrate. Recently, Liu et al. [119] proposed a microwave heating method to control growth of long ZnO nanowire arrays, which can avoid the growth timeout and retain the reactants concentration in dynamic equilibrium.

In nanoseed-mediated microwave heating growth method, initially, a ZnO thin layer is deposited on fluorine doped tin oxide glass substrate by pulsed-laser deposition technique (PLD). Then, the ZnO deposited substrate is dipped into a beaker including zinc nitrate hexahydrate and hexamethylenetetramine aqueous solution and heated with a microwave oven with frequency 2.45 GHz, followed by injection into the two Teflon tubes. The third Teflon tube is used for pumping the extra solution on the surface of the solution to keep the solution volume during the experimental process. The heating process is performed for 0.5–5.0 h with a power setting at 640 W. After finishing the reaction, the substrate is brought out from the growth solution, followed by washing several times with DI water, and dried in an air environment. The obtained results show that the length of the nanowires increases linearly with growth time. Also, it was found that ZnO nanowires were produced with a length 10 μm after growing for 2–3 h and the growth rate is 58–78 nm/min.

## 3.4 TWO-DIMENSIONAL NANOMATERIALS

### 3.4.1 SYNTHESIS OF NANOFILMS

#### 3.4.1.1 Chemical Precipitation with Subsequent Thermal Treatment

The synthesis of nanostructured oxide films is conducted via chemical precipitation with subsequent thermal treatment of the obtained product. This method is the most general because of mixing the initial reagents at the ionic-molecular level; this subsequently gives an opportunity to obtain oxide nanocrystalline powders of the given compositions with a high dispersion (from 1 to 100 nm) at sufficiently low temperatures ($\leq 600°C$) [120,121].

The solutions of cerium and zirconium nitrates and aqueous ammonia solution are used as the initial reagents and precipitator, respectively, to synthesize the $CeO_2$–$ZrO_2$ powders. After precipitation, the gel precipitates, including $Ce(OH)_3$ and $ZrO(OH)_2$, are filtered and then exposed

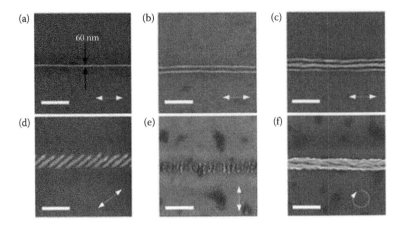

**FIGURE 3.37** SEM images of nanowires grown from horizontally polarized laser radiation show: (a) single, (b) double, and (c) triple nanowires, and the single wire has a width of 60 nanometers. SEM images demonstrate the polarization dependence of nanowire patterns for, (d) linear polarization 45 degrees from the scanning direction, at (e) linear polarization 90 degrees from the scanning direction, and for, (f) circular polarization. (Adapted from J.I. Mitchell et al., Scientific Reports, *Nature*, 4, 3908–3912.)

to fast freezing at –25°C for 24 h. Thermolysis of the synthesized powders based on $CeO_2$ are studied by DTA (Figure 3.37). In Figure 3.37a of the precipitated gels without freezing, two endothermic effects are recorded in 165°C (the removal of adsorption and crystallization water) and 360°C (the removal of hydroxyl groups). Figure 3.37b shows reduction of the water content by the freezing of the coprecipitated gels in the crystallohydrates of the amorphous hydroxides [122] (Figure 3.38).

### 3.4.1.2 Chemical Bath Technique

Deposition of CdS nanofilms have been deposited on well cleaned glass substrates. For deposition of nanofilms of CdS, an aqueous solution of 0.02 M cadmium chloride anhydrous has been taken in a glass beaker to add drop-wise ammonia solution (25%). Thereafter, a solution of TX-100 and double distilled water is added to the previous solution. The solution has been stirred for 15 min to obtain a clear homogeneous solution. After all, an aqueous solution of 0.01 M thiourea is added under mixing condition and constant bath temperature of 343 K and pH of the solution at 11.60. The resulted films with pale yellow, uniform and good adherence to the substrate have been found. Many gas bubbles have been formed in the bath through the chemical reaction. Nonionic surfactant TX-100 has been used to remove the gas bubble formation during the reaction. The chemical reactions leading to the layer formation may be represented as follows:

$$[Cd(NH_3)n]^{2+} \rightleftharpoons Cd^{2+} + nNH_3, n = 1,2,3,\ldots 6 \tag{3.4}$$

$$CS(NH_2)_2 + 2OH^- \rightarrow CH_2N_2 + 2H_2O + S^{2-} \tag{3.5}$$

$$[Cd(NH_3)_n]^{2+} + CS(NH_2)_2 + 2OH^- \rightarrow Cds_{film} + CN_2H_2 + nNH_3 + 2H_2O \tag{3.6}$$

### 3.4.1.3 Molecular Layer Deposition

Continuous and uniform carbon nanofilms are synthesized by the pyrolysis of molecular layer deposition (MLD) formed-polyimide films [123]. The cycle numbers are a good parameter to easily control the film thickness at the nanometer scale.

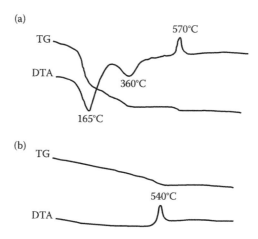

**FIGURE 3.38** DTA and TG curves of coprecipitated powders in $ZrO_2$–$CeO_2$ system (a) without freezing the gel and (b) after freezing the coprecipitated gel at –20°C. (Adapted from S. Kumar, P. Sharma, V. Sharma, *Journal of Applied Physics*, 111, 2012, 043519-6.)

In this method, the deposition of polyimide was done in a closed type, hot-wall ALD reactor. Ethylenediamine (EDA) and 1,2,4,5-benzenetetracarboxylic anhydride (PMDA) precursors with $N_2$ carrying gas was followed by transferring into a quartz tube furnace and annealing at 600°C for 2 h under protecting $H_2$/Ar gas flow at normal pressure to produce continuous and uniform carbon nanofilms.

At first, spherical gold nanoparticles solution is dropped on a Cu grid supported with $Al_2O_3$ layer to deposit with polyimide films, after air-drying. Au nanoparticles coated with continuous and uniform carbon nanofilms are finally produced by annealing the above-mentioned procedure. After the deposition of polyimide on anodic aluminum oxide template with a 200 nm pore diameter, the sample is annealed at 600°C for 2 h under protecting $H_2$/Ar gas, and then immersed into 1 M NaOH aqueous solution at 45°C for 2 h to remove the anodic aluminum oxide template.

### 3.4.2 Nanocoatings

#### 3.4.2.1 Hydrothermal Method Combined with a Mild Ultrasonic

A straightforward hydrothermal method based on a mild ultrasonic sonication is used to fabricate a magnetically recyclable thin-layer $MnO_2$ nanosheet-coated $Fe_3O_4$ nanocomposite [124]. This method provides $MnO_2$/$Fe_3O_4$ nanocomposite with good stability and photocatalytic efficiency to degrade methylene blue under UV–vis light irradiation and also acid resistance and stable recyclability. The ferromagnetic properties of the produced $MnO_2$/$Fe_3O_4$ nanocomposite have been verified by magnetic measurements. This property can be useful to separate by simply applying an external magnetic field after the photocatalytic reaction. The obtained nanocoating can be used in water treatment, dye pollutants degradation, and environmental cleaning.

#### 3.4.2.2 Graphene-Based Nanocoatings

Several methods of graphene synthesis have been discussed by starting with micromechanical exfoliation (scotch tape method), a very easy method for isolating graphene from graphite. However, there are some limitations in this procedure. It uses adhesive tape to exfoliate patterned to generated highly oriented pyrolytic graphite (HOPG) [125].

The tape is then folded to obtain a few layered graphene sheets. At the end, the thin flakes are attached to the film that can be transferred to a suitable substrate such as silicon oxide [126]. Mechanical exfoliation is the simplest way to prepare micro-size graphene flakes for fundamental research purposes [127]. The advantage of this method is production of large size and unmodified graphene sheets. Its disadvantage is due to the very small scale of production. Geim and Novoselov used this method to isolate graphene for the very first time in 2004 [125].

The second method is the epitaxial growth from silicon carbide (SiC). In this method, SiC is heated at high temperatures under vacuum to produce graphene. In this time, silicon atoms sublimate from the substrate to rearrange themselves into graphene layers. The epitaxial graphene grown on the Si substrate shows the excellent crystal quality and large coverage.

The third method is chemical vapor depositions (CVD) in which substrates like copper or nickel are used as catalysts. In this method, the substrates are placed into a furnace and heated at low vacuum to 1100°C. One of the advantages of this method is mass production of high quality graphene.

The last method is a chemical method (Hummer's method), which is the main method for the synthesis of graphene oxide and reduced graphene oxide. In this method, graphite flakes are oxidized into graphene oxide in the presence of sulfuric acid, hydrochloric acid, and potassium permanganate, and then graphene oxide is later reduced by a strong reducing agent known as hydrazine hydrate.

## 3.5   THREE-DIMENSIONAL MATERIALS

Three-dimensional or bulk nanomaterials are materials that are not limited to the nanoscale in any dimension. These materials have three dimensions beyond 100 nm, but may comprise one or more types of any kind of nanomaterials that are in close contact with the matrix. The most important examples of three-dimensional nanomaterials contain nanoporous, nanocomposites, and extremely intercalated multilayered materials such as clay and graphene oxide.

Synthesis of nanocomposite and multilayered particles are discussed in detail in other parts of this book; thus, in the following, we will briefly explain some facts about the preparation of nanoporous materials.

Nanoporous materials are classified into three groups by IUPAC:

- Microporous materials: 0.2–2 nm
- Mesoporous materials: 2–50 nm
- Macroporous materials: 50–1000 nm

Microporous and mesoporous materials, such as zeolites and mesoporous silica, have obtained great interest in the field of drug delivery and imaging. Zeolites are microporous, crystalline aluminosilicates or silicates with pore sizes typically ranging from 5 to 10 Å [128]. This material could be produced by various templates that lead to forming zeolites with distinguished pore sizes and connectivity. It should be noted that synthetic zeolites have some advantages in comparison to natural zeolites; for example, they can be produced with higher purity and also their morphology can be adjusted with controlling synthesis conditions [129].

The most significant method for preparation of zeolites is the hydrothermal technique [130]. In this technique, silicoaluminate as an appropriate precursor is put in a sealed reaction container, typically a Teflon-lined autoclave. In the following, a usual procedure for hydrothermal preparation of zeolite is described [131].

1. Amorphous precursors of silica and alumina are dissolved in a basic medium. Usually, alkali ion hydroxides, which are well known as mineralizing agents, are applied to obtain essential high pH values.
2. The aqueous solution is heated in a sealed reactor or in a preheated oven.
3. The reactants remain amorphous for a long time, and after passing the onset temperature producing the first zeolites crystals is observed.
4. Finally, all amorphous precursors dissolved in the solution convert to a nearly identical amount of zeolite crystals and could be obtained by purification, washing, and drying.

In a common example of synthesis of zeolite by the hydrothermal method, sodium metasilicate—well known as water glass—and sodium aluminate are used as precursors. By adding aluminate to silicate, a precursor make gel would be obtained in which both aluminate and silicate oligomer chains exist. This is known as the precrystalline phase, which features depend on various factors such as silicon/aluminum ratio, solution pH value, nature of reactants, and purity of the precursors. Heating the mixture causes depolymerization of chains to form smaller units.

Finally, in the presence of the above-mentioned mineralizing agent the crystalline zeolites, which comprise Si–O–Al bonds, could be produced. In this method, no considerable enthalpy change is observed due to a similar bond type in precursor oxides (Si–O and Al–O bonds) and produced zeolite (Si–O–Al bonds). For this reason, the total enthalpy change for thermal preparation of zeolite is generally negligible. Therefore, preparation of zeolite is mainly controlled by the kinetics of the total procedure [132–134].

An outstanding review by Cundy and Cox [135] on the synthesis of zeolites and its crystallization mechanism is very useful.

## 3.6 CONCLUSION

Nanotechnology is an interdisciplinary research area that has gained wide attention worldwide in the past few years. The nanomaterials have been categorized into four groups, including zero-, one-, two-, and three-dimensional. General and specific synthesis methods for preparation of different nanomaterials have been considered; however, emphasis was mostly on chemical methods such as microwave-assisted synthesis, sonochemical synthesis, sol–gel method, aerosol method, emulsion method, microfluidic method, laser ablation method, arc discharge method, chemical bath technique, and so on. It can be concluded that zero- and one-dimensional nanomaterials are priority groups, respectively. Nevertheless, it could be anticipated that two- and three-dimensional nanomaterials will find more attention in the near future.

## REFERENCES

1. R.W. Siegel, G.L. Trigg, Nanophase materials, In: G.L. Trigg, (ed.), *Encyclopedia of Applied Physics*, Weinheim: VCH, 11, 1994, pp. 1–27.
2. M.P. Pileni, Semiconductor nanocrystals. In: K.J. Klabunde, (ed.), *Nanoscale Materials in Chemistry*, John Wiley & Sons, Inc., New York, USA, 2001, pp. 61–84.
3. E. Reverchon, R. Adami, Nanomaterials and supercritical fluids, *The Journal of Supercritical Fluids*, 37, 2006, 1–22.
4. I. Rahman, V. Padavettan, Synthesis of silica nanoparticles by sol–gel: Size-dependent properties, surface modification, and applications in silica-polymer nanocomposites: A review, *Journal of Nanomaterials*, 1, 2012, 1–15.
5. J. Zheng, P.R. Nicovich, R.M. Dickson, Highly fluorescent noble-metal quantum dot, *Annual Review of Physical Chemistry*, 58, 2007, 409–431.
6. B. Han, E. Wang, DNA-templated fluorescent silver nanoclusters, *Analytical and Bioanalytical Chemistry*, 402, 2012, 129–138.
7. S. Choi, R.M. Dickson, J. Yu, Developing luminescent silver nanodots for biological applications, *Chemical Society Reviews*, 41, 2012, 1867–1891.
8. G. Li, R. Jin, Atomically precise gold nanoclusters as new model catalysts, *Account of Chemical Research*, 46, 2013, 1749–1758.
9. X. Wang, J. Zhuang, Q. Peng, L. Yadong, A general strategy for nanocrystal synthesis, *Nature*, 437, 2005, 121–124.
10. Y. Zhu, H. Qian, M. Zhu, R. Jin, Thiolate-protected aun nanoclusters as catalysts for selective oxidation and hydrogenation processes, *Advanced Materials*, 22, 2010, 1915–1920.
11. J. Zhong, J. Qu, F. Ye, C. Wang, L. Meng, J. Yang, The bis(p-sulfonatophenyl)phenylphosphine-assisted synthesis and phase transfer of ultrafine gold nanoclusters, *Journal of Colloid and Interface Science*, 361, 2011, 59–63.

12. A. Mathew, P.R. Sajanlal, T. Pradeep, A fifteen atom silver cluster confined in bovine serum albumin, *Journal of Materials Chemistry*, 21, 2011, 11205–11212.
13. L. Shang, S.J. Dong, G.U. Nienhaus, Ultra-small fluorescent metal nanoclusters: Synthesis and biological applications, *Nano Today Journal*, 6, 2011, 401–418.
14. M. Cui, Y. Zhao, Q. Song, Synthesis, optical properties and applications of ultra-small luminescent gold nanoclusters, *Trends in Analytical Chemistry*, 57, 2014, 73–82.
15. R.J. Giguere, T.L. Bray, S.M. Duncan, Application of commercial microwave ovens to organic synthesis, *Tetrahedron Letter*, 27, 1986, 4945–4948.
16. R. Gedye, F. Smith, K. Westaway, H. Ali, L. Baldisera, L. Laberge, J. Rousell, The use of microwave ovens for rapid organic synthesis, *Tetrahedron Letter*, 27, 1986, 279–282.
17. A. Beeri, E. Berman, R. Vishkautsan, Y. Mazur, Reactions of H atoms produced by microwave discharge with olefins in acetone and toluene, *Journal of American Chemical Society*, 108, 1986, 6413–6414.
18. Y. Yue, T.Y. Liu, H.W Li, Z. Liub, Y. Wu, Microwave-assisted synthesis of BSA-protected small gold nanoclusters and their fluorescence-enhanced sensing of silver(I) ions, *Nanoscale Journal*, 4, 2012, 2251–2254
19. W.Y. Chen, J.Y. Lin, W.J. Chen, L.Y. Luo, E.W.G. Diau, Y.C. Chen, Functional gold nanoclusters as antimicrobial agents for antibiotic-resistant bacteria, *Nanomedicine Journal*, 5, 2010, 755–764.
20. Sh Liu, F. Lua, J.J Zhu, Highly fluorescent Ag nanoclusters: Microwave-assisted green synthesis and $Cr^{3+}$ sensing, *Chemical Communications Journal*, 47, 2011, 2661–2663.
21. L. Yan, Y.Q. Cai, B.Z. Zheng, H.Y. Yuan, Y. Guo, D. Xiao, M.M.F. Choi, Microwave-assisted synthesis of BSA-stabilized and HSA-protected gold nanoclusters with red emission, *Journal of Material Chemistry*, 22, 2012, 1000–1005.
22. L. Shang, L.X. Yang, F. Stockmar, R. Popescu, V. Trouillet, M. Bruns, D. Gerthsen, G.U. Nienhaus, Microwave-assisted rapid synthesis of luminescent gold nanoclusters for sensing $Hg^{2+}$ in living cells using fluorescence imaging, *Nanoscale Journal*, 4, 2012, 4155–4160.
23. T. Liu, L, Zhang, H, Song, Z. Wang, Y. Lv, Sonochemical synthesis of Ag nanoclusters: Electro generated chemiluminescence determination of dopamine, *The Journal of Biological and Chemical Luminescence*, 28, 2013, 530–535.
24. Y.G. Sun, B. Mayers, Y.N. Xia, Transformation of silver nanospheres into nanobelts and triangular nanoplates through a thermal process, *Nano Letter*, 3, 2003, 675–679.
25. Xu Hangxun, S. Kenneth Suslick, Sonochemical synthesis of highly fluorescent nanoclusters, *American Chemical Society*, 4, 2010, 3209–3214.
26. I. Diez, M. Pusa, S. Kulmala, H, Jiang, A, Walther, A. S. Goldmann, Color tunability and electrochemiluminescence of silver, nanoclusters, *Angewandte Chemie International*, 48, 2009, 2122–2125.
27. L. Shang, S.J. Dong, Facile preparation of water-soluble fluorescent silver nanoclusters using a polyelectrolyte template, *Chemical Communications*, 9, 2008, 1088–1090.
28. M.A.H. Muhammed, P.K. Verma, S.K. Pal, A. Retnakumari, M. Koyakutty, S. Nair, T. Pradeep, Luminescent quantum clusters of gold in bulk by albumin induced core etching of nanoparticles: Metal ion sensing, metal-enhanced luminescence, and biolabeling, *Chemistry – A European Journal*, 16, 2010, 10103–10112.
29. Y. Negishi, Y. Takasugi, S. Sato, H. Yao, K. Kimura, T. Tsukuda, Magic-numbered $Au_n$ clusters protected by glutathione monolayers (n = 18, 21, 5, 28, 32, and 39): Isolation and spectroscopic characterization, *Journal of the American Chemical Society*, 126, 2004, 6518–6519.
30. T.G. Schaaff, R.L. Whetten, Giant gold-glutathione cluster compounds: Intense optical activity in metal-based transitions, *The Journal of Physical Chemistry B*, 104, 2000, 2630–2641.
31. S.H. Chen, H. Yao, K. Kimura, Reversible transference of Au nanoparticles across the water and toluene interface: A Langmuir type adsorption mechanism, *Langmuir*, 17, 2001, 733–739.
32. Y. Negishi, K. Nobusada, T. Tsukuda, Glutathione-protected gold clusters revisited: Bridging the gap between gold(I)-thiolate complexes and thiolate protected gold nanocrystals, *Journal of the American Chemical Society*, 127, 2005, 5261–5270.
33. A.P. Alivisatos, Semiconductor clusters, nanocrystals, and quantum dots, *Science, New Series*, 271, 1996, 933–937.
34. C.J. Murphy, J.L. Coffer, Quantum dots: A primer, *Journal of Applied Spectroscopy*, 56, 2002, 16A–27A.
35. E. Petryayeva, W.R. Algar, IL. Medintz, Quantum dots in bioanalysis: A review of applications a cross various platforms for fluorescence spectroscopy and imaging, *Journal of Applied Spectroscopy*, 67, 2013, 215–52.
36. C.B. Murray, D.J. Noms, M.G. Bawendi, Synthesis and characterization of nearly monodisperse cde (e = s, se, te) semiconductor nanocrystallites, *Journal of the American Chemical Society*, 115, 1993, 8706–8715.

37. B.O. Dabbousi, J. Rodriguez-Viejo, V. Mikulec, Frederic, J.R. Heine, H. Mattoussi, R. Ober, K.F. Jensen, M.G. Bawendi, (CdSe)ZnS core–shell quantum dots: Synthesis and optical and structural characterization of a size series of highly luminescent materials, *The Journal of Physical Chemistry B*, 101, 1997, 9463–9475.

38. X. Peng, M.C. Schlamp, A.V. Kadavanich, A.P. Alivisatos, Epitaxial growth of highly luminescent CdSe/CdS core/shell nanocrystals with photostability and electronic accessibility, *Journal of the American Chemical Society*, 119, 1997, 7019–7029.

39. M. Costa-Fernandez, R. Pereiro, A. Sanz-Medel, The use of luminescent quantum dots for optical sensing, *Trends in Analytical Chemistry*, 25, 2006, 207–218.

40. I.L. Medintz, H.T. Uyeda, E.R. Goldman, H. Mattoussi, Quantum dot bioconjugates for imaging, labelling and sensing, *Nature Materials*, 4, 2005, 435–446.

41. D. Bera, L. Qian, T.K. Tseng, P.H. Holloway, Quantum dots and their multimodal applications: A review, *Journal of Materials*, 3, 2010, 2260–2345.

42. R. D. Tilley, Synthesis and applications of nanoparticles and quantum dots, *Chemistry in New Zealand Journal*, 1, 2008, 146–150.

43. F.E. Kruisal, H. Fissana, A. Peleda, Synthesis of nanoparticles in the gas phase for electronic, optical and magnetic applications – A review, *Journal of Aerosol Science*, 29, 1998, 511–535.

44. A. Henglein, Small-particle research: Physicochemical properties of extremely small colloidal metal and semiconductor particles, *Chemical Reviews*, 89, 1989, 1861–1873.

45. M. Hayat, *Colloidal Gold: Principles, Methods and Applications*, Academic Press, San Diego, London, 1989.

46. J. Turkevich, J. Hillier, P.C. Stevenson, A study of the nucleation and growth processes in the synthesis of colloidal gold, *Discussions of the Faraday Society*, 11, 1951, 55–75.

47. M. Brust, M. Walker, D. Bethell, D.J. Schiffrin, R. Whyman, Synthesis of thiol-derivatised gold nanoparticles in a two-phase liquid–liquid system, *Journal of the Chemical Society, Chemical Communications*, 7, 1994, 801–802.

48. M.C. Daniel, D. Astruc, Gold nanoparticles: Assembly, supramolecular chemistry, quantum-size-related properties, and applications toward biology, catalysis, and nanotechnology, *Journal of Chemical Review*, 104, 2004, 293–346.

49. A. Frattini, N. Pellegri, D. Nicastro, O.D. Sanctis, Effect of amine groups in the synthesis of Ag nanoparticles using aminosilanes, *Materials Chemistry and Physics*, 94, 2005, 148–152.

50. X. Dong, Ji. Xiaohui, Wu. Hongli, L. Zhao, Li. Jun, Y. Wensheng, Shape control of silver nanoparticles by stepwise citrate reduction, *The Journal of Physical Chemistry C*, 113, 2009, 6573–6576.

51. J.P. Abid, A.W. Wark, P.F. Brevetb, H.H. Giraulta, Preparation of silver nanoparticles in solution from a silver salt by laser irradiation, *Chemical Communications Journal*, 7, 2002, 792–793.

52. J. Belloni, M. Mostafavi, H. Remita, J.-L. Marignier, M.-O. Delcourt, Radiation-induced synthesis of mono- and multi-metallic clusters and nanocolloids, *New Journal of Chemistry*, 11, 1998, 1239–1255.

53. S. Eustis, H.Y. Hsu, M.A. El-Sayed, Gold nanoparticle formation from photochemical reduction of $Au^{3+}$ by continuous excitation in colloidal solutions: A proposed molecular mechanism, *The Journal of Physical Chemistry B*, 109, 2005, 4811–4815.

54. A.K. Mittal, Y. Chisti, U. Ch. Banerjee, Synthesis of metallic nanoparticles using plant extracts, *Biotechnology Advances Journal*, 31, 2013, 346–356.

55. S. Rajesh, D.P. Raja, J.M. Rathi, K. Sahayaraj, Biosynthesis of silver nanoparticles using *Ulva fasciata* (Delile) ethyl acetate extract and its activity against *Xanthomonas campestris* pv. Malvacearum, *Journal of Biopesticides*, 5, 2012, 119–128.

56. J.A. Dahl, B.L.S. Maddux, J.E. Hutchison, Toward greener nanosynthesis, *Chemical Review*, 107, 2007, 2228–2269.

57. J.E. Hutchison, Greener nanoscience: A proactive approach to advancing applications and reducing implications of nanotechnology, *ACS Nano*, 2, 2008, 395–402.

58. P. Raveendran, J. Fu, S.L. Wallen, Completely green synthesis and stabilization of metal nanoparticles, *Journal of the American Chemical Society*, 125, 2003, 13940–13941.

59. A. Moreno-Vega, T. Gomez-Quintero, R. Nunez-Anita, L. Acosta-Torres, V. Castaño, Polymeric and ceramic nanoparticles in biomedical applications, *Hindawi Publishing Corporation Journal of Nanotechnology*, 1, 2012, 1–10.

60. I. Rahman, V. Padavettan, Synthesis of silica nanoparticles by sol–gel: Size-dependent properties, surface modification, and applications in silica–polymer nanocomposites: A review, *Hindawi Publishing Corporation Journal of Nanomaterials*, 1, 2012, 1–15.

61. K. Klabunde, J. Stark, O. Koper, C. Mohs, D. Park, Sh. Decker, Y. Jiang, I. Lagadic, D. Zhang, Nanocrystals as stoichiometric reagents with unique surface chemistry, *The Journal of Physical Chemistry*, 100, 1996, 12142–12153.
62. L.L. Hench, J.K. West, The sol–gel process, *Chemical Reviews*, 90, 1990, 33–72.
63. W. Stober, A. Fink, E. Bohn, Controlled growth of monodisperse silica spheres in the micron size range, *Journal of Colloid and Interface Science*, 26, 1968, 62–69.
64. H. Liu, G. Ning, Zh. Gan, Y. Lin, A simple procedure to prepare spherical α-alumina powders, *Materials Research Bulletin*, 44, 2009, 785–788.
65. P. Padmaja, P.K. Pillai, K.G.K. Warrier, Adsorption isotherm and pore characteristics of nano alumina derived from sol–gel boehmite, *Journal of Porous Materials*, 11, 2004, 147–155.
66. S. Granado, V. Ragel, Influence of alfa-Al$_2$O$_3$ morphology and particle size on drug release from ceramic/polymer composites, *Journal of Materials Chemistry*, 7, 1997, 1581–1585.
67. H. Iida, K. Takayanagi, T. Nakanishi, T. Osaka, Synthesis of Fe$_3$O$_4$ nanoparticles with various sizes and magnetic properties by controlled hydrolysis, *J. Colloid Interface Science*, 314, 2007, 274–280.
68. M.R. Ganjali, M. Hosseini, M. Khobi, Sh. Farahani, M. Shaban, F. Faridbod, A. Shafiee, P. Norouzi, A novel europium-sensitive fluorescent nano-chemo sensor based on new functionalized magnetic core–shell Fe$_3$O4@SiO$_2$ nanoparticles, *Talanta*, 115, 2013, 271–276.
69. F. E. Kruisal, H. Fissana, A. Peleda, Synthesis of nanoparticles in the gas phase for electronic, optical and magnetic applications – A review, *Journal of Aerosol Science*, 29, 1998, 511–535.
70. A.B.D., Nandiyanto, K. Okuyama, Progress in developing spray-drying methods for the production of controlled morphology particles: from the nanometer to submicrometer size ranges, *Advanced Powder Technology*, 22, 2011, 1–19.
71. J. Poostforooshan, S. Rennecke, M. Gensch, S. Beuermann, G.P. Brunotte, G. Ziegmann, A.P. Weber, Aerosol process for the *in situ* coating of nanoparticles with a polymer shell, *Aerosol Science and Technology*, 48, 2014, 1111–1122.
72. Y. Tamsilian, A. Ramazani S.A., Producing nanostructure of polymeric core–shell to intelligent control solubility of hydrophilic polymer during polymer flooding process, MSc. Thesis, Sharif University of Technology, 2011.
73. Y. Tamsilian, A. Ramazani S.A., M. Shaban, Sh. Ayatollahi, R. Tomovska, High molecular weight polyacrylamide nanoparticles prepared by inverse emulsion polymerization: Reaction conditions-properties relationships, *Colloid and Polymer Science*, 294, 2016, 513–525.
74. Y. Tamsilian, A. Ramazani S.A., Enhanced oil recovery performance and time-dependent role of polymeric core–shell nanoemulsion, *Scientia Iranica F*, 21, 2014, 1174–1178.
75. Y. Tamsilian, A. Ramazani S.A., Producing nanostructure of polymeric core–shell to intelligent control solubility of hydrophilic polymer during polymer flooding process, U.S. Patent, US20140187451, United States, 2014.
76. Y. Tamsilian, A. Ramazani S.A., Smart polymer flooding process with nanoscale core–shell structure, *The International Conference on Nanotechnology: Fundamentals and Applications (ICNFA 2013)*, Canada, 2013.
77. Y. Tamsilian, A. Ramazani S.A., Smart polymer flooding process, U.S. Patent, US20150148269, United States, 2015.
78. Y. Tamsilian, A. Ramazani S.A., Sh. Ayatollahi, Smart enhanced oil recovery process with thermoviscosifying polymer nanoparticles, Ph.D. Thesis, Sharif University of Technology, 2014.
79. S.F. Chin, A. Azman, S.C. Pang, Size controlled synthesis of starch nanoparticles by a microemulsion method, *Journal of Nanomaterials*, 7, 2014.
80. U.S. Khan, N.S. Khattak, A. Rahman, F. Khan, Optimal method for preparation of magnetite nanoparticles, *Journal of the Chemical Society of Pakistan*, 33, 2011, 628–633.
81. J. Wang, W. Chen, J. Sun, Ch. Liu, Q. Yin, L. Zhang, Y. Xianyu, X. Shi, G. Hu, X. Jiang, A microfluidic tubing method and its application for controlled synthesis of polymeric nanoparticles, *Lab on Chip*, 14, 2014, 1673–1677.
82. B. Ku, H. Seo, B.G. Chung, Synthesis and characterization of thermoresponsive polymeric nanoparticles, *BioChip Journal*, 8, 2014, 8–14.
83. L. Wei, T. Tang, B. Huang, Synthesis and characterization of polyethylene/clay–silica nanocomposites: A montmorillonite/silica-hybrid-supported catalyst and *in situ* polymerization, *Journal of Polymer Science Part A: Polymer Chemistry*, 42, 2004, 941–949.
84. T.J. Pinnavaia, G.W. Beall, *Polymer–Clay Nanocomposites*, Wiley, New York, 97, 2000.
85. D. Lee, H. Kim, K. Yoon, K.E. Min, K.H. Seo, S.K. Noh, Polyethylene/MMT nanocomposites prepared by *in situ* polymerization using supported catalyst systems, *Science and Technology of Advanced Materials*, 6, 2005, 457–462.

86. R.D. Farahani, A. Ramazani S.A., Melt preparation and investigation of properties of toughened Polyamide 66 with SEBS-g-MA and their nanocomposites, *Materials & Design*, 29, 2008, 105–111.

87. H. Baniasadi, A. Ramazani S.A., Sh. Mashayekhan, F. Ghaderinezhad, Preparation of conductive polyaniline/graphene nanocomposites via *in situ* emulsion polymerization and product characterization, *Synthetic Metals*, 196, 2014, 199–205.

88. H. Baniasadi, A. Ramazani S.A., S. Javan Nikkhah, Investigation of *in situ* prepared polypropylene/clay nanocomposites properties and comparing to melt blending method, *Materials and Design*, 31, 2010, 76–84.

89. A. Ramazani S.A., F. Tavakolzadeh, Preparation of polyethylene/layered silicate nanocomposites using *in situ* polymerization approach, *Macromolecular Symposia*, 274, 2008, 65–71.

90. K.M. Molapo, P.M. Ndangili, R.F. Ajayi, G. Mbambisa, S.M. Mailu, N. Njomo, M. Masikini, P. Baker, E.I. Iwuoha, Review paper electronics of conjugated polymers (I): Polyaniline, *International Journal of Electrochemical Science*, 7, 2012, 11859–11875.

91. S. Sinha, S. Bhadra, D. Khastgir, Effect of dopant type on the properties of polyaniline, *Journal of Applied Polymer Science*, 112, 2009, 3135–3140.

92. F. Yilmaz, Polyaniline: Synthesis, characterization, solution properties and composites, Ph.D. Thesis, 2007.

93. I. Harada, Y. Furukawa, F. Ueda, Vibrational spectra and structure of polyaniline and related compounds, *Synthetic Metals*, 29, 1989, 303–312.

94. T. Thanpitcha, A. Sirivat, A.M. Jamieson, R. Rujiravanit, Preparation, and characterization of polyaniline/chitosan blend film, *Carbohydrate Polymers*, 64, 2006, 560–568.

95. M.S. Dresselhaus, G. Dresselhaus, P.C. Eklund, *Science of Fullerenes and Carbon Nanotubes*. Academic Press, San Diego, 1996.

96. H. Dai, *Nanotube Growth and Characterization*, Springer, Berlin, 2001, pp. 29–53.

97. J. Kong, H. Soh, A. Cassell, C.F. Uate, H. Dai, Synthesis of individual single-walled carbon nanotubes on patterned silicon wafers, *Nature*, 395, 1998, 878–881.

98. S. Fan, M. Chapline, N. Franklin, T. Tombler, A. Cassell, Self-oriented regular arrays of carbon nanotubes and their field emission properties, *Science*, 283, 1999, 512–514.

99. H. Dai, Carbon nanotubes: Synthesis, integration, and properties, *Accounts of Chemical Research*, 2002, 35, 1035–1044.

100. A. Cassell, N. Franklin, T. Tombler, E. Chan, J. Han, Directed growth of free-standing single-walled carbon nanotubes, *Journal of the American Chemical Society*, 121, 1999, 7975–7976.

101. N. Franklin, H. Dai, An enhance chemical vapor deposition method to extensive single-walled nanotube networks with directionality, *Advanced Materials*, 12, 2000, 890–894.

102. Y. Zhang, A. Chan, J. Cao, Q. Wang, W. Kim, Electric field-directed growth of aligned single-walled carbon nanotubes, *Applied Physics Letters*, 79, 2001, 3155–3157.

103. J.R. Sanchez-Valencia, Th. Dienel1, O. Grouning1, I. Shorubalko, A. Mueller, M. Jansen, K Amsharov, P. Ruffieux, R. Fasel, Controlled synthesis of single-chirality carbon nanotubes, *Nature*, 512, 2014, 61–64.

104. K. Ohta, T. Nishizawa, T. Nishiguchi, R. Shimizu, Y. Hattori, Sh. Inoue, M. Katayama, K. Mizu-Uchic, T. Konoc, Synthesis of carbon nanotubes by microwave heating: Influence of diameter of catalytic Ni nanoparticles on diameter of CNTs, *Journal of Materials Chemistry A*, 2, 2014, 2, 2773–2780.

105. S. Arepalli, Laser ablation process for single-walled carbon nanotube production, *Journal of Nanoscience and Nanotechnology*, 4, 2004, 317–325.

106. A. Thess, R. Lee, P. Nikolaev et al., Crystalline ropes of metallic carbon nanotubes, *Science*, 1996, 273, 483–487.

107. M. Wilson, K. Kannangara, G. Smith, M. *Simmons, Nanotechnology: Basic Science and Emerging Technologies*, Chapman & Hall/CRC, Boca Raton, FL, 2002.

108. B. Hornbostel, M. Haluska, J. Cech, U. Dettlaff, S. Roth, Arc discharge and laser ablation synthesis of single walled carbon nanotubes, *Carbon Nanotubes*, Springer, Netherlands, 2006.

109. H. Omachi, T. Nakayama, E. Takahashi, Y. Segawa, K. Itami, Initiation of carbon nanotube growth by well-defined carbon nanorings, *Nature Chemistry*, 5, 2013, 572–576.

110. G. Mogilevsky, O. Hartman, E.D. Emmons, A. Balboa, J.B. DeCoste, B.J. Schindler, I. Iordanov, Ch.J. Karwacki, Bottom-up synthesis of anatase nanoparticles with graphene domains, *ACS Applied Materials & Interfaces*, 6, 2014, 10638–10648.

111. M. Mohammadikish, Green synthesis of nanorod Ni(salen) coordination complexes using a simple hydrothermal method, *CrystEngComm*, 16, 2014, 8020–8026.

112. B. Nikoobakht, M.A. El-Sayed, Preparation and growth mechanism of gold nanorods (NRs) using seed-mediated growth method, *Chemistry of Materials*, 15, 2003, 1957–1962.

113. X. Xu, Y. Zhao, X. Xue, Sh. Huo, F. Chen, G. Zou, X.J. Liang, Seedless synthesis of high aspect ratio gold nanorods with high yield, *Journal of Materials Chemistry A*, 2, 2013, 3528–3535.

114. K.A. Kozek, K.M. Kozek, W.C. Wu, S.R. Mishra, J.B. Tracy, Large-scale synthesis of gold nanorods through continuous secondary growth, *Chemistry of Materials*, 25, 2013, 4537–4544.

115. W.I. Park, D.H. Kim, S.W. Jung, G.C. Yi, Metal organic vapor-phase epitaxial growth of vertically well-aligned ZnO nanorods, *Applied Physics Letters*, 80, 2002, 4232–4234.

116. G.Ch. Yi, Ch. Wang, W. Park, ZnO nanorods: Synthesis, characterization and applications, *Semiconductor Science and Technology*, 20, 2005, S22–S34,

117. J.I. Mitchell, N. Zhou, W. Nam, L.M. Traverso, X. Xu, Sub-diffraction laser synthesis of silicon nanowires, Scientific Reports, *Nature*, 4, 3908–3912.

118. S. Kim, Y. Lee, A. Gu, Ch.You, K. Oh, S. Lee, Y. Im, Synthesis of vertically conformal zno/cuo core–shell nanowire arrays by electrophoresis-assisted electroless deposition, *The Journal of Physical Chemistry C*, 118, 2014, 7377–7385.

119. L. Liu, K. Hong, X. Ge, D. Liu, M. Xu, Controllable and rapid synthesis of long ZnO nanowire arrays for dye-sensitized solar cells, *The Journal of Physical Chemistry C*, 118, 2014, 15551–15555.

120. T.I. Panova, M. Yu. Arsent'ev, L.V. Morozova, I.A. Drozdova, Synthesis and investigation of the structure of ceramic nanopowders in the $ZrO_2$–$CeO_2$– $Al_2O_3$ system, *Glass Physics and Chemistry*, 36, 2010, 470–477.

121. M. Yu, Arsent'ev, M.V. Kalinina, P.A. Tikhonov, L.V. Morozova, A.S. Kovalenko, N. Koval'ko, I.I. Khlamov, O.A. Shilova, Synthesis and study of sensor oxide nanofilms in a $ZrO_2$–$CeO_2$system, *Glass Physics and Chemistry*, 40, 2014, 362–366.

122. S. Kumar, P. Sharma, V. Sharma, CdS nanofilms: Synthesis and the role of annealing on structural and optical properties, *Journal of Applied Physics*, 111, 2012, 043519-6.

123. P. Yang, G. Wang, Zh. Gao, H. Chen, Y. Wang, Y. Qin, Uniform and conformal carbon nanofilms produced based on molecular layer deposition, *Materials*, 6, 2013, 5602–5612.

124. L. Zhang, J. Lian, L. Wu, Zh. Duan, J. Jiang, L. Zhao, Synthesis of a thin-layer $MnO_2$ nanosheet-coated $Fe_3O_4$ nanocomposite as a magnetically separable photocatalyst, *Langmuir*, 30, 2014, 7006–7013.

125. K.S. Novoselov, A.K. Geim, S.V. Morozov, D. Jiang, Y. Zhang, Electric field effect in atomically thin carbon films, *Science*, 306, 2004, 666–669.

126. D. Jinhong, H. Cheng, The fabrication, properties, and uses of graphene/polymer composites, *Macromolecular Chemistry and Physics*, 213, 2012, 1060–1077.

127. X. Du, I. Skachko, A. Barker, E.Y. Andrei, Approaching ballistic transport in suspended graphene, *Nature Nanotechnology*, 3, 2008, 491–495.

128. D.W. Breck, *Zeolite Molecular Sieves: Structure, Chemistry, and Use*, Wiley, New York, 1974.

129. C.S. Cundy, P.A. Cox, The hydrothermal synthesis of zeolites: History and development from the earliest days to the present time, *Chemical Reviews*, 103, 2003, 663–701.

130. A.B. Murcia, Ordered porous nanomaterials: The merit of small-article review, *Hindawi Publishing Corporation ISRN Nanotechnology*, 1, 2013, 1–29.

131. A. Corma, M.E. Davis, Issues in the synthesis of crystalline molecular sieves: Towards the crystallization of low frame work density structures, *Chem Phys Chem*, 5, 2004, 304–313.

132. R.A. van Santen, G. Ooms, C.J.J. denouden, B.W. van Beest, M.F.M. Post, Computational studies of zeolite framework stability, *Zeolite Synthesis*, American Chemical Society, New York, USA, 1989, pp. 617–633.

133. A. Navrotsky, I. Petrovic, Y. Hu, C.Y. Chen, M.E. Davis, Little energetic limitation to microporous and mesoporous materials, *Microporous Materials*, 4, 1995, 95–98.

134. P.M. Piccione, S. Yang, A. Navrotsky, M.E. Davis, Thermodynamics of pure-silica molecular sieve synthesis, *Journal of Physical Chemistry B*, 106, 2002, 3629–3638.

135. C.S. Cundy, P.A. Cox, The hydrothermal synthesis of zeolites: Precursors, intermediates and reaction mechanism, *Microporous and Mesoporous Materials*, 82, 2005, 1–78.

# 4 Optical Properties of Nanomaterials

*Veronica Marchante Rodriguez and Hrushikesh A. Abhyankar*

## CONTENTS

## 4.1 INTRODUCTION

Over the last few decades, nanoparticles and nanomaterials have attracted the interest of researchers due to their unique properties when compared to macroscopic (bulk materials) and atomic (molecules) levels. In terms of optical properties, some nanoparticles and nanomaterials present good characteristics that make them suitable for applications like imaging, sensing, coloration, etc. This has favored the research on the implementation of nanomaterials for optical applications, as well as the development of new technologies or adjustment of existing ones.

This chapter briefly introduces the physical phenomena resulting from the interaction of light-matter and the characterization techniques. This chapter also contains a detailed explanation about the application of nanomaterials based on their optical properties, focusing on relevant fields like sensing, solar cells, photocatalysis, colorants, etc.

## 4.2 BASIC CONCEPTS IN OPTICS

### 4.2.1 INTERACTION LIGHT-MATTER

When an electromagnetic radiation (like light) interacts with matter, if the response is proportional to the intensity and/or properties of the radiation, the matter exhibits linear optic properties. On the other hand, when the response is not a linear function of the electromagnetic radiation/field the

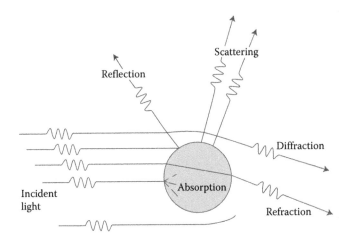

**FIGURE 4.1** Physical phenomena from the interaction light-matter. (Adapted from K. Okuyama, Laser light scattering, in *Aerosol Particles,* Okuyama Group, August 2002. http://aerosols.wustl.edu/AAARworkshop08/materials/Okuyama/.)

material exhibits nonlinear optical properties. In other words, nonlinear optical behavior appears when the response of the media (dielectric polarization) is nonlinear to the incident electromagnetic field. Nonlinear properties usually appear under high intensity electromagnetic field. The following section discusses this phenomenon in detail.

### 4.2.2 LINEAR OPTICAL PROPERTIES

The physical phenomena that result from the interaction between light and matter with linear optical behavior can be summarized in Figure 4.1. These phenomena are reflection and refraction, absorption and scattering, and diffraction.

#### 4.2.2.1 Reflection and Refraction of Light

Reflection of light occurs when the photons of a light beam are sent back to the hemisphere of the incident light. When the reflection takes place on smooth surfaces, it is called specular reflection, while on rough surfaces it is called scatter reflection. On the other hand, refraction refers to the changes in the light beam direction when it goes through a media with different speed of light. Then, the refractive index "n" is defined as the ratio between the speed of light in the media and the speed of light in a vacuum. Some media are able to attenuate electromagnetic waves, and they are said to have complex refractive index. In general, both phenomena, reflection and refraction, are described using Snell's law: $n_1 \cdot \sin \theta_1 = n_2 \cdot \sin \theta_2$ (including media with complex refractive index) (Figure 4.2) [2].

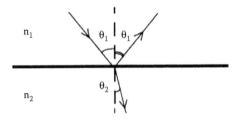

**FIGURE 4.2** Reflection and refraction of a light beam at a refractive surface. (Adapted from G. Sharma and R. Bala, *Digital Color Imaging Handbook.* CRC Press, 2002.)

### 4.2.2.2 Absorption and Scattering of Light

All the processes in which the intensity of the light is reduced are related to the absorption of light; for example, the transformation of the radiation energy of the light in any other type of energy, like heat. The most extensive theory about the light absorption of materials is the Beer–Lambert–Bouguer law. According to this, the variation of the intensity of a collimated light beam $d\phi$ when it goes through a medium of thickness dx with a concentration c of dispersed particles is proportional to the concentration c, to the initial intensity of the light beam $\phi$ and to the thickness of the medium dx (Figure 4.3):

$$d\phi = -\varepsilon(\lambda) \cdot c \cdot \phi \cdot \ln(10)dx$$

$$\phi(X) = \exp\left[-X \cdot c \cdot \varepsilon(\lambda)\ln 10\right] \cdot \phi(0) = 10^{-[X \cdot c \cdot \varepsilon(\lambda)]} \cdot \phi(0)$$

In which $\lambda$(nm) is the wavelength of the light beam, $\varepsilon(\lambda)$ (m²/mol) is the molar absorption coefficient.

Transmittance or transmittance spectrum $T(\lambda)$ is defined as the rate between the light intensity after $\phi(X)$ and before $\phi(0)$ going through the medium:

$$T(\lambda) = \phi(X) / \phi(0) = \exp[-X \cdot c \cdot \varepsilon(\lambda)\ln 10] = 10^{-[X \cdot c \cdot \varepsilon(\lambda)]}$$

When $T(\lambda) = 1$ the medium is considered transparent (the medium does not absorb or interact with light), while $T(\lambda) = 0$ corresponds to an opaque medium (the light beam cannot go through the medium). When $0 \leq T(\lambda) \leq 1$, then the medium is called translucent. Usually, Lambert–Beer law is expressed in logarithmic terms:

$$D(\lambda) = -\log T(\lambda)$$

In which $D(\lambda)$ is the absorbance of the medium. Analogously when $D(\lambda) = 0$ the medium is transparent, and when $D(\lambda) = \infty$ the medium is opaque. The absorbance $D(\lambda)$ is generally considered as an additive property when the particles do not interact. Then for a mixture of substances, the total absorption will be given by the summation of the absorption of each substance, $D(\lambda) = \Sigma_i D_i(\lambda)$, while the transmittance will be the product of the transmittance of each substance, $T(\lambda) = \Pi_i T_i(\lambda)$.

Lambert–Beer law for diffuse light beams is the following:

$$d\phi = -2 \cdot \varepsilon(\lambda) \cdot c \cdot \phi \cdot \ln(10)dx$$

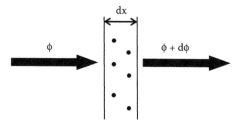

**FIGURE 4.3** Light absorption of a medium with thickness dx and concentration c. (Adapted from G. Sharma and R. Bala, *Digital Color Imaging Handbook*. CRC Press, 2002.)

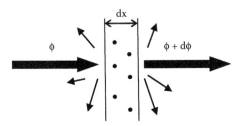

**FIGURE 4.4** Light scattering in medium of thickness dx with dispersed particle concentration c. (Adapted from G. Sharma and R. Bala, *Digital Color Imaging Handbook.* CRC Press, 2002.)

Scattering of light refers to the processes that deviate photons in different directions (Figure 4.4). In heterogeneous media, this phenomenon is related to differences in the refraction index, like in a matrix with small particle size filler. The variation of the flux intensity of a collimated light beam $d\phi$ when it goes through a medium of thickness dx is given by:

$$d\phi = -\beta(\lambda) \cdot \phi \cdot dx$$

$$\beta(\lambda) = \sigma(\lambda) \cdot c \cdot \ln(10)$$

In which $\lambda$ (nm) is the wavelength, $\beta(\lambda)$ is the scattering coefficient of the medium, and $\sigma(\lambda)$ $(m^2/mol)$ is the molar scattering extinction coefficient (light scattering caused by a mol of spheric particles of radius r). Therefore, both phenomena (scattering and absorption of light) contribute in the total extinction coefficient of a particle $\varepsilon_T(\lambda)$:$\varepsilon_T(\lambda) = \varepsilon(\lambda) + \sigma(\lambda)$.

Using these physical properties, the parameters that describe the optical behavior of materials (absorption coefficient $K(\lambda)$ and scattering coefficient $S(\lambda)$) can be calculated:

$$K = 2 \cdot \ln 10 \cdot \varepsilon(\lambda) \cdot c \quad S = 2 \cdot \ln 10 \cdot \frac{\sigma(\lambda)}{2} \cdot c$$

### 4.2.2.3  Diffraction of Light

The diffraction of light is a phenomenon that occurs when a light beam is deviated from its direction when it encounters an obstacle (opening, particle, barrier, etc.). This phenomenon is only noticeable when the wavelength and the size of the obstacle are comparable, that is, similar magnitude order.

In light–matter interactions, Mie theory is one of the most extensive and accepted theories to describe the light diffraction phenomenon that occurs when a light beam interacts with a single particle. Detailed explanation and resolution of Maxwell's equations for Mie theory and for the limit theories can be found in a general bibliography about optics. Briefly, the hypothesis applied in Mie theory is as follows:

- The incident electromagnetic field is plane and monochromatic.
- The particle is spherical, isotropic, and homogeneous.
- The medium surrounding the particle is homogeneous, not conductive and not magnetic.

For the case of a spherical, isotropic, and homogeneous particle, the equations that describe the light scattering according to Mie theory are

$$E^s = -\frac{e^{ikd}}{ikd} \mathbf{S} E^0 \rightarrow \begin{bmatrix} E_1^S \\ E_2^S \end{bmatrix} = -\frac{e^{ikd}}{ikd} \begin{bmatrix} S_2 & 0 \\ 0 & S_1 \end{bmatrix} \begin{bmatrix} E_1^0 \\ E_2^0 \end{bmatrix}$$

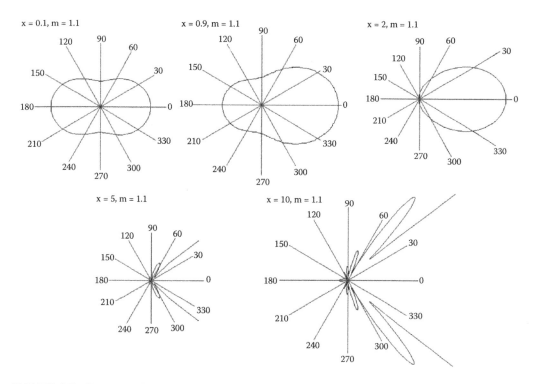

**FIGURE 4.5** Representation of the response of the diffraction for a particle when the ratio of refractive indexes is 1.1 (m = 1.1) for several values of x.

where **S** is the amplified scattering matrix, "1" and "2" represent the polarization states of the incident field parallel and perpendicular to the plane of dispersion, respectively, "0" indicates incident wave (or electromagnetic field), subindex "S" indicates scattered wave, and "E" is the vector that represents the electric field. Figures 4.5 and 4.6 represent the solutions to the Mie theory for the diffraction of light by a particle, for different ratios between particle size and wavelength of incident light ("$x = \pi D/\lambda$" where D is the particle diameter and $\lambda$ is the wavelength), in two situations: when the ratio between the refractive index of the particle and in the medium ("m"= particle refractive index/medium) is very low (m = 1.1, similar refractive index in the particle and in the medium) (Figure 4.5), or very high (m = ∞, particle refractive index is very high compared to the medium) (Figure 4.6).

For one single particle, limit theories of MIE scattering theory ("$x = \pi D/\lambda$" where D is the particle diameter and $\lambda$ the wavelength; and "m" is the ratio between refractive index in the particle/ material and in the media) are presented in Figure 4.7.

### 4.2.3 NONLINEAR OPTICAL PROPERTIES

Nonlinear optical properties are those related to the atomic response of dielectric materials to electromagnetic fields of high intensity. The result is the emission of a modified electromagnetic field (polarization) with different characteristics (phase, frequency, and amplitude) from the incident field (light beam). When the resulting electromagnetic field (polarization) has double the frequency of the incident field (light beam), the process is called "second harmonic generation" (SHG). This phenomenon occurs commonly in nonsymmetric crystalline materials.

For electromagnetic fields with low intensity, the induced polarization is usually directly proportional to the incident light beam. This gives origin to the linear optical properties. When the

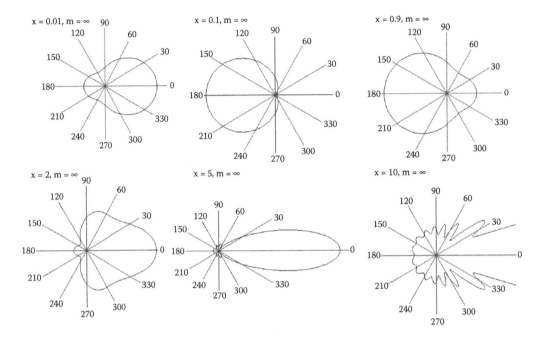

**FIGURE 4.6** Representation of the response of the diffraction for a particle when the ratio of refractive indexes is high (m = ∞) for several values of x.

intensity of the incident electromagnetic field ($\vec{E}$) is high (compared to the atomic dipolar forces), the polarization ($\vec{P}$) is given by [3]:

$$\vec{P} = \varepsilon_0 \gamma^{(1)}\vec{E} + \gamma^{(2)}\vec{E}\vec{E} + \gamma^{(3)}\vec{E}\vec{E}\vec{E} + \cdots$$

where $\varepsilon_0$ is the permittivity of the free space, and $\gamma^{(1)}$ is the linear susceptibility of the medium (related to linear optical properties like refractive index, absorption, scattering, etc.), and the other $\gamma^{(i)}$ are the nonlinear susceptibilities of the medium. $\gamma^{(2)}$ is related to SHG, frequency mixing, and parametric generation, and $\gamma^{(3)}$ is related to the "third harmonic generation" (THG), stimulated

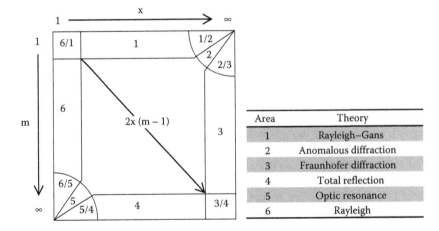

**FIGURE 4.7** Representation of the limit theories for the MIE theory and the range in which they applied.

Raman scattering, etc. A more detailed explanation about the mathematical description of SHG and THG can be found in the literature.

## 4.3 OPTICAL CHARACTERIZATION OF NANOMATERIALS

The techniques used for the optical characterization of materials are based on the study of the response of the materials when they interact with an incident light beam. In general terms, spectroscopy is the technique used to measure the spectrum of the material, and then quantify the amount of energy absorbed, scattered, diffracted, etc. There are several optical spectroscopic techniques, depending on the properties to evaluate/study and the type of material (solid, liquid, and gas). The most common ones are UV–visible, infrared (IR), and Raman spectroscopy. However, there are other techniques like nonlinear spectroscopy and dynamic light scattering (DLS) that give information about properties at the atomic and molecular levels. These techniques are summarized in Table 4.1 [4].

Almost all these techniques are based on the same working/designing principles. The elements of these systems are a source of light, a filter to obtain a monochromatic light beam (one single wavelength), a sample, a detector to measure/capture the light after interacting with the sample, and a recorder to analyze the signal (Figure 4.8). The difference among the spectroscopic techniques relies on the range of the spectrum to analyze. This affects the source of light used. Some systems

**TABLE 4.1**
**Summary of Spectroscopic Techniques for Characterization of Materials**

| Energy Exchange between Photons and Matter | | |
| --- | --- | --- |
| **Type of Energy Transfer** | **Region of Electromagnetic Spectrum** | **Spectroscopic Technique** |
| Absorption | γ-ray | Mossbauer spectroscopy |
| | X-ray | X-ray absorption spectroscopy |
| | UV/vis | UV/vis spectroscopy |
| | | Atomic absorption spectroscopy |
| | IR | Infrared spectroscopy |
| | | Raman spectroscopy |
| | Microwave | Microwave spectroscopy |
| | Radio wave | Electron spin resonance spectroscopy |
| | | Nuclear magnetic resonance spectroscopy |
| Emission (thermal excitation) | UV/vis | Atomic emission spectroscopy |
| Photoluminescence | X-ray | X-ray fluorescence |
| | UV/vis | Fluorescence spectroscopy |
| | | Phosphorescence spectroscopy |
| | | Atomic fluorescence spectroscopy |
| Chemiluminescence | UV/vis | Chemiluminescence spectroscopy |
| No Energy Exchange between Photons and Matter | | |
| **Type of Interaction** | **Region of Electromagnetic Spectrum** | **Spectroscopic Technique** |
| Diffraction | X-ray | X-ray diffraction |
| Refraction | UV/vis | Refractometry |
| Scattering | UV/vis | Nephelometry |
| | | Turbidimetry |
| Dispersion | UV/vis | Optical rotary dispersion |

*Source:* Adapted from D. Harvey, Spectroscopic methods, in *Analytical Chemistry 2.0*, 2009, p. 124.

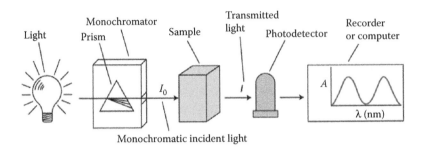

**FIGURE 4.8** Schematic representation of the elements and process of an spectrophotometer. (Adapted from L. Taiz and E. Zeiger, Topic 7.1 principles of spectrophotometry, in *Plant Physiology*, 5th ed. Online, Sinauer, 2010.)

split the incident light beam and compare the nonaltered light beam with the beam that interacts with the sample. In other cases, an integrating sphere is implemented in the system to obtain a diffused source of light (in which the flux is equally divided in all directions) instead of using a unidirectional (collimated) light beam. This is applied for obtaining information about the optical response of materials under ambient conditions, like reflection, absorption, and scattering of light. Techniques like dynamic light scattering (DLS), based on Rayleigh scattering, are used mainly to determine particle size distribution in particle dispersions in liquid or gas medium.

All these techniques have been modified and/or adapted in order to accommodate for the characterization of nanomaterials and nanoparticles, but they operate under the same principles.

## 4.4  NANOMATERIALS IN OPTICAL APPLICATIONS

The word "nanoparticle" (NP) represents a wide variety of materials and shapes. One generic classification of nanoparticles is based on the shape (Figure 4.9). Within each type, we have to consider different composition base. In some cases, the final shape of the nanoparticle depends on the synthesis process, and then the nanomaterial will fall in one or another category. Nanomaterials are applied in a great number of sectors (Figure 4.10), and research is being conducted to extend and/or improve their application in others. Applications based on optical properties of nanomaterials include optical detector, LASER, sensor, imaging, phosphor, display, solar cell, photocatalysis, photoelectrochemistry, biomedicine, colorants, etc. [6]. Some of these applications are explained in this section.

### 4.4.1  NANOMATERIALS IN SOLAR CELL, PHOTOCATALYSIS, AND PHOTOELECTROCHEMISTRY

Obtaining electricity from an unlimited source of energy, like sunlight, is a promising technology to replace those depending on fossil fuel. Photocatalysis and photoelectrochemistry applications and solar cells are all based on this principle. Apart from the production of electricity, they can also be used in obtaining $H_2$ through water splitting process, purification of water, air, and soil, chemical reactions, etc. Briefly, the steps that take place in the photovoltaic process are absorption of sunlight photons by semiconducting materials, excitation of electrons as a result of the sunlight energy absorption, transference of excited electrons in the media and/or material system towards an electrode, and storage or usage of the electron flow (electricity).

The photovoltaic effect and photovoltaic cells have been known since the end of the nineteenth century. However, it was after the 1970s that this technology experienced a boost. With the aim of developing a highly efficient and competitive technology to substitute fossil fuel, a wide variety of materials were investigated. Particularly for solar cell application, the most representative semiconducting material is silicon, classified according to crystallinity and crystal size (monocrystalline,

Quantum dots        Liposomes        Iron oxide NPs        Carbon nanotubes

Gold NPs        Polymeric NPs        Dendrimers        Micro- and nanobubbles

Upconverting NPs        Iron-platinum NPs        Nanoclusters        Functionalized NPs

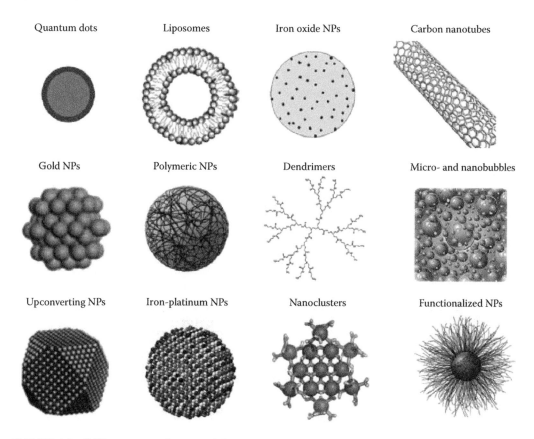

**FIGURE 4.9** Different types of nanoparticles (NPs). (Adapted from F. Re, R. Moresco, and M. Masserini, *J. Phys. Appl. Phys.*, 45(7), 073001, Feb. 2012.)

polycrystalline, and ribbon). But also, thin films of materials like active material—CdTe, CIGS, amorphous Si, GaAs—between two panes of glass are used. New research trends for solar cell application are considering materials like perovskite, liquid inks, doped rare earth materials, light-absorbing dyes, quantum dots, and organic/polymer solar cells. Other materials, like $TiO_2$, present exceptional light absorption properties, which make them suitable as photocatalysts for water splitting.

The main objective for new developments is to increase the efficiency of the photovoltaic materials. Nanomaterials have been studied for this purpose, and apparently quantum dots and combinations of nanomaterials and light-absorbing dyes are promising alternatives for these applications.

### 4.4.1.1 Combination of Nanomaterials and Light-Absorbing Dyes

Nanoparticles in combination with light-absorbing dyes are applied in "dye-sensitized solar cell" (DSSC). In general, DSSCs are arranged in sandwich structures (Figure 4.11), with the same basic elements of an electrochemical cell [9,10]: photoelectrode (anode), made of conductive glass coated by a semiconducting material (mesoporous $TiO_2$ or ZnO nanoparticles) film in which a light-absorbing dye (Ru bipyridyl complexes) is adsorbed; counter-electrode or cathode, made also of conductive glass but coated by a catalyst thin film (palladium, graphite); and electrolyte, an ionic solution (typically $I^-/I_3^-$), to facilitate the transference of electrons and charge.

Briefly, the physical-chemical processes that occur in a DSSC can be described as follows: the dye absorbs photons (light harvesting) and gets into an excited state. Then, the dye transfers electrons to the semiconducting material, getting to an oxidized state. These electrons flow from

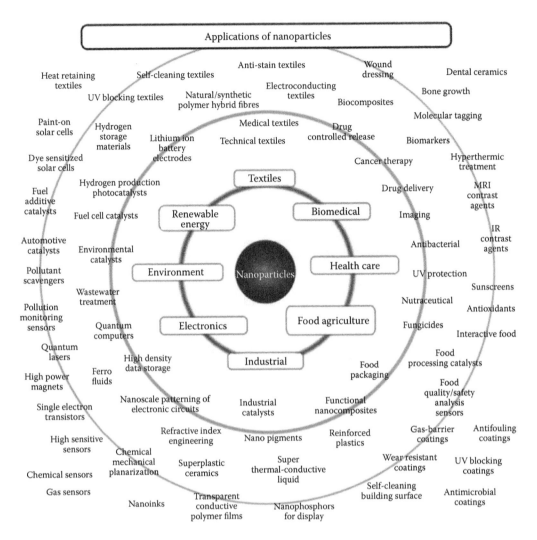

**FIGURE 4.10** Commercial applications of inorganic NPs. (Adapted from T., Tsuzuki, *Int. J. Nanotechnol.*, 6(5/6), 567, 2009.)

the photoelectrode (semiconducting material) to the counter-electrode through an external circuit (Figure 4.12). On the other hand, the oxidized dye is regenerated by the electrolyte, which becomes oxidized. Finally, the oxidized electrolyte is reduced in the counter-electrode using electrons from the external circuit.

The main advantage of using semiconducting materials like $TiO_2$ and $ZnO$ is the high stability and durability in solution, but their ability to absorb light is low because they have wide band gaps (Figure 4.13) [11]. The light-absorbing dyes adsorbed on the surface of the semiconducting material act as initiators: they absorb the light photons and transfer the electrons to the semiconducting material, starting the photovoltaic process. $TiO_2$ and $ZnO$ NPs increase the efficiency of this type of system, as they provide high surface area for the interaction with the dye (from 40 to 150 m²/g) [12,13,10]. Mesoporous nanocrystalline $TiO_2$ has a sponge-like structure that allows adsorbing a great amount of dye, scattering light and the diffusion of the electrolyte to regenerate the dye. As a result, the light harvesting efficiency (LHE) can increase to 90%–100% for the dye maximum wavelength absorption. It has been proved that the light-to-electricity conversion efficiency of a DSSC can reach 10.4% [10].

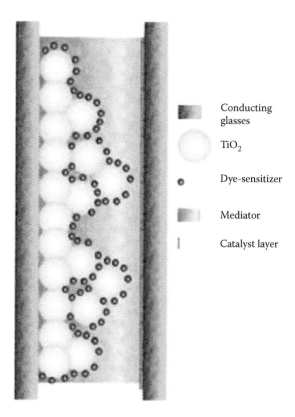

Conducting glasses

TiO$_2$

Dye-sensitizer

Mediator

Catalyst layer

**FIGURE 4.11** **(See color insert.)** Schematic arrangement of a dye-sensitized solar cell. (With kind permission from Springer Science+Business Media: *Nanoenergy: Nanotechnology Applied for Energy Production*, 2012, de Souza, F. L. and Leite, E. R.)

#### 4.4.1.2 Quantum Dots

Following the same structure as DSSC, solar cells can be manufactured replacing the light-absorbing dye by quantum dots (quantum dot solar cells, QDSC) [14]. In this case, TiO$_2$ and ZnO NPs are doped with semiconducting nanoparticles with low band gap to permit the absorption of light in the visible-IR range of the spectrum. Some examples of these quantum dots are cadmium, antimony, or lead salts (CdS, CdSe, Sb$_2$S$_3$, and PbS). In some cases, researchers have reported energy conversion efficiencies up to 8.55% for QDSC [15].

Doped TiO$_2$ and ZnO NPs are also very useful in other applications, like in photocatalysis. They can be used as photocatalysts for the degradation of contaminants in the purification of water and air. For this sort of application, the NPs selected as doping agents are in general noble metal ions, like silver (Ag), gold (Au), platinum (Pt), or palladium (Pd); nonmetal ions like nitrogen (N), carbon (C), fluorine (F), phosphorus (P), or sulfur (S); or a combination of both [16]. Noble metallic ions absorb light in the visible range due to the surface plasmon resonance (SPR) effect (oscillating motion of electrons on metal surfaces excited by an electromagnetic field). These doping elements extend the light absorption of the TiO$_2$ or ZnO NPs from the UV region to the visible region. But also they prevent the charge recombination of the photoexcited electron–hole pairs. On the other hand, doping TiO$_2$ or ZnO NPs with nonmetallic ions creates localized wavelength absorption in the visible region. Apparently, nitrogen is the doping agent that exhibits higher enhancement [16–18]. A promising alternative is using both types of dopants simultaneously. Few studies have been conducted in this topic, but particularly TiO$_2$ nanoparticles doped with Au and N (Au–N/TiO$_2$ NPs) showed higher visible light absorption and catalyst activity than TiO$_2$ NPs doped with each of the doping agents (Au/TiO$_2$ NPs or N/TiO$_2$ NPs) [16,19,20].

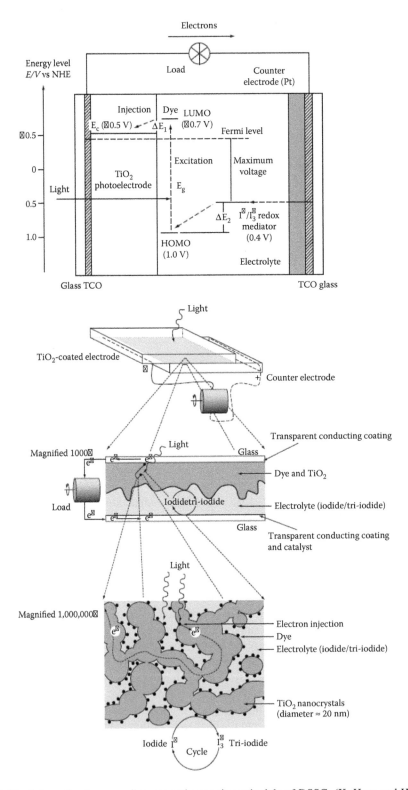

**FIGURE 4.12** Schematic of energy diagram and operating principle of DSSC. (K. Hara and H. Arakawa: Dye-sensitized solar cells. In *Handbook of Photovoltaic Science and Engineering*. A. Luque and S. Hegedus, Eds. 2003. 663–700. Copyright Wiley-VCH Verlag GmbH & Co. KGaA. Reproduced with permission.)

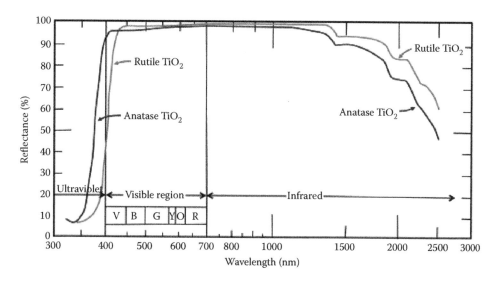

**FIGURE 4.13** Typical reflectance spectra of $TiO_2$. In the visible range, 400–700 nm, the reflection of $TiO_2$ is close to 100% what indicates that the absorption is close to 0%. (Adapted from Y. Lan, Y. Lu, and Z. Ren, *Nano Energy*, 2(5), 1031–1045, Sep. 2013.)

Figure 4.15 shows some examples of the variation in the absorbance spectra of $TiO_2$ and ZnO after being doped. Pure $TiO_2$ and ZnO present very low absorbance in the visible range (400–700 nm), but very high in the UV range (200–380 nm). Doping with metal particles (Fe, Au, and Pt) increases the absorbance in the visible region, demonstrated by the shoulders and peaks that the samples present in that region.

Some examples of successful application of doped $TiO_2$ and ZnO NPs for the removal of contaminants from water and/or air are the degradation of up to 95% of p-nitrophenol using $TiO_2$ nanoparticles doped with $Fe^{3+}$ [24], and the degradation of 60% and 80% methyl orange in aqueous solution using $TiO_2$ and ZnO NPs doped with N, respectively [22].

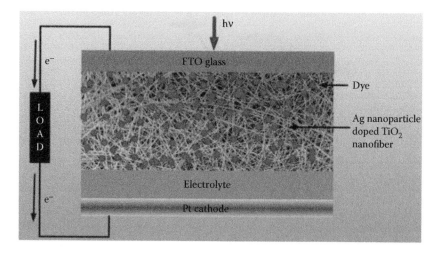

**FIGURE 4.14** (**See color insert.**) Scheme of a solar cell combining quantum dots and dye-sensitizers. (Adapted from J. Li et al., *Chem. Phys. Lett.*, 514(1–3), 141–145, Sep. 2011.)

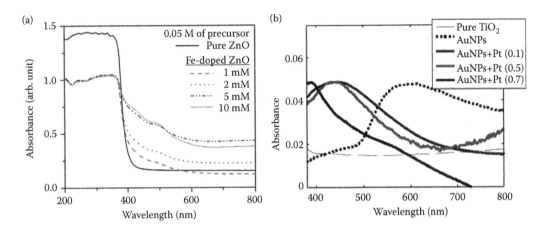

**FIGURE 4.15** Examples of the variation of the absorbance spectra of doped $TiO_2$ and ZnO. (Adapted from M. Silambarasan, S. Saravanan, and T. Soga, *Phys. E: Low Dimens. Syst. Nanostruct.*, 71, 109–116, Jul. 2015; S. Sun et al., *Ceram. Int.*, 39(5), 5197–5203, July 2013; J. P. Campos-López et al., *Mater. Sci. Semicond. Process.*, 15(4), 421–427, Aug. 2012.)

### 4.4.2 NANOMATERIALS IN OPTICAL DETECTOR AND SENSORS

The evolution of technology toward small, compact, and more efficient devices is the driving force in the development of nanotechnology in the sensor field. Some nanomaterials, especially inorganic and metallic ones, exhibit very good properties like high conductivity, magnetism, chemical inertness, etc., that makes them suitable for electronic and sensor applications [25]. A sensor is a device that detects a signal (typically a physical stimulus) and responds to it. Analogically, a nanosensor is a device with at least one component in the nanosize range able to detect a signal (optical, electronic, electrical, chemical, etc.) and respond to it (like transforming it into another signal that carries the information). Figure 4.16 shows a generic representation for the modification of nanoparticles (NPs) according to their final purpose. By attaching/bonding different agents, NPs can be functionalized

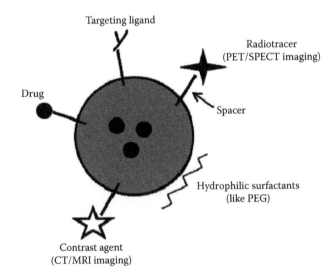

**FIGURE 4.16** Multifunctionalized NP for different applications (molecular imaging, drug delivery, targeting, and stealth) depending on the type of surfactant. (Adapted from F. Re, R. Moresco, and M. Masserini, *J. Phys. Appl. Phys.*, 45(7), 073001, Feb. 2012.)

and tailored for specific applications. Nanosensors can be classified depending on the detected signal or application field (Figure 4.17). But also, there are classifications based on the type and shape of the nanomaterial they are made of, like cantilever, nanotubes, and nanowires.

In this section we will focus on the nanomaterials applied to sensors due to their optical properties. Some of the main types of optical nanosensors are [28]

- *Electro-optical sensors:* They convert an optical signal (properties of light-wavelength, intensity, polarization, etc. or changes in light, using measurements of absorbance, reflectance, luminesce, etc.) into an electronic signal. These kinds of sensors are applied in LCD for screen devices like mobile phones, TVs, and notebooks [26].
- *Photoluminescent nanosensors:* They use changes in the luminescence properties of the nanomaterial, like inhibition (Si nanocrystals) or exhibition (nanoparticles of metal oxides or quantum dots), as a result of physical or chemical alteration of the surface.
- *Nanostructured surface plasmon resonance sensors:* Metallic nanoparticles, like gold and silver, exhibit surface plasmon resonance. Then the variation of aggregation state can lead to changes in the absorbance properties, which can be measured.
- *Fiber-optic nanosensors:* These sensors are mostly based on the alteration on light intensity or property (color, refractive index, etc.) produced by the substance to be detected [29,30]. The main advantages of fiber optic chemical sensors are the safety in the detection of explosive substances, insensitivity to electromagnetic fields, and the capability of carrying a huge amount of information [31].

Some nanomaterials change their optical properties in the presence of certain substances or as a result of changes in environment conditions. These materials can be applied as detectors. There are mainly three optical properties that are easy to observe and/or measure: absorbance, photoluminescence, and chemiluminescence [32].

Changes in the absorbance of the nanomaterials can be triggered by humidity, gases, aggregation state, etc., and it implies a change in color. Some potential applications for this type of sensor are detection of glucose using gold-nanoparticle-based colorimetric sensor [33], screening of antibodies in clinical diagnosis, and detection of gases, such as detection of CO using Au/NiO thin film or humidity [34,35].

The detectors and sensors based on photoluminescence of nanomaterials are based mainly on the inhibition of the photoluminescence ("quenching") of silicon nanocrystals. The variation in the photoluminescence is caused by chemical modification of the surface, which can be chemical absorption or adsorption (physisorption). A potential application is the detection of explosives in mines (due to the photoluminescence quenching effect of nitrotoluenes in Si nanocrystals). Other nanomaterials that can find application in this type of sensor are quantum

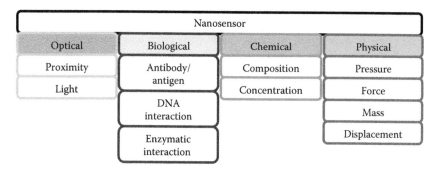

**FIGURE 4.17** Generic classification of nanosensors based on application and signal detected. (Adapted from T. A. Faisal, Nanosensors basics, design and applications, 15:53:44 UTC, January 2015. http://es.slideshare.net/ taifalawsi1/nanosensors-basics-design-and-applications. [Accessed May 5, 2011]; S. Agrawal and R. Prajapati, *Int. J. Pharm. Sci. Nanotechnol.*, 4(4), 1528–1535, 2012.)

**TABLE 4.2**

**Nanomaterials Applied to the Detection of Explosives**

| Type of Detection Method | Principle | Nanoplatform | Explosive Detected |
|---|---|---|---|
| Photoluminescence | Change in the photoluminescence of sensor element in response to an analyte | Quantum dots Nanoparticles | TNT, RDX, HMX, ammonium nitrate |
| Fiber optic-based | Fiber optical sensors rely on changes in frequency or intensity of electromagnetic radiation | Nanoparticles Nanowires Quantum dots | DNT, DNB |
| LASER-induced breakdown | Determination of explosive materials composition by detecting the surface plasma generated using an optical probe | Quantum dots Nanoparticles | TNT, RDX |
| Surface enhanced Raman scattering (SERS) | Identification of Raman signal from trace amounts of analyte molecules when they are absorbed on an activated metal surface or SERS substrate | Quantum dots Nanoparticles Nanostructures | TNT, RDX, PETN |
| Tera-Hertz (THz) detection | THz explosive sensors are based on differential absorption of a sample region illuminated with THz radiations of two frequencies, chosen for a specific explosive, and to maximize the contrast between presence and absence of an explosive | Carbon nanotubes Quantum dots Graphene | PETN, RDX, HMX, TNT, TDX |

*Source:* Adapted from Nanotechnology sensors for the detection of trace explosives. [Online]. http://www.nanowerk.com/ spotlight/spotid=28691.php [Accessed May 7, 2015].

dots (CdS doped with $Cd^{2+}$, $Cu^{+/2+}$, $Zn^{2+}$, etc.) and fluorescence dyes supported on nanoporous aluminum oxide and polymer nanobeads [32]. The opposite effect, presence of luminescence effect after chemical reaction between gases and solid surfaces (chemiluminescence) can also be used for sensing applications. Nanoparticles of MgO, $ZrO_2$, $TiO_2$, $Al_2O_3$, $Y_2O_3$, and $SrCO_3$ exhibit luminescence after reacting with organic gases, and some sensor prototypes have been tested based on this principle [32].

As it has been explained, optical nanosensors have a wide variety of applications. The most promising ones are in the explosive detection (Table 4.2), in virology, chemical vapor sensing, food safety, and screen devices.

### 4.4.3 NANOCOLORANTS AND NANOPIGMENTS

Some nanomaterials have received attention due to their color and potential color-based applications. In the color industry, colorant refers to any substance that gives color when applied to a substrate. One of the criteria to classify colorants is in relation to their miscibility with the substrate: when colorant and substrate are miscible and the colorant can be diluted, it is considered a dye, while pigments are immiscible with the substrate and they get dispersed in the substrate, forming a different phase. Using the same terminology, nanocolorants and nanopigments can be distinguished. In general, the term nanocolorant is applied to solutions or dispersions of nanoparticles with color properties, while the word nanopigment is applied mainly for solid state and powder.

One of the most peculiar effects is the change in color when properties of the nanoparticle (like shape and size) change. Nanocolorants are one of the most typical examples. One popular kind of nanocolorant is the one based on metallic nanoparticles synthesized in different ways and media, so the final size and shape of the nanoparticle varies. The variation in shape and size affects the color of the colloidal dispersion and/or solution of the nanocolorant. This effect

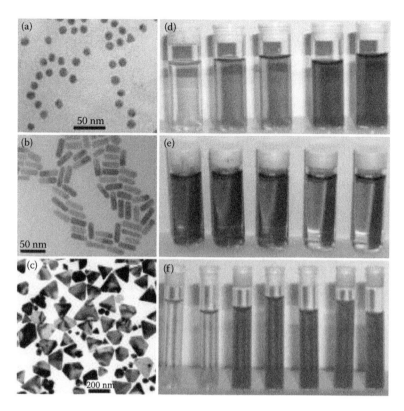

**FIGURE 4.18** **(See color insert.)** TEM images of gold nanospheres (a), gold nanorods (b), and silver nanoprisms (c); and visual appearance of colloidal dispersions of AuAg alloy nanoparticles with increasing Au concentration (d), Au nanorods of increasing aspect ratio (e), and Ag nanoprisms with increasing lateral size (f). (Adapted from M. Liz-Marzán, *Mater. Today*, 7(2), 26–31, Feb. 2004.)

is known as surface plasmon resonance (related to the frequency at which electrons oscillate in response to an electric field or incident electromagnetic radiation). Only metals with free electrons exhibit this effect in the visible region of the spectrum [37]. Some examples are shown in Figures 4.18 and 4.19. Theoretically, these nanocolorants could provide a wide gamut of color. However, the application of this type of nanocolorant has not been developed yet, thus limiting its potential commercialization. On the other hand, this type of nanocolorant has potential application in the sensor sector.

As mentioned before, the term "nanopigments" is generally applied to nanomaterials in solid state that can be applied as colorants. The nanoclay-based pigments (NCP) (Figure 4.20) are organically modified nanoclay in which all or part of the cationic exchange capacity (CEC) of the nanoclay is exchanged with an organic dye and, in some cases, with organic surfactants, like quaternary

**FIGURE 4.19** **(See color insert.)** Variation of color of silver nanoparticles with different dilutions. (Adapted from A. Chhatre et al., *Colloids Surf. Physicochem. Eng. Asp.*, 404, 83–92, June 2012.)

**FIGURE 4.20** **(See color insert.)** Nanoclay-based pigments. (Adapted from M. Pomares, *Diario Informacion*, 2, Apr. 2009.)

ammonium [39,40]. Generally, organically modified nanoclays are a particular type of nanofiller for polymer. Their remarkable importance is derived from the fact that with low content (around 2–5 wt%) they can confer great improvement in the polymer properties [41–44]. Therefore, nanoclay-based pigments can be applied as colorants for polymers, acting at the same time as reinforcement fillers. There are a great number of studies about the incorporation of dyes in clays and the properties of the clay-dye systems [45–57] and also, some studies have been carried out to assess the influence of the NCP in thermoplastic polymers [58–61]. Moreover, they can be used to produce coatings, paints or to color other substrates. Others studies have been done to study the synthesis of this type of nanoclay-based pigment with different dyes [62,63], observing that these nanopigments can significantly improve the thermal and UV stability of some substrates [63]. Despite their great potential, this type of nanopigment has not been commercialized yet. Then, the development of this type of pigment is limited to the laboratory scale.

Traditionally, titanium dioxide ($TiO_2$) and zinc oxide (ZnO) have been used as pigments in many sectors (paints, plastics, toothpastes, cosmetics, pharmaceutic products, etc.) because of their white color, opacity, and durability [11,65,66]. With the development of the nanotechnology, the properties of titanium dioxide nanoparticles ($TiO_2$ NP) and zinc oxide nanoparticles (ZnO NP) have been enhanced (brightness, light absorption, etc.) compared to the bulk material. Then, $TiO_2$ and ZnO NPs are also used as nanopigments for certain applications, particularly in cosmetics for sunlight protection, due to their ability to absorb UV radiation (Figure 4.12). This has been translated in an increase in effectiveness in the cosmetic sector for sunscreen products. Nevertheless, there is controversy about the use of these nanopigments. Currently, organizations like the Scientific Committee on Consumer Safety (SCCS) and Food and Drug Administration (FDA) have confirmed the safety of using $TiO_2$ and ZnO NPs at concentration levels up to 25 wt.% and have regulated their application as additives for food and cosmetic products in the United States and Europe [67–69].

### 4.4.4 NANOMATERIALS IN NONLINEAR OPTIC APPLICATIONS [3]

Some nanomaterials present nonlinear optical properties, like second harmonic generation (SHG) or third harmonic generation (THG). Nanowires (ZnO, GaN, and $KNbO_3$), noble metallic nanoparticles, nanocrystals ($Fe(IO_3)_3$, KTiOPO, and $BiTiO_3$), and some QDs (core/shell CdTe/CdS), etc. are some examples of nanomaterials that exhibit SHG signals. These signals can be used for imaging applications, like tracking nanoparticles inside cells and organisms. In fact, a harmonic holographic microscope has been developed for 3D imaging of nanocrystals in cells without scanning (Figure 4.21) [70]. These nanocrystals emit a stable coherent signal suitable for long-term observations. Other advantages are the fact that they do not present photo bleaching, the excitation wavelength can be varied, and the time response is fast.

In addition, noble metallic nanoparticles with SPR can be used as enhancement agents for THG. Semiconducting nanomaterials (ZnO think films, CdSe QDs, and $Fe_3O_4$ NPs) also exhibit THG signals that can be used for label-free imaging of nanostructures in cells and tissues.

## 4.5 CONCLUSIONS AND FUTURE TRENDS

After more than 20 years of research, scientists continue to develop new and/or improved applications, products, and techniques in the nanomaterials, nanoscience, and nanotechnology fields. In this chapter, the applications and characterization techniques for nanomaterials focusing exclusively on the optical properties have been presented. Titanium dioxide, metal quantum dots, and organically modified nanoclays are some of the nanomaterials that exhibit excellent optical properties for practical applications like solar cells, sensors, coatings, etc. However, researchers are constantly investigating new (nano) materials.

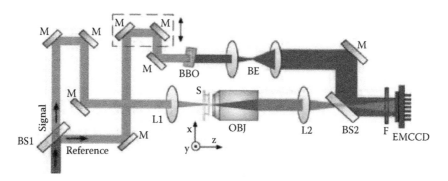

**FIGURE 4.21** $H_2$ microscope experimental setup. BS1 and BS2, beam splitters; M, mirror; L1 and L2, lens; S, sample; OBJ, microscope objective; BE, beam expander; F, band-pass filter centered at 400 nm. BS1 splits the laser into signal and reference beams. In the signal arm, L1 slightly focuses the excitation beam into the sample with SHRIMPs. OBJ and L2 form a 4F imaging system to collect and optically magnify the SHG image of SHRIMP. The EMCCD is placed away from the 4F imaging plane. A band-pass filter is placed in front of the EMCCD to remove the excitation from the SHG signal. The reference beam goes through a translation stage and a BBO crystal so that the coherent reference SHG laser pulses are generated and can be temporally and spatially overlapped with the signal on the EMCCD. The signal and reference beams are combined collinearly by BS2 and therefore an on-axis digital hologram is recorded on the EMCCD. (Adapted from C.-L. Hsieh et al., *Opt. Express*, 17(4), 2880, Feb. 2009.)

For photoelectrochemical applications, new materials are being developed with the aim of increasing the efficiency in the energy conversion, like graphene honeycomb-like structures to replace platinum in dye-sensitized solar cells, copper indium selenide sulfide quantum dots, as they are nontoxic, graphene and molybdenum diselenide solar cells, graphene coated with zinc oxide nanowires, silver nanocubes scattered over a thin gold layer, etc. [71]. Nevertheless, Sablon et al. [72] reported an increase of 50% in the conversion efficiency of solar cells (mainly due to the enhancement in IR energy harvesting) when using InAs/GaAs QDs, compared to GaAs solar cell.

In photocatalysis, recent developments involve increasing the light harvesting efficiency (i.e., increasing the wavelength absorption range) based on doped $TiO_2$ nanoparticles, like N/Au-$TiO_2$ NPs [16] or Fe-doped $TiO_2$ NPs [24], which can be used for the degradation of organic compounds, like nitrophenols.

Over the last few decades, carbon nanotubes (CNTs) have received a lot of attention/interest due to their excellent properties. However, toxicological studies have shown that pure/pristine CNTs can pose a potential risk for human health [73,74]. In addition, since the discovery of a method for extracting graphene from graphite by Geim and Novoselov in 2003 [75], scientific interest has moved to graphene (flake-like nanoparticles of carbon). Graphene has excellent physicochemical properties, and currently several investigations are being conducted to determine the toxicology of graphene, as well as derivatives like graphene oxide [76–79]. It is expected that in the near future, new nanomaterials based on (functionalized) graphene, as well as new technologies for the application and implementation of nanomaterials will be developed. A very interesting example is the harmonic holographic microscope (Figure 4.21) for tracking nanoparticles in cells, which can be used in biomedicine and diagnosis.

## REFERENCES

1. K. Okuyama, Laser light scattering, in *Aerosol Particles*, Okuyama Group, August 2002. http://aerosols.wustl.edu/AAARworkshop08/materials/Okuyama/ [Accessed May 1, 2015].
2. G. Sharma and R. Bala, *Digital Color Imaging Handbook*. Boca Raton, London, New York, Washington D.C.: CRC Press, 2002.
3. S. Suresh and D. Arivuoli, Nanomaterials for nonlinear optical (NLO) applications: A review, *Rev. Adv. Mater. Sci.*, 30(3), 243–253, 2012.
4. D. Harvey, Spectroscopic methods, in *Analytical Chemistry 2.0*, 2009, p. 124. http://www.asdlib.org/onlineArticles/ecourseware/Analytical%20Chemistry%202.0/Text_Files.html
5. L. Taiz and E. Zeiger, Topic 7.1 principles of spectrophotometry, in *Plant Physiology*, 5th ed. Online, Sinauer, 2010. http://5e.plantphys.net/index.php
6. J. Z. Zhang, Introduction, in *Optical Properties and Spectroscopy of Nanomaterials*, World Scientific Publishing, 2009, p. 400. Ebook (online). http://www.worldscientific.com/nanosci/7093.html [Accessed: April 28, 2015].
7. F. Re, R. Moresco, and M. Masserini, Nanoparticles for neuroimaging, *J. Phys. Appl. Phys.*, 45(7), 073001, Feb. 2012.
8. T. Tsuzuki, Commercial scale production of inorganic nanoparticles, *Int. J. Nanotechnol.*, 6(5/6), 567, 2009.
9. F. L. de Souza and E. R. Leite, *Nanoenergy: Nanotechnology Applied for Energy Production*. Spinger, Heidelbarg, New York, Dordrecht, London: Springer Science & Business Media, 2013.
10. K. Hara and H. Arakawa, Dye-sensitized solar cells, in *Handbook of Photovoltaic Science and Engineering*, A. Luque and S. Hegedus, Eds. John Wiley & Sons, Ltd., 2003, pp. 663–700. http://onlinelibrary.wiley.com/doi/10.1002/0470014008.ch15/pdf. http://onlinelibrary.wiley.com/doi/10.1002/0470014008.ch15/pdf [Accessed May 21, 2015].
11. Y. Lan, Y. Lu, and Z. Ren, Mini review on photocatalysis of titanium dioxide nanoparticles and their solar applications, *Nano Energy*, 2(5), 1031–1045, Sep. 2013.
12. P. Das, D. Sengupta, U. Kasinadhuni, B. Mondal, and K. Mukherjee, Nano-crystalline thin and nano-particulate thick $TiO_2$ layer: Cost effective sequential deposition and study on dye sensitized solar cell characteristics, *Mater. Res. Bull.*, 66, 32–38, June 2015.

13. C. Li, Y. Luo, X. Guo, D. Li, J. Mi, L. Sø, P. Hald, Q. Meng, and B. B. Iversen, Mesoporous $TiO_2$ aggregate photoanode with high specific surface area and strong light scattering for dye-sensitized solar cells, *J. Solid State Chem.*, 196, 504–510, Dec. 2012.

14. J. Li, X. Chen, N. Ai, J. Hao, Q. Chen, S. Strauf, and Y. Shi, Silver nanoparticle doped $TiO_2$ nanofiber dye sensitized solar cells, *Chem. Phys. Lett.*, 514(1–3), 141–145, Sep. 2011.

15. C.-H. M. Chuang, P. R. Brown, V. Bulović, and M. G. Bawendi, Improved performance and stability in quantum dot solar cells through band alignment engineering, *Nat. Mater.*, 13(8), 796–801, Aug. 2014.

16. S. Bouhadoun, C. Guillard, F. Dapozze, S. Singh, D. Amans, J. Bouclé, and N. Herlin-Boime, One step synthesis of N-doped and Au-loaded $TiO_2$ nanoparticles by laser pyrolysis: Application in photocatalysis, *Appl. Catal. B: Environ.*, 174–175, 367–375, Sep. 2015.

17. R. Asahi, T. Morikawa, T. Ohwaki, K. Aoki, and Y. Taga, Visible-light photocatalysis in nitrogen-doped titanium oxides, *Science*, 293(5528), 269–271, 2001.

18. S. Sato, R. Nakamura, and S. Abe, Visible-light sensitization of $TiO_2$ photocatalysts by wet-method N doping, *Appl. Catal. Gen.*, 284(1–2), 131–137, Apr. 2005.

19. B. Tian, C. Li, F. Gu, and H. Jiang, Synergetic effects of nitrogen doping and Au loading on enhancing the visible-light photocatalytic activity of nano-$TiO_2$, *Catal. Commun.*, 10(6), 925–929, Feb. 2009.

20. Y. Wu, H. Liu, J. Zhang, and F. Chen, Enhanced photocatalytic activity of nitrogen-doped titania by deposited with gold, *J. Phys. Chem. C*, 113(33), 14689–14695, 2009.

21. M. Silambarasan, S. Saravanan, and T. Soga, Effect of Fe-doping on the structural, morphological and optical properties of ZnO nanoparticles synthesized by solution combustion process, *Phys. E: Low Dimens. Syst. Nanostruct.*, 71, 109–116, Jul. 2015.

22. S. Sun, X. Chang, X. Li, and Z. Li, Synthesis of N-doped ZnO nanoparticles with improved photocatalytical activity, *Ceram. Int.*, 39(5), 5197–5203, July 2013.

23. J. P. Campos-López, C. Torres-Torres, M. Trejo-Valdez, D. Torres-Torres, G. Urriolagoitia-Sosa, L. H. Hernández-Gómez, and G. Urriolagoitia-Calderón, Optical absorptive response of platinum doped $TiO_2$ transparent thin films with Au nanoparticles, *Mater. Sci. Semicond. Process.*, 15(4), 421–427, Aug. 2012.

24. S. Sood, A. Umar, S. K. Mehta, and S. K. Kansal, Highly effective Fe-doped $TiO_2$ nanoparticles photocatalysts for visible-light driven photocatalytic degradation of toxic organic compounds, *J. Colloid Interface Sci.*, 450, 213–223, July 2015.

25. I. Capek, Nanosensors based on metal and composite nanoparticles and nanomaterials, in *Nanoscience and Nanotechnology*, Encyclopedia of Life Support Systems (EOLSS). http://www.eolss.net/sample-chapters/c05/e6-152-25-00.pdf [Accessed May 5, 2015].

26. T. A. Faisal, Nanosensors basics, design and applications, 15:53:44 UTC, January 2015. http://es.slideshare.net/taifalawsi1/nanosensors-basics-design-and-applications [Accessed May 5, 2011].

27. S. Agrawal and R. Prajapati, Nanosensors and their pharmaceutical applications: A review, *Int. J. Pharm. Sci. Nanotechnol.*, 4(4), 1528–1535, 2012.

28. A. Cusano, F. J. Arregui, M. Giordano, and A. Cutolo, *Optochemical Nanosensors*. Boca Raton, London, New York: CRC Press, 2012. https://www.crcpress.com/Optochemical-Nanosensors/Cusano-Arregui-Giordano-Cutolo/9781439854891.

29. C. Elosúa, C. Bariáin, I. R. Matías, F. J. Arregui, A. Luquin, and M. Laguna, Volatile alcoholic compounds fibre optic nanosensor, *Sens. Actuators B: Chem.*, 115(1), 444–449, May 2006.

30. B. Renganathan and A. R. Ganesan, Fiber optic gas sensor with nanocrystalline ZnO, *Opt. Fiber Technol.*, 20(1), 48–52, Jan. 2014.

31. M. Consales, A. Cutolo, M. Penza, P. Aversa, M. Giordano, and A. Cusano, Fiber optic chemical nanosensors based on engineered single-walled carbon nanotubes, *J. Sens.*, 2008, e936074, Sep. 2008.

32. J. Shi, J. Zhu, X. Zhang, W. R. G. Baeyens, and A. M. García-Campaña, Recent developments in nanomaterial optical sensors, *TrAC Trends Anal. Chem.*, 23(5), 351–360, May 2004.

33. A. Gole, A. Kumar, S. Phadtare, A. B. Mandale, and M. Sastry, Glucose induced in-situ reduction of chloroaurate ions entrapped in a fatty amine film: formation of gold nanoparticle–lipid composites, *Phys. Chem. Commun.*, 4(19), 92–95, Jan. 2001.

34. M. Ando, T. Kobayashi, and M. Haruta, Optical CO detection by use of CuO/Au composite films, *Sens. Actuators B: Chem.*, 25(1–3), 851–853, Apr. 1995.

35. M. Ando, Y. Sato, S. Tamura, and T. Kobayashi, Optical humidity sensitivity of plasma-oxidized nickel oxide films, *Solid State Ion.*, 121(1–4), 307–311, June 1999.

36. Nanotechnology sensors for the detection of trace explosives. [Online]. http://www.nanowerk.com/spotlight/spotid=28691.php [Accessed May 7, 2015].

37. L. M. Liz-Marzán, Nanometals: Formation and color, *Mater. Today*, 7(2), 26–31, Feb. 2004.

38. A. Chhatre, P. Solasa, S. Sakle, R. Thaokar, and A. Mehra, Color and surface plasmon effects in nanoparticle systems: Case of silver nanoparticles prepared by microemulsion route, *Colloids Surf. Physicochem. Eng. Asp.*, 404, 83–92, June 2012.

39. M. I. Beltrán, V. Benavente, V. Marchante, and A. Marcilla, The influence of surfactant loading level in a montmorillonite on the thermal, mechanical and rheological properties of EVA nanocomposites, *Appl. Clay Sci.*, 83–84, 153–161, Oct. 2013.

40. M. I. Beltrán, V. Benavente, V. Marchante, H. Dema, and A. Marcilla, Characterisation of montmorillonites simultaneously modified with an organic dye and an ammonium salt at different dye/salt ratios. Properties of these modified montmorillonites EVA nanocomposites, *Appl. Clay Sci.*, 97–98, 43–52, Aug. 2014.

41. S. Pavlidou and C. D. Papaspyrides, A review on polymer–layered silicate nanocomposites, *Prog. Polym. Sci.*, 33(12), 1119–1198, Dec. 2008.

42. Q. H. Zeng, A. B. Yu, G. Q. (Max) Lu, and D. R. Paul, Clay-based polymer nanocomposites: Research and commercial development, *J. Nanosci. Nanotechnol.*, 5(10), 1574–1592, 2005.

43. S. Livi, J. Duchet-Rumeau, T.-N. Pham, and J.-F. Gérard, A comparative study on different ionic liquids used as surfactants: Effect on thermal and mechanical properties of high-density polyethylene nanocomposites, *J. Colloid Interface Sci.*, 349(1), 424–433, Sep. 2010.

44. S. Sinha Ray and M. Okamoto, Polymer/layered silicate nanocomposites: A review from preparation to processing, *Prog. Polym. Sci.*, 28(11), 1539–1641, Nov. 2003.

45. K. Bergmann and C. T. O'Konski, A spectroscopic study of methylene blue monomer, dimer, and complexes with montmorillonite, *J. Phys. Chem.*, 67(10), 2169–2177, Oct. 1963.

46. J. Bujdák, V. Martínez Martínez, F. López Arbeloa, and N. Iyi, Spectral properties of rhodamine 3B adsorbed on the surface of montmorillonites with variable layer charge, *Langmuir*, 23(4), 1851–1859, Feb. 2007.

47. J. Bujdák, N. Iyi, Y. Kaneko, A. Czímerová, and R. Sasai, Molecular arrangement of rhodamine 6G cations in the films of layered silicates: The effect of the layer charge, *Phys. Chem. Chem. Phys.*, 5(20), 4680, 2003.

48. F. Gessner, C. C. Schmitt, and M. G. Neumann, Time-dependent spectrophotometric study of the interaction of basic dyes with clays. I. methylene blue and neutral red on montmorillonite and hectorite, *Langmuir*, 10(10), 3749–3753, Oct. 1994.

49. Z. Grauer, A. B. Malter, S. Yariv, and D. Avnir, Sorption of rhodamine B by montmorillonite and laponite, *Colloids Surf.*, 25(1), 41–65, July 1987.

50. Z. Klika, H. Weissmannová, P. Čapková, and M. Pospíšil, The rhodamine B intercalation of montmorillonite, *J. Colloid Interface Sci.*, 275(1), 243–250, July 2004.

51. A. Landau, A. Zaban, I. Lapides, and S. Yariv, Montmorillonite treated with Rhodamine-6G mechanochemically and in aqueous suspensions, *J. Therm. Anal. Calorim.*, 70(1), 103–113, Aug. 2002.

52. F. López Arbeloa, R. Chaudhuri, T. Arbeloa López, and I. López Arbeloa, Aggregation of rhodamine 3B adsorbed in wyoming montmorillonite aqueous suspensions, *J. Colloid Interface Sci.*, 246(2), 281–287, Feb. 2002.

53. P. Monvisade and P. Siriphannon, Chitosan intercalated montmorillonite: Preparation, characterization and cationic dye adsorption, *Appl. Clay Sci.*, 42(3–4), 427–431, Jan. 2009.

54. P. Čapková, P. Malý, M. Pospíšil, Z. Klika, H. Weissmannová, and Z. Weiss, Effect of surface and interlayer structure on the fluorescence of rhodamine B–montmorillonite: Modeling and experiment, *J. Colloid Interface Sci.*, 277(1), 128–137, Sep. 2004.

55. M. Pospíšil, P. Čapková, H. Weissmannová, Z. Klika, M. Trchová, M. Chmielová, and Z. Weiss, Structure analysis of montmorillonite intercalated with rhodamine B: Modeling and experiment, *J. Mol. Model.*, 9(1), 39–46, Feb. 2003.

56. M. J. Tapia Estevez, F. Lopez Arbeloa, T. Lopez Arbeloa, and I. Lopez Arbeloa, Absorption and fluorescence properties of Rhodamine 6G adsorbed on aqueous suspensions of Wyoming montmorillonite, *Langmuir*, 9(12), 3629–3634, Dec. 1993.

57. C.-C. Wang, L.-C. Juang, T.-C. Hsu, C.-K. Lee, J.-F. Lee, and F.-C. Huang, Adsorption of basic dyes onto montmorillonite, *J. Colloid Interface Sci.*, 273(1), 80–86, May 2004.

58. S. Raha, I. Ivanov, N. H. Quazi, and S. N. Bhattacharya, Photo-stability of rhodamine-B/montmorillonite nanopigments in polypropylene matrix, *Appl. Clay Sci.*, 42(3–4), 661–666, Jan. 2009.

59. V. Marchante, F. M. Martínez-Verdú, M. I. B. Rico, and A. M. Gomis, Mechanical, thermal and colorimetric properties of LLDPE coloured with a blue nanopigment and conventional blue pigments, *Pigment Resin Technol.*, 41(5), 263–269, Sep. 2012.

60. V. Marchante, A. Marcilla, V. Benavente, F. M. Martinez-Verdu, and M. I. Beltran, Linear low-density polyethylene colored with a Nanoclay-based pigment: Morphology and mechanical, thermal and colorimetric properties, *J. Appl. Polym. Sci.*, 129(5), 2716–2726, 2013.

61. V. Marchante, V. Benavente, A. Marcilla, F. M. Martinez-Verdu, and M. I. Beltran, EVA/nanoclay-based pigments composites: Morphology, rheology, and mechanical, thermal and colorimetric properties, *J. Appl. Polym. Sci.*, 130(4), 2987–2994, 2013.

62. S. Raha, N. Quazi, I. Ivanov, and S. Bhattacharya, Dye/clay intercalated nanopigments using commercially available non-ionic dye, *Dyes Pigments*, 93(1–3), 1512–1518, Apr. 2012.

63. J. Sivathasan, Preparation of clay-dye pigment and its dispersion in polymers, Master's Thesis, RMIT University, Melbourne, Australia, 2007.

64. M. Pomares, Color inteligente, *Diario Informacion*, April 21, 2009. http://medias.diarioinformacion. com/suplementos/2009-04-30_SUP_2009-04-22_19_57_05_paraninfo.pdf.

65. A. Weir, P. Westerhoff, L. Fabricius, and N. von Goetz, Titanium dioxide nanoparticles in food and personal care products, *Environ. Sci. Technol.*, 46(4), 2242–2250, Feb. 2012.

66. G. Buxbaum and G. Pfaff, *Industrial Inorganic Pigments*, 3rd ed. Weinheim: Wiley-VCH Verlag GmbH & Co. KGaA, 2005.

67. CFR - Code of Federal Regulations Title 21. [Online]. http://www.accessdata.fda.gov/scripts/cdrh/ cfdocs/cfcfr/CFRSearch.cfm?fr=73.575 [Accessed May 5, 2015].

68. CFR - Code of Federal Regulations Title 21. [Online]. http://www.accessdata.fda.gov/scripts/cdrh/ cfdocs/cfcfr/CFRSearch.cfm?fr=352.10 [Accessed May 5, 2015].

69. EUR-Lex - 31976L0768 - EN, Official Journal L 262, 27/09/1976 P. 0169 - 0200; Greek special edition: Chapter 13 Volume 4 P. 0145; Spanish special edition: Chapter 15 Volume 1 P. 0206; Portuguese special edition Chapter 15 Volume 1 P. 0206; Finnish special edition: Chapter 13 Volume 5 P. 0198; Swedish special edition: Chapter 13 Volume 5 P. 0198. [Online]. http://eur-lex.europa.eu/legal-content/EN/TXT/ HTML/?uri=CELEX:31976L0768&from=EN [Accessed May 5, 2015].

70. C.-L. Hsieh, R. Grange, Y. Pu, and D. Psaltis, Three-dimensional harmonic holographic microscopy using nanoparticles as probes for cell imaging, *Opt. Express*, 17(4), 2880, Feb. 2009.

71. E. Boysen, Nanotechnology in Solar Cells, *UnderstandingNano.com*, 2014. [Online]. http://www.under-standingnano.com/solarcells.html.

72. K. A. Sablon, J. W. Little, V. Mitin, A. Sergeev, N. Vagidov, and K. Reinhardt, Strong enhancement of solar cell efficiency due to quantum dots with built-in charge, *Nano Lett.*, 11(6), 2311–2317, June 2011.

73. S. Y. Madani, A. Mandel, and A. M. Seifalian, A concise review of carbon nanotube's toxicology, *Nano Rev.*, 4, Dec. 2013. doi:10.3402/nano.v4i0.21521.

74. C. Lam, J. T. James, R. McCluskey, S. Arepalli, and R. L. Hunter, A review of carbon nanotube toxicity and assessment of potential occupational and environmental health risks, *Crit. Rev. Toxicol.*, 36(3), 189–217, Jan. 2006.

75. K. S. Novoselov, A. K. Geim, S. V. Morozov, D. Jiang, Y. Zhang, S. V. Dubonos, I. V. Grigorieva, and A. A. Firsov, Electric field effect in atomically thin carbon films, *Science*, 306(5696), 666–669, Oct. 2004.

76. S. Xu, Z. Zhang, and M. Chu, Long-term toxicity of reduced graphene oxide nanosheets: Effects on female mouse reproductive ability and offspring development, *Biomaterials*, 54, 188–200, June 2015.

77. L. Mu, Y. Gao, and X. Hu, l-Cysteine: A biocompatible, breathable and beneficial coating for graphene oxide, *Biomaterials*, 52, 301–311, June 2015.

78. H. Mao, W. Chen, S. Laurent, C. Thirifays, C. Burtea, F. Rezaee, and M. Mahmoudi, Hard corona composition and cellular toxicities of the graphene sheets, *Colloids Surf. B: Biointerfaces*, 109, 212–218, Sep. 2013.

79. S. Li, X. Pan, L. K. Wallis, Z. Fan, Z. Chen, and S. A. Diamond, Comparison of $TiO_2$ nanoparticle and graphene–$TiO_2$ nanoparticle composite phototoxicity to *Daphnia magna* and *Oryzias latipes*, *Chemosphere*, 112, 62–69, Oct. 2014.

# 5 Microscopy of Nanomaterials

*Mazeyar Parvinzadeh Gashti, Farbod Alimohammadi,*
*Amir Kiumarsi, Wojciech Nogala, Zhun Xu,*
*William J. Eldridge, and Adam Wax*

## CONTENTS

## 5.1 INTRODUCTION

The initial idea of nanotechnology was proposed in 1959 by Richard Feynman while the exact term "nanotechnology" was used by Norio Taniguchi in 1974. Feynman discussed this idea during his lecture entitled: "There's Plenty of Room at the Bottom" [1]. When the dimensions of the solid materials become very small, they present different physical and chemical properties compared to those of the same material in larger size [2]. Based on definition, a product is considered a nanomaterial if at least one of its dimensions falls within the 1–100 nm range. Nanomaterials are typically metals, ceramics, polymers, organic materials, or composites [3]. The novel properties of nanomaterials could be due to their extremely large surface area and the high surface-to-volume ratio of these compounds compared to similar conventional materials [4]. Nanomaterials include but are not limited to nanoparticles, nanocomposites, nanocrystals, nanoclusters, nanofibers, nanotubes, nanofilms, nanowires, and nanorods. There has been tremendous progress in the field of nanotechnology in recent decades [1]. Nanotechnology is growing fast because nanomaterials are well incorporated into many aspects of our lives [5].

A precise definition for *nanotechnology* is given by Meyer as: "The manipulation, precision placement, measurement, modeling or manufacture of sub-100 nanometer scale matter" [6].

The definition of nanomaterial was recently given by the European Commission as: "A natural, incidental or manufactured material containing particles, in an unbound state or as an aggregate or as an agglomerate and where, for 50% or more of the particles in the number size distribution, one or more external dimensions is in the size range 1–100 nm" [7].

Unique properties of nanomaterials have attracted interest from government, private enterprises, and scientific researchers. This, in turn, has resulted in exploring the unique enhanced properties of nanomaterials in many different fields [4,8–12]. There has been intensive scientific research on the applications of nanomaterials in a vast area including but not limited to new sources of energy and storage (i.e., solar cells, fuel cells, and batteries), electrics and electronics, optical and optoelectronics, sensors, catalysts, mechanical, construction, composite materials, bioengineering, biomedical, pharmaceuticals, food, cosmetics, and other life science applications [1,5,13].

To study nanomaterials and introduce new applications, the particle characteristics of these materials must be accurately defined in order to ensure that results are reproducible [3,5,14]. These properties are mainly shape, size, surface area and properties, dispersion, and physical and chemical properties [5,15–18]. The analytical process must however be simple, inexpensive, reproducible, and reliable.

Visualization at such a small scale has always been an important step toward understanding these materials. Microscopy is a very powerful and reliable research technique that helps scientists in the field of nanotechnology by providing valuable information regarding surface information, morphology, size, and shape of the particles. The electron microscope, for instance, captures images of the nanoparticle's surface and provides reliable information of their size and shape [19–22].

Some of the most widely used techniques that characterize nanomaterials are scanning electron microscopy (SEM), scanning tunneling microscopy (STM), atomic force microscopy (AFM), scanning electrochemical microscopy (SECM), neutron diffraction, x-ray scattering, x-ray fluorescence spectrometry, x-ray diffraction (XRD), and transmission electron microscopy (TEM) [23,24]. Some of these techniques such as high-resolution transmission electron microscopy (HRTEM) and energy dispersive x-ray spectroscopy have been particularly developed to characterize the composition, size, morphology, crystal structure, and orientation of nanoparticles [19].

Nanocharacterization is a new field in nanotechnology that refers to the techniques and methods used to characterize nanomaterials. This field of research and study is gaining importance and microscopy is considered the most important technique in this field [25–27].

## 5.2  SCANNING ELECTRON MICROSCOPY (SEM)

### 5.2.1  BACKGROUND

For research in the nanoparticle field, it is necessary to observe the morphology and particle size of samples. Nowadays, the scanning electron microscope (SEM) is one of the most widely used instruments to characterize and analyze nanoparticles and nanostructures [28]. SEM probes the surface morphology with a beam of electrons and the achieved image provides a three-dimensional appearance that can be useful to investigate the surface structure [28–32].

Image formation in SEM is based on the signals that achieved from primary electrons and sample interaction. Interactions between sample and primary electrons excite and emit various signals including secondary electrons, transmitted electrons, backscattered electrons, auger electrons, and x-ray continuum (Figure 5.1) [33,34].

Electrons of the sample atoms are exited during the ionization, which generates secondary electrons, the conventional electrons which are applied in SEM. Secondary electrons present topographic contrast in the SEM with good resolution including the visualization of the surface texture and roughness [35].

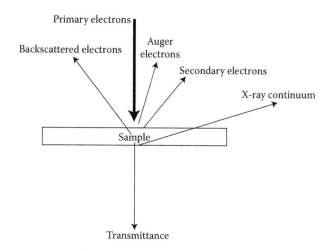

**FIGURE 5.1**  Interaction between primary electron beam and the sample in SEM.

## 5.2.2  PURPOSE, ADVANTAGES, AND LIMITATIONS

The manipulation on nanoparticles is a powerful method to survey and analyze the mobility of nanoparticles on the solid surface and it opens a new window to a deeper understanding on nanomechanics and nanotribology. The scanning tunneling microscope (STM) and atomic force microscope (AFM) have caused superior capabilities to fabricate and survey nanostructures. By employing the local action of a sharp tip, which can be controlled above a sample surface with subnanometer precision, several approaches to structure fabrication have been made [36].

The friction properties of nanoparticles via atomic force microscopy have been reported. However, a real-time manipulation technique inside a SEM was recently applied for tribological studies of nanoparticles (Figure 5.2).

This method enables a real-time visual feedback in a signal line scan regarding the trajectory of particles. However, there are still some limitations of manipulation inside a SEM. Slow processes can be visualized and fast processes can be noticed because the scanning rate of the electron is restricted to a few hertz. Meanwhile, the electron beam is capable of causing a considerable amount of energy, which can lead to partial melting of the sample or an electrostatic charging [37].

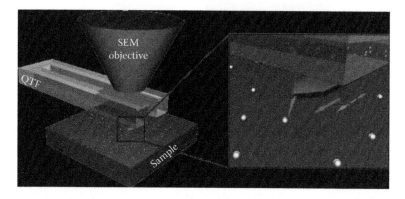

**FIGURE 5.2**  Schematics of the manipulation experiments by an SEM. (Reprinted with permission from B. Polyakov et al. *Beilstein Journal of Nanotechnology* 5, 2014: 133–140.)

SEM and TEM (discussed later in this chapter) are the conventional methods for direct imaging and dimensional measurements of nanoparticles. However, TEMs typically attain a higher resolution than SEM because of electron energies above 100 KeV, and TEMs are more expensive in comparison to the SEMs. Therefore, TEM is achieved by applying an ordinary SEM with moderate electron energy equipped by a proper transmission electron detector in order to transmit particles in the nanometer size range. This type of microscope is named TSEM to discriminate it from conventional TEMs. It has been shown that SEM in transmission mode determines the size and form of particle with high sensitivity, and this technique has been used to image silica, gold, and latex nanoparticles [38].

The ZnO nanoparticles were capped by oleate, and the polyethylene-like organic components covered the nanoparticles by a facile atmospheric cold plasma polymerization of n-octane. Figure 5.3 presents the SEM, the bright-field TSEM, and the dark-field TSEM images of a covered agglomerate particle at electron acceleration voltage of 20 KV. The SEM picture (Figure 5.3a) shows the surface of agglomerated particles capped by the polymer. The bright-field TSEM picture (Figure 5.3b) shows dark agglomerate due to the scattering of electrons to higher angles than the acceptance angle of the bright-field detector. Figure 5.3c represents the agglomerate particles better than the previous pictures. The higher magnification, dark-field TSEM picture in Figure 5.3d confirms the presence of the organic layer on the nanoparticles (NPs) agglomerate and make it possible to evaluate the thickness of the organic phase on the particle (between 10 and 20 nm) [39].

Environmental scanning electron microscopy (ESEM) is a typically new version of SEM, and allows a wet sample to be imaged without probably damaging the specimen through the use of partial water vapor pressure in the microscope sample chamber. Prior sample preparation is not essential and the sample can be observed in its natural state, which is the key advantage over conventional SEM. This can be attained by permitting the presence of a gas in the sample chamber, instead of the normal high vacuum requirements of SEM. Imaging samples in liquids provides the capability to image colloids and nanoobjects in the liquid phase [40].

ESEM has been used to investigate the surface morphology and particle size. Water permeability resistant property and microstructure of concrete containing nano-SiO$_2$ have been investigated by ESEM. Microstructures of normal and nano-SiO$_2$ concrete were confirmed using

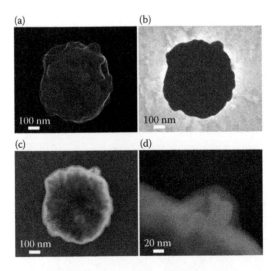

**FIGURE 5.3** (a) SEM, (b) bright-field TSEM, (c) dark-field TSEM, and (d) high magnification dark-field TSEM. (Reprinted with permission from F. Fanelli, A. M. Mastrangelo, F. Fracassi. Aerosol-assisted atmospheric cold plasma deposition and characterization of superhydrophobic organic–inorganic nanocomposite thin films. *Langmuir* 30, 857–865. Copyright 2014 American Chemical Society.)

**FIGURE 5.4** (a) Microstructure of normal concrete (curing time of 28 days), (b) microstructure of normal concrete (curing time of 180 days), (c) microstructure of nano-SiO$_2$ concrete (curing time of 28 days). (Reprinted from *Cement and Concrete Research*, Vol 35, T. Ji, Preliminary study on the water permeability and microstructure of concrete incorporating nano-SiO$_2$, 1943–1947, Copyright 2005, with permission from Elsevier.)

ESEM. For this purpose, the samples were prepared from the concrete cubes at different curing ages (Figure 5.4), and then soaked in the isopropyl alcohol to stop hydration by removing free water. It can be observed which microstructure of the concrete containing nano-SiO$_2$ is more uniform compared to the normal concrete [41].

## 5.3 TRANSMISSION ELECTRON MICROSCOPE (TEM)

### 5.3.1 BACKGROUND

The shape, distribution, and particle size of nanomaterials were studied via TEM. TEM images provide information about the atomic-resolution lattice as well as physical and chemical information at a 1 nm resolution or higher, and provide direct investigation and identification of a single crystal [42].

In a STEM, the image is provided by scanning a focused beam over the sample and collecting transmitted electrons by detector. In other words, STEM is one type of TEM which is equipped with additional devices such as scanning coils and detectors. Meanwhile, images are the result of raster scanning the subnanometer probe over the surface and collecting electrons pixel by pixel that are transmitted through the sample [43–45].

On the other hand, STEM-in-SEM is one method to improve resolution and overcome the limitations of the SEM. In this method, the beam targets the small area and scans over the sample, and the image is extracted by mapping some signal intensity synchronously with the scan. A STEM system used to a standard SEM is generally designated as "low-voltage STEM."

### 5.3.2 Purpose, Advantages, and Limitations

Observation and analysis of charged-induced nanoparticles dynamic in solution have been investigated by STEM. Therefore, STEM has been used to image platinum nanoparticles on an insulating membrane, and the membrane is one of two electron-transparent windows separating a liquid solution from the microscope vacuum. Figure 5.5 shows a time series of STEM images. At first, nanoparticles are immobile but after several seconds of imaging they start to move. By increasing the exposure time, the nanoparticles leave the center of the field of view. Also, it has been shown on receiving a dose of ~$10^4$ e/nm$^2$, initially embedded nanoparticles start to motion along trajectories, and by increasing the dose rates the particle motion dramatically increased. It has been elucidated that, even under mild imaging conditions, the *in situ* electron microscopy of aqueous environments can cause electrophoretic charging phenomena that manipulate the dynamics of nanoparticles. Ultimately, it seems that the nanoparticles attain charge as a direct outcome of being subjected to the imaging beam [46].

Furthermore, the TEM technique has been developed to investigate the growth mechanism and the particle characterization using *in situ* liquid transmission method. In this method, the final morphology of nanostructures can be observed during particle growth in real time with subnanometer spatial resolution. Figure 5.6 shows the lead sulfide nanoparticles growth under different chemical composition while the relative ratio of chemical components has changed. It was reported that changing the chemical composition affected the nanoparticle's formation; the shape and size of nanoparticles formed, and also resulted in a shift from nanoparticle form to the evolution of flower-shaped nanoparticles [47].

High-resolution transmission electron microscope (HRTEM) is an imaging mode of the TEM that provides the direct looking at the atomic structure of the samples. HRTEM images are produced by the interference of the electron beams rather than the absorbance of them. This is a powerful technique for direct imaging of the projected shape of a single nanoparticle when the particle size is small.

HRTEM gives higher resolution images that are widely and effectively applied for analyzing crystal structure and lattice space in nanoparticles on an atomic scale. Figure 5.7 presents the SEM and HRTEM images of the iron oxide nanocubes and mesocrystals. Figure 5.7a–c shows the SEM

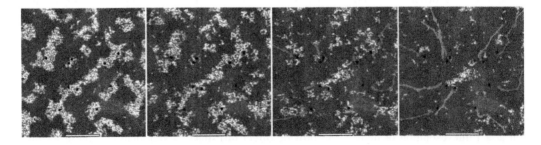

**FIGURE 5.5** (**See color insert.**) Time series of STEM images taken with 350 nm × 350 nm from the center of the field of view. The images are provided ~7 s apart, by the passage of time to the right. The trajectories of 10 particles are presented by the green tracks, with starting points assigned by red dots and the shade of green increasing among the frames. (Reprinted with permission from E.R. White et al. Charged nanoparticle dynamics in water induced by scanning transmission electron microscopy. *Langmuir* 28, 3695–3698. Copyright 2012 American Chemical Society.)

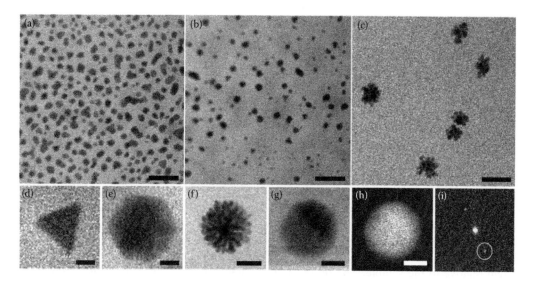

**FIGURE 5.6** Nanoparticle growth under varying chemical composition. *In situ* bright-field scanning TEM images of 2:1 (a), 1:1 (b), and 1:1.25 (c) (lead acetate:thioacetamid) trigonal (d), hexagonal (e), flower-like (f), and spherical (g). Lattice fringes for the (220) plane of PbS at 0.21 nm resolution can be observed in (g) and (h) and Bragg reflections circled in (i). Scale bars represent 100 nm (a–c), 12.5 nm (d and e), 25 nm (f), and 2.5 nm (g and h). (Reprinted with permission from J.E. Evans et al. Controlled growth of nanoparticles from solution with *in situ* liquid transmission electron microscopy. *Nano Letter* 11, 2809–2813. Copyright 2011 American Chemical Society.)

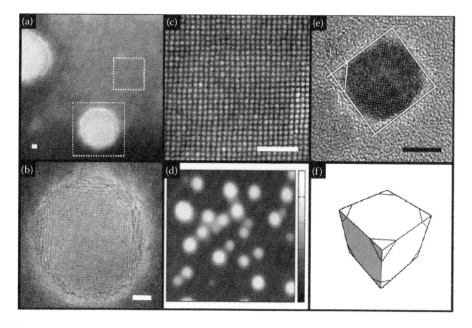

**FIGURE 5.7** Iron oxide nanocube mesocrystals, (a–c) Scanning electron microscopy images of the (a) overall and (b and c) structures of the iron oxide nanocubes on Ge substrate. Scale bars represent 100 nm. (d) Atomic force microscopy image of a $10 \, \mu m \times 10 \, \mu m$ area of the same sample, (e) high-resolution TEM photograph of nanocubes, and (f) schematic of the truncated cube. (Reprinted with permission from S. Disch et al. Shape induced symmetry in self-assembled mesocrystals of iron oxide nanocubes. *Nano Letter* 11, 1651–1656. Copyright 2011 American Chemical Society.)

images with information about the detailed structure of iron oxides nanocubes on Ge substrate with the scale bars 100 nm. Figure 5.7e and f represents the HRTEM images of nanocubes at 800x magnification and schematic of the truncated cube, respectively. HRTEM images represent that the cubes' faces are ceased on the {100}, and the corners and edges correspond to the {111} and {110} planes of the spinal structure [48].

## 5.4 SCANNING ELECTROCHEMICAL MICROSCOPY (SECM)

### 5.4.1 BACKGROUND

Scanning electrochemical microscopy (SECM) [49–51] is one of the most important methods used in nanoparticulate catalyst research. Analytically, SECM allows mapping the topography as well as the lateral variations of the specific (electro) chemical activity, measuring local concentrations of reactants or products of heterogeneous reactions, and investigating heterogeneous kinetics [51]. The sample for SECM can be insulator, conductor, or semiconductor. The central element of an SECM is the ultramicroelectrode (UME) probe, which is moved by a positioning system relative to the sample. The tip-to-sample distance is kept longer than in scanning tunneling microscopy in order to prevent tunneling current interference in the chemical signal. The UME acting as a sensor (or a source of nanoparticle precursor) is connected to a (bi)potentiostat together with the reference electrode (RE) and the counter electrode (CE) establishing a three-electrode cell (Figure 5.8).

### 5.4.2 PURPOSE

Contrary to other scanning probe techniques that are mainly used to investigate the structure of nanoparticles and their distribution, SECM provides information on the usability of materials as a catalyst for various specific chemical reactions. Among them are NPs catalyzed reactions which occur in devices for energy conversion and storage such as hydrogen oxidation (fuel cell anode reaction) [52], oxygen reduction (ORR, (bio)fuel cell cathode reaction) [53–66], hydrogen evolution [67–69], and water oxidation [70–72]. ORR catalyzed by NPs may lead to undesired $H_2O_2$ generation, which can be probed by SECM [53,65]. Simultaneous SECM imaging of various microsamples of nanoparticulate catalysts within one scanning area is a helpful tool in engineering multicomponent catalysts [54,55,65,66,72] and also investigating dependencies between catalytic activities of NPs and their size [66], structure [60], and shape [61]. This approach assures the same analytical

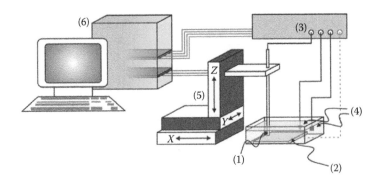

**FIGURE 5.8** Schematic setup of SECM. (1) UME, (2) sample, (3) (bi)potentiostat, (4) reference and counter electrodes, (5) positioning system, and (6) control PC. (K. Powers et al.: Research strategies for safety evaluation of nanomaterials. Part VI. Characterization of nanoscale particles for toxicological evaluation. *Toxicological Sciences.* 2006. 90. 296–303. Copyright Wiley-VCH Verlag GmbH & Co. KGaA. Reproduced with permission.)

conditions for each microsample and allows quick study of a large number of samples with only little wastage of materials. SECM is able to measure lateral conductivity of NP films or composites deposited on insulating solid supports [73,74], as well as on soft liquid–gas interfaces [75]. Liquid–liquid interfaces in SECM setups were applied as a locus of NP synthesis [76] or as an interface for electron transfer rate measurements between NPs dispersed in the organic phase and an aqueous oxidant [77,78]. Heterogeneous electron transfer rate constants through the thiol monolayers covering gold clusters [79] and AgNP [80] were also measured with SECM.

SECM imaging of NPs found application in forensic sciences. Latent finger marks printed on dark surfaces developed with NPs are often invisible by optical methods contrary to SECM imaging [81,82].

Detection of individual NPs captured at a nanopore was done by monitoring the faradaic current related to electro-oxidation of redox-active molecules on an SECM tip positioned at the opening of the pore [83]. NP capture results in a decrease of the transport rate of an electroactive substance through the pore, thus the tip current decreases. A single NP deposited on carbon fiber electrode (CFE) was detected in an SECM experiment by moving the electrode from air to an electrolyte containing $Fe^{3+}$ ions [84]. At certain conditions the electrocatalytic reduction of $Fe^{3+}$ to $Fe^{2+}$ occurs exclusively on Pt. Kwon and Bard studied diffusion-controlled collisions of single NP to SECM tip by monitoring SECM current transients [70]. This study provides information about the nature of NP interactions, that is, type of collisions (elastic or adsorptive) and deactivation.

Recently, special interest in SECM has been gained in the study of biological applications of NPs, such as electrical wiring of enzymes [56,59,60,85], enzyme immobilization for enhanced detection of DNA hybridization [86–88], and to visualize proteins immobilized on solid polymeric membrane [89]. Shao and coworkers employed SECM in studying interactions between Hela cells and AgNPs [90].

Besides analytical applications, SECM is a tool for localized synthesis of metallic NPs [76,84,85,91–97] and local deposition of previously synthesized NPs [98–102]. Individual nanoparticles were prepared by SECM approaching precursor solution/air interface with CFE [84].

### 5.4.3 Examples of NPs Characterized by SECM

A wide variety of NPs have been studied with SECM for the purposes mentioned earlier, which are summarized in Table 5.1 based on their compositions. The majority of them are gold, silver, and platinum NPs probably due to their overwhelming popularity in different applications and research areas compared to other types of NPs and their catalytic activities.

**TABLE 5.1**

**Types of Nanoparticles Studied by SECM**

| NPs Type | References | NPs Type | References |
|---|---|---|---|
| Ag | [55,57,75,76,80–82,86,89,90,92,95,96,100] | Pd-W | [64] |
| Au | [54,61,62,66,73,74,77–82,85,86,92–95,97–99,101–103] | Polystyrene | [83] |
| C | [52,56,58,59,104] | Pt | [52–55,57,60,63,67,84,105] |
| CdTe | [92] | Rh | [54] |
| Composite | [52,54,55,57–59,63,64,68,74,80,101,106] | $RuO_2$ | [72] |
| $IrO_x$ | [70] | Si | [107] |
| Ni | [84] | $SiO_2$ | [87,88] |
| Pb | [108] | $TaO_2$ | [109] |
| Pd | [53,68,69] | $TiO_2$ | [72] |
| Pd–Au | [65] | $WO_3$ | [71] |
| Pd–Pt | [65] | | |

### 5.4.4 Advantages and Limitations

The lateral resolution of SECM is limited by the size of the probe and is usually worse than scanning probe microscopy (SPM) techniques, such as AFM and STM, but instead large areas can be analyzed. Application of the probe size depends on the rate of reaction on the sample. Normalized rate constant, $\kappa = k_0 a / D$, where $k_0$—heterogeneous rate constant, $a$—probe radius, and $D$—diffusion coefficient, is usually measurable within a range of 0.01–10. Sluggish reactions require bigger probes (lower resolution) than fast sample processes. Individual NPs modified with fast redox moiety terminated flexible chains can be distinguished by combined AFM-SECM nanoelectrode probe (Figure 5.9) [103]. Despite the resolution limitation, SECM remains an invaluable supplementary tool in NPs' characterization, especially in mapping and evaluation of catalytic activity of NPs [104,105,107–109].

## 5.5 PHOTOACOUSTIC MICROSCOPY (PA)

### 5.5.1 Background

Photoacoustic effect occurs when a short-pulsed laser beam, or an intensity-modulated laser beam, irradiates the target surface. Absorption of light results in a rapid thermoelastic expansion, and further conversion to an ultrasound emission, which is referred to as photoacoustic waves [110,111]. Since its demonstration in 1881, the photoacoustic effect has been applied to imaging and spectroscopy in material science, biology science, and medicine [112,113]. Photoacoustic (PA) imaging combines the advantages of both high sensitivity in optical imaging and low scattering in ultrasonic imaging. Besides, the nonionizing irradiation in PA imaging is not hazardous to tissues in contrast to the ionizing x-ray based imaging such as CT and micro-CT [114]. Many optical absorbers have proven useful in PA imaging, such as hemoglobin [115], melanoma [116], water [117,118], lipid [119], and various natural and artificial contrast agents [120–123].

### 5.5.2 Purposes of Application of Photoacoustic Imaging

Recently, nanomaterials have been massively applied to PA imaging, owing to their strong optical absorption in both visible and infrared spectrum [91,112,124,125]. Various conjugates and

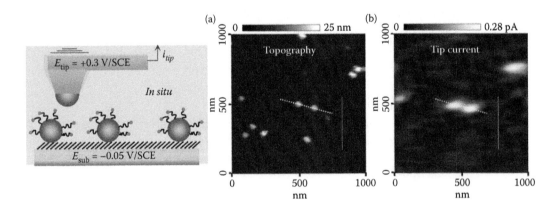

**FIGURE 5.9   (See color insert.)** AFM-SECM tapping mode imaging of a gold surface bearing a random array of 20 nm Fc-PEGylated AuNP. Simultaneously acquired topography (a) and tip current images (b). (Reprinted with permission from A. Srivatsan et al. Gold nanocage-photosensitizer conjugates for dual-modal image-guided enhanced photodynamic therapy. *Theranostics* 4(2), 163. Copyright 2014 American Chemical Society.)

composites have been developed from two types of basic nanomaterials: carbon-based nanomaterials and metal-based nanomaterials.

One typical cylindrical form of the carbon-based nanomaterials, also called single-walled carbon nanotubes (SWNTs), improves the signal contrast by strengthening the PA signal over the entire wavelength range from 740 to 820 nm [125], and such a broad absorption spectrum for SWNTs also makes it flexible in wavelength option. Besides, the linear correlation between PA signal and their concentration within the specific range [124] makes the quantitative analysis possible. Apart from the advantages in optics, the capability of coupling the peptides [124], coating with the polymer films [112], and incorporating to the polymer scaffold [114] also broaden their PA applications in molecular imaging and tissue engineering.

Among various metal-based nanomaterials, gold nanoparticles, nanocages, and nanoshells [126,127] are most widely used in PA imaging. The laser-induced nanobubble formation from their accumulation in tumors [128–130] causes nonlinear enhancement of photoacoustic effects [131,132]. Another important feature of the gold nanomaterials that enables enhanced contrast is their unique tunability of localized surface plasmon resonance (LSPR) [129,131] via their size, shape, and composition adjustment. The optical spectrum related to the amount of chemical medium, the sensitivity related to the shapes of their nanostructures, along with their enhancement ability and biocompatibility make gold nanostructures an effective contrast agent in PA imaging that is otherwise a challenge to see inside the tissue.

### 5.5.3 EXAMPLES OF NANOPARTICLES CHARACTERIZED BY PHOTOACOUSTIC MICROSCOPY

Cai et al. [114] applied both optical-resolution photoacoustic microscopy (OR-PAM) and acoustic-resolution photoacoustic microscopy (AR-PAM) to observe the scaffolds incorporating SWNTs in simulated physiological environments (Figure 5.10).

It was demonstrated that multiscale PAM is not only suitable for SWNTs-incorporated polymeric scaffolds imaging in biological tissues, but it can also provide the quantitative information on porosity, pore sizes, and degradation effects during tissue regeneration [133].

Cai and Wu [134] further developed a "green synthesis" of much smaller (~10 nm) and solvent-free carbon nanoparticles called luminescent carbon nanoparticles (OCN). The rapid particle relocation and clearance of OCN makes the real-time PA imaging possible on sentinel lymph nodes (SLN). The rapid lymphatic transport and the small size of OCN offers the surgical convenience, lower cost, and great potential in stable large-scale commercial manufacturing (Figure 5.11).

Taking the advantages of LSPR peak tunability and noncovalent conjugation of Au nanostructures to drug molecules, Srivatsan et al. [110] developed AuNC–NPPH conjugates as a multifunctional agent in enhanced photodynamic therapy, monitored by PA imaging in near-infrared (NIR) region (Figure 5.12).

The experiment results showed a ~740% increase of PA amplitude in the mouse spleen after injection of AuNC at a dose of only 10 pmol, which indicated the AuNC is a highly sensitive exogenous contrast agent for PA imaging.

Other than the generally used size and shape control methods, Liu et al. [135] developed the semiconductor–metal heterodimer nanoparticles. The absorbance spectrum is flattened by the combination of plasmonic metal nanoparticles with heavily doped semiconductor nanoparticles, which allows multimodal imaging, including deep tissue photoacoustic imaging in vivo.

In addition to centimeter-large tumors [128–130], cerebral cortex [91], SLN [134], and gastrointestinal tract [136], nanomaterials were also applied to quantitative visualization of micrometer-sized cells and tissues [106]. The nanoparticle-loaded cell imaging by PAM provided local spatial distribution of molecular markers, which showed great potential in validating the effectiveness of molecular targeting strategies (Figure 5.13).

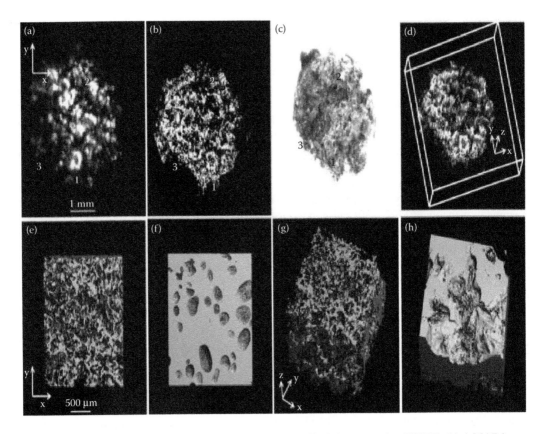

**FIGURE 5.10** PAM and micro-CT images of the PLGA scaffolds incorporating SWNTs. (a) A MAP image of the scaffold by AR-PAM. (b) A MAP image of the scaffold by OR-PAM. (c) An optical microscope image of the scaffold. The common features that can be identified from the images have matching numbers. (d) A 3D depiction of the OR-PAM image. (e) A micro-CT MAP image of the scaffold in dry surrounding. (f) A micro-CT MAP image of the scaffold in fetal bovine serum. (g) A 3D depiction of the micro-CT image in dry surroundings. (h) A 3D depiction of the micro-CT image in fetal bovine serum. Micro-CT, microcomputed tomography; 3D, three dimensional; MAP, maximum amplitude projection. (Adapted from X. Cai, B.S. Paratala, S. Hu et al. *Tissue Engineering Part C: Methods* 18(4), 2011: 310–317.)

### 5.5.4 Advantages and Limitations

Nanomaterials serving as an important class of optical contrast agent in PA imaging are well suited for various biomedical applications across a wide range of length scales. The strong optical absorption in the NIR spectrum region improves the SNR in deep tissue imaging, and the broad absorption spectrum achieved by various tuning techniques enables multimodality imaging with high contrast. Yet more efforts are still being put in order to make PA imaging with nanomaterials better clinically translatable, which is warranted for addressing material stability, healthcare cost, patient inconvenience, and small risks [134].

## 5.6 HYPERSPECTRAL MICROSCOPY

### 5.6.1 Background and Application

The strong dependence of the plasmonic properties of nanoparticles on wavelengths creates a natural connection to hyperspectral microscopy where illumination at a series of wavelengths is used to assess spectral properties. Figure 5.14 shows an example of a hyperspectral microscopy

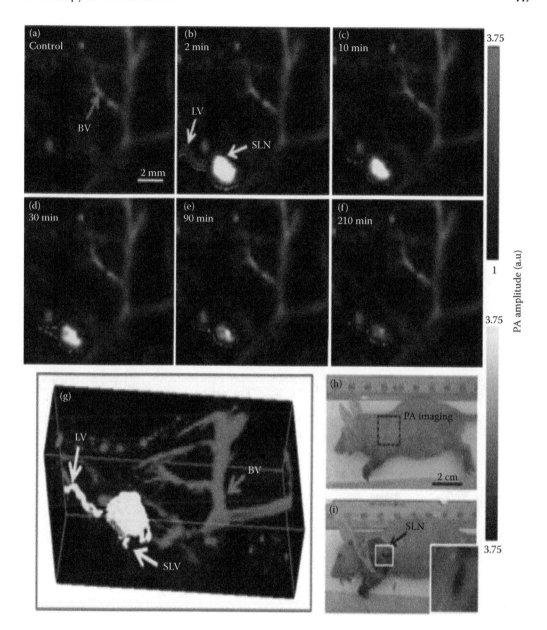

**FIGURE 5.11** (**See color insert.**) Noninvasive real-time *in vivo* PA imaging of SLN in nude mouse: For all PA images, the laser was tuned to 650 nm wavelength. (a) Control PA image acquired before OCN injection. Red parts represent optical absorption from blood vessels (BV); (b) PA image acquired immediately (2 min) after the OCN injection, blood vessel (BV), lymph vessel (LV), and sentinel lymph node are marked with arrows, and the SLN is visible in (b–e), however, the contrast is much weaker after 210 min postinjection in (f). (g) 3D depiction of the SLN and BVs immediately after OCN particles injection, (h) photograph of the nude mouse before taking the PA images. The scanning region is marked with a black dotted square. (i) Photograph of the mouse with the skin removed after PA imaging, accumulation of dark-colored OCN particles are visible in the lymph node. (Adapted from L. Wu, X. Cai, K. Nelson et al. *Nano Research* 6(5), 2013: 312–325.)

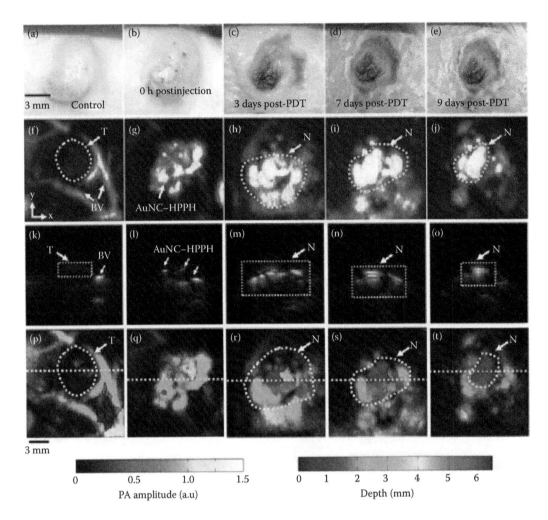

**FIGURE 5.12** **(See color insert.)** (a–e) Photograph of the tumor taken before and after PDT with an injection of AuNC–HPPH at various time points up to 9 days. PA images acquired (f) before intratumoral injection of AuNC–HPPH, (g) after injection, and (h–j) 3, 7, and 9 days posttreatment. (k–o) Depth-resolved PA B-scan images cut along the dotted lines in (f–j), respectively. (p–t) Depth-encoded PA images of (f–j), respectively. BV, blood vessels; T, tumor boundary; and N, tumor necrotic region. (Adapted from A. Srivatsan et al. *Theranostics* 4(2), 2014: 163.)

system. In Figure 5.14a, an acousto-optic tunable filter (AOTF) was employed in the illumination path [137,138] to enable the illumination wavelength to be swept in time. In this scheme, the swept source is driven by a super continuum laser source, offering high intensity illumination across a wide spectral range, and coupled to a microscope platform using a custom dark-field illumination scheme [137]. As implemented, this dark-field microspectroscopy system enables imaging of changes in plasmonic resonance peaks.

Briefly, our epi-illumination dark-field microspectroscopy system consists of collimated white light input into an axicon to form a ring of light. This hollow ring is focused onto the back-focal plane of an objective, which in turn focuses the cone of light onto the sample. Nanoparticles backscatter this light toward the objective, which includes an iris that blocks the back-reflected illumination ring from being imaged onto the CCD. As an alternative to using the AOTF on the illumination side, the approach shown in Figure 5.14b can be implemented, where the full spectrum from a white light source is incident on the sample and the scattered light is filtered using a tunable filter

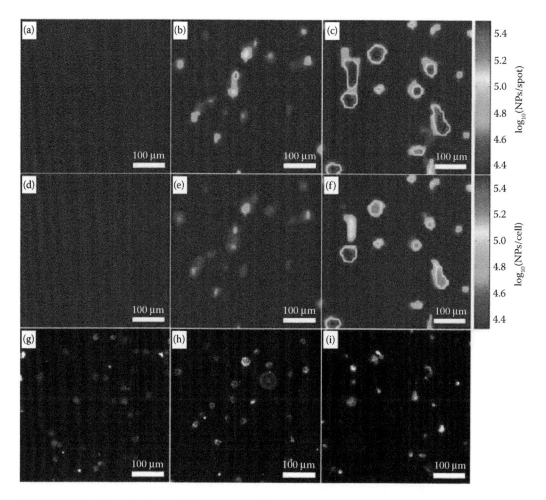

**FIGURE 5.13** **(See color insert.)** Quantitative PA imaging results (a–f) and coregistered dark-field micros-copy (g–i) of J774A.1 cells incubated with and without NPs. Quantitative PA images scaled to the number of NPs per focal spot (a–c) and number of NPs per cell (d–f) are shown. Images of cells incubated without NPs (a, d, and g), with $1.9 \times 10^{12}$ NPs per milliliter of cell culture media (b, e, and h), and with $3.7 \times 10^{12}$ NPs per milliliter of cell culture media (c, f, and i) are shown. (Adapted from J.R. Cook, W. Frey, S. Emelianov. *ACS Nano* 7(2), 2013: 1272–1280.)

on the detection side. In either scheme, by synchronizing the AOTF sweep with the CCD exposure, a hyperspectral image cube may be generated. The scattering intensity can then be obtained as a function of wavelength, revealing the spectral properties of individual nanoparticles.

To illustrate the utility of hyperspectral imaging of plasmonic nanoparticles, we examined their use as molecular contrast agents for cellular imaging. Plasmonic nanoparticles can readily be func-tionalized with DNA, peptides, and antibodies of interest to target specific molecules [133,138–142]. Immunolabeled nanoparticles can be used to obtain biologically relevant information when bound to cells, which highly express surface receptors of interest. For example, anti-epidermal growth factor receptor (anti-EGFR) functionalized gold nanospheres have been used to identify and study cells, which overexpress this receptor [133,139–142]. Using the hyperspectral microscopy system described above, spectral data were analyzed to assess the refractive indices (RI) of particular cellular structures of skin cancer cells (A431). By assessing the shift in the scattering spectrum of functionalized gold nanospheres using hyperspectral microscopy, local refractive indices within attoliter volumes, including that of the cellular membrane, were estimated with an uncertainty of

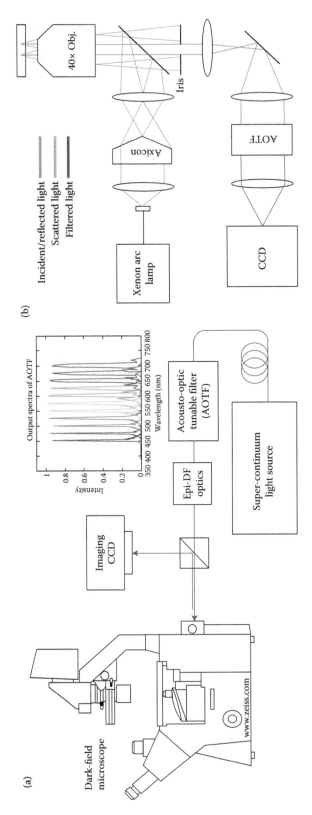

**FIGURE 5.14** (**See color insert.**) (a) Block diagram of the hyperspectral dark-field system. Inset: Example output spectra of white light as filtered by the AOTF. (Adapted from K. Seekell et al. *Methods* 56.2, 2012: 310–316; K. Seekell et al. *Biomedical Optics Express* 4(11), 2013: 2284–2295.) (b) Schematic of the epi-illumination and detection trains of the dark-field microspectroscopy system showing the optical paths of the incident/reflected illuminating light (gray), the backscattered light (green), and the filtered light (red). Note that the AOTF can be placed in either the illumination (a) or detection (b) path.

0.02 RI units [139]. Furthermore, the level of EGFR expression as well as differences in refractive indices between multiple EGFR-positive cancer cell lines was shown to impact the mean scattering peak of nanoparticle tagged cells [143]. For increasing EGFR expression, the scattering intensity was shown to be larger and red-shifted from the scattering peak of unbound functionalized gold nanospheres due to shifts in the local refractive index. Moreover, plasmonic nanoparticle tags were shown to be performing equally as well as fluorescent EGFR tags while having the additional benefit of increased biocompatibility and increased optical stability (no photo bleaching) [143].

The versatility of hyperspectral imaging can be seen through implementation of spectrally multiplexed molecular imaging of multiple cellular receptors using different nanoparticle species [138]. As shown in Figure 5.15 below, different nanoparticle geometries and compositions produce distinct scattering spectra; for example, gold nanorods scatter in the red to near-infrared region, silver nanospheres scatter in the blue to blue-green optical region, while gold spheres predominantly scatter light in the green optical region. The spectral distinction between these species through the use of hyperspectral microscopy allows simultaneous molecular imaging of multiple receptors. Scattering in distinct spectral windows (Figure 5.15e and f) provides a simple way to assess expression of multiple receptors. Such information can be used to reveal information regarding cell phenotype and receptor signaling pathways [138].

Another application of the hyperspectral dark-field microspectroscopy system for cellular study was to identify spacing between receptor locations on the plasma membranes of live cells via the plasmonic coupling phenomenon [144]. When two adjacent nanoparticles have an interparticle gap less than the individual nanoparticle diameter, plasmonic coupling occurs, leading to a red-shift in the scattering spectra observed from the two neighboring dimerized nanoparticles. This red-shift is dictated by the interparticle gap, surrounding refractive index, the particle diameter, and the antibody coating thickness. Therefore, given sufficient knowledge of the surrounding environment and the nanoparticle geometry, the distance between adjacent EGF receptors can be calculated by analyzing the degree of plasmonic

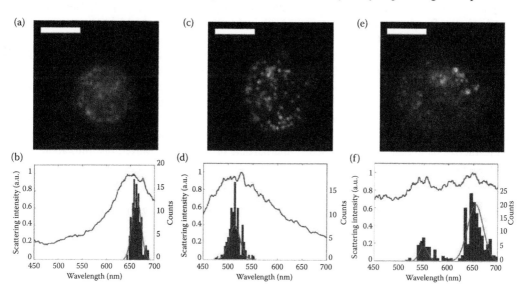

**FIGURE 5.15** **(See color insert.)** (a) Dark-field image of MDA-MB-468 cell tagged with anti-EGFR functionalized GNRs. (b) Example scattering spectrum and distribution of peak scattering wavelengths (Gaussian fit with peak of $664.0 \pm 9.3$ nm) of MDA-MB-468 cells tagged with anti-EGFR GNRs. (c) A549 cell tagged with anti-IGF-1R silver nanospheres. (d) Example scattering spectrum and distribution of peak scattering wavelengths (Gaussian fit with peak $520.8 \pm 11.3$ nm) of A549 cells tagged with anti-IGF-1R silver nanospheres. (e) A549 cell tagged with both anti-EGFR GNRs and anti-HER-2 gold nanospheres (f). Example scattering profile with peak scattering distribution showing two peaks due to the two nanoparticle species employed in (e).

coupling via hyperspectral microscopy. By measuring the spectra of plasmonic coupled NP pairs, the spacing between adjacent HER-2 receptors on SK-BR-3 cells was determined to be 10.4 nm [144].

Hyperspectral imaging can be used to examine plasmonic coupling in more detail. For example, the effect of the incident light polarization on the peak wavelength of scattering can be thoroughly mapped using this approach [145]. Consider a pair of nanospheres that are close enough to induce plasmonic coupling. Two axes can be used to describe the pair: a long and a short axis. Incident polarization parallel to the short axis will result in a scattering profile similar to a single nanosphere. However, if the polarization is aligned parallel to the long axis, the scattering spectrum is red-shifted, indicative of plasmonic coupling [145]. Therefore, by simply adding a linear polarizer to the hyperspectral imaging system, the scattering spectra associated with the long and short axes of the coupled NPs can be determined and, consequently, used to determine interparticle separation without the need for sample manipulation. In this example, hyperspectral microscopy can be applied to obtain nanoscale information far beyond the diffraction limited resolution of optical imaging.

Finally, hyperspectral imaging can take advantage of the spectral tunability of gold nanorods (GNRs), which offer a plasmonic peak that can be varied across a large spectral window from 600 nm to well over 1 micron [146,147]. The ability to shift the plasmonic peak of GNRs by adjusting their aspect ratio allows for an even larger number of spectrally distinct NP species, and, perhaps more importantly, allows GNRs to be used for tissue imaging by tuning the resonance to the optical therapeutic window (between 700 and 900 nm) where intrinsic absorption due to water and hemoglobin in human tissue is lowest [146,147]. For example, using an adapted hyperspectral imaging system to switch between different wavelength windows enables discrimination of diseased tissues by observing binding of functionalized GNRs to target receptors. In a recent study of an orthotropic brain tumor model, EGFR-overexpressing malignant glioma cells (GBM-270) were identified by binding of anti-EGFR labeled GNRs. By using tumor cells that were transfected to also express green fluorescent protein (GFP) so that they could serve as a positive marker for malignancy, the GNRs' molecular specificity and spatial accuracy could be assessed [148]. Here the spectral selectivity of hyperspectral imaging was exploited to discriminate the highly scattering blood component at 550 nm from GNR contrast with its peak scattering wavelength of 620 nm. Using this system, binary predictions for the presence of cancer were enabled, indicating significant agreement (kappa statistic of 0.75) between GFP and GNR positive regions [148].

In summary, hyperspectral dark-field microscopy provides many benefits for nanoparticle imaging, especially in the context of biological systems. Nanoparticles offer biocompatibility and stable optical properties not offered by fluorophores. Furthermore, single nanoparticle scattering spectra and changes in their spectra reveal biological functional parameters including intracellular location, local refractive index, interparticle spacing, and particle orientation; however, one limitation is the need for either *a priori* information or experimental validation of RI values, NP size, etc. in order to accurately and precisely calculates these biological parameters [139,143–145]. In addition, hyperspectral microscopy allows for the spectral multiplexing of NP molecular contrast agents for multiple receptors tracking of cells through the use of different NP species with distinct and separated spectral scattering signatures [138,139,148]. Thus, hyperspectral imaging can be a power tool for the analysis and classification of plasmonic nanoparticles, especially when used for biological applications.

## 5.7 SUMMARY AND CONCLUSION

Using microscopy techniques to study the structure of nanomaterials in different applications opens up several perspectives:

1. Upgrading and promoting available functions to microscopy tools are required for enhancing the quality of imaging the nanoparticles and understanding their functionality and applications.
2. Developing other kinds of microscopy is needed by combing different tools with the microscopes to open new horizons in characterization of nanomaterials.

Therefore, the progress in microscopic tools can overshadow nanocoatings. For example, introducing the fourth dimension electron microscopy and scanning transmission electron microscopy can provide the door to observation of different structural and morphological phenomena of nanomaterials, nanostructures, and nanoparticles surfaces.

## REFERENCES

1. L. Zhang, T.J. Webster. Nanotechnology and nanomaterials: Promises for improved tissue regeneration. *Nano Today* 4, 2009: 66–80.
2. P.J.A. Borm, D. Robbins, S. Haubold et al. The potential risks of nanomaterials: A review carried out for ECETOC. *Particle and Fibre Toxicology* 3, 2006: 1–35.
3. The Royal Society and Royal Academy of Engineering, Nanoscience and Nanotechnologies: Opportunities and Uncertainties. RS policy document 35/06, 2006.
4. H. Liu, T.J. Webster. Nanomedicine for implants: A review of studies and necessary experimental tools. *Biomaterials* 28, 2007: 354–369.
5. R.C. Murdock, L. Braydich-Stolle, A.M. Schrand et al. Characterization of nanomaterial dispersion in solution prior to *in vitro* exposure using dynamic light scattering technique. *Toxicological Sciences* 101, 2008: 239–253.
6. M. Meyer, O. Kuusi. Nanotechnology: Generalizations in an interdisciplinary field of science and technology. *International Journal for Philosophy of Chemistry* 10, 2002: 153–168.
7. W. Tanthapanichakoon, J. Chaichanawong, S. Ratchahat. Review of nanomaterials R&D in 21st century in Thailand with highlight on nanoparticle technology. *Advanced Powder Technology* 25, 2014: 189–194.
8. X.Y. Qin, J.G. Kim, J.S. Lee. Synthesis and magnetic properties of nanostructured g-Ni–Fe alloys. *Nanostructal Materials* 11, 1999: 259–270.
9. M. Ferrari. Cancer nanotechnology: Opportunities and challenges. *Nature Reviews Cancer* 5, 2005: 161–171.
10. J.K. Vasir, M.K. Reddy, V.D. Labhasetwar. Nanosystems in drug targeting: Opportunities and challenges. *Current Nanoscience* 1, 2005: 47–64.
11. T.J. Webster, R.W. Siegel, R. Bizios. Osteoblast adhesion on nanophase ceramics. *Biomaterials* 20, 1999: 1221–1227.
12. T.J. Webster, C. Ergun, R.H. Doremus. Enhanced functions of osteoblasts on nanophase ceramics. *Biomaterials* 21, 2000: 1803–1810.
13. S. Ringer, K. Ratinac. On the role of characterization in the design of interfaces in nanoscale materials technology. *Microscopy Microanalysis* 10, 2004: 324–335.
14. E. Hood. Nanotechnology: Looking as we leap. *Environmental Health Perspectives* 112, 2004: A740–A749.
15. J. Bucher, S. Masten, B. Moudgil et al. Developing experimental approaches for the evaluation of toxicological interactions of nanoscale materials. *Final Workshop Report* 3–4, 2004: 1–37.
16. G. Oberdorster, A. Maynard, K. Donaldson et al. Principles for characterizing the potential human health effects from exposure to nanomaterials: Elements of a screening strategy. *Particle and Fibre Toxicology* 2, 2005: 8.
17. G. Oberdorster, E. Oberdorster, J. Oberdorster. Nanotoxicology: An emerging discipline evolving from studies of ultrafine particles. *Environmental Health Perspectives* 113, 2005: 823–839.
18. K. Powers, S. Brown, V. Krishna et al. Research strategies for safety evaluation of nanomaterials. Part VI. Characterization of nanoscale particles for toxicological evaluation. *Toxicological Sciences* 90, 2006: 296–303.
19. C. Mao. Introduction: Nanomaterials characterization using microscopy. *Microscopy Research and Technique* 64, 2004: 345–346.
20. C.B. Mao, C.E. Flynn, A. Hayhurst et al. Viral assembly of oriented quantum dot nanowires. *Proceedings of the National Academy of Sciences of the United States of America* 100, 2003: 6946–6951.
21. X. Peng, U. Manna, W. Yang et al. Shape control of CdSe nanocrystals. *Nature* 404, 2000: 59–61.
22. F.X. Redl, K.S. Cho, C.B. Murray et al. Three-dimensional binary superlattices of magnetic nanocrystals and semiconductor quantum dots. *Nature* 423, 2003: 968–971.
23. C.B. Mao, D.J. Solis, B.D. Reiss et al. Virus-based toolkit for the directed synthesis of magnetic and semiconducting nanowires. *Science* 303, 2004: 213–217.
24. C.B. Murray, C.R. Kagan, M.G. Bawendi. Synthesis and characterization of monodisperse nanocrystals and close-packed nanocrystal assemblies. *Annual Review of Materials Research* 30, 2000: 545–610.

25. T. Kauffeldt, H. Lakner, M. Lohmann et al. Nanocharacterization of small particles produced in a gas. *Nanostructured Materials* 6, 1995: 365–368.

26. L.L. Kazmerski. Atomic imaging, atomic processing and nanocharacterization of $CuInSe_2$ using proximal probe techniques. *Japanese Journal of Applied Physics Part 1* 32, 1993: 25–34.

27. T. Suzuki, K. Nagatani, K. Hirano et al. AFM/ MFM hybrid nanocharacterization of martensitic transformation and degradation for Fe–Pd shape memory alloy. *Proceedings of SPIE—The International Society for Optical Engineering* 5045, 2003: 63–72.

28. C.T.K.H. Stadtländer. Scanning electron microscopy and transmission electron microscopy of mollicutes: Challenges and opportunities. Editors: A. Méndez-Vilas, J. Díaz, *Modern Research and Educational Topics in Microscopy*. Formatex, Spain, 2007.

29. A. Bogner, P.H. Jouneau, G. Thollet et al. A history of scanning electron microscopy developments: Towards "wet-STEM" imaging. *Micron* 38, 2007: 390–401.

30. K.A. Marx. Introduction: Enabling nanomanufacturing using microscopy techniques, microscopy research and technique, doi: 10.1002/jemt.20472.

31. M. Parvinzadeh Gashti, F. Alimohammadi, J. Hulliger et al. Microscopic methods to study the structure of scaffolds in bone tissue engineering: A brief review. Editor: A. Méndez-Vilas, *Current Microscopy Contributions to Advances in Science and Technology*, Vol. 1, Formatex Research Center, Spain, 2012, 625–638.

32. M. Parvinzadeh Gashti, F. Alimohammadi, G. Song et al. Characterization of nanocomposite coatings on textiles: A brief review on Microscopic technology. Editor: A. Méndez-Vilas, *Current Microscopy Contributions to Advances in Science and Technology*, Vol. 2, Formatex Research Center, Spain, 2012, 1424–1437.

33. C.E. Hall. *Introduction to Electron Microscopy*. McGraw-Hill Publishing Co, USA, 1996.

34. B. Fultz, J.M. Howe. *Transmission Electron Microscopy and Diffractometry of Materials*. Springer, Heidelberg, 4th ed., 2013.

35. H. Seiler. Secondary electron emission in the scanning electron microscope. *Journal of Applied Physics* 54, 1983: 1–18, doi: 10.1063/1.332840.

36. T. Junno, K. Deppert, L. Montelius et al. Controlled manipulation of nanoparticles with an atomic force microscope. *Applied Physics Letters* 66, 1995: 3627–3629.

37. B. Polyakov, S. Vlassov, L.M. Dorogin et al. Manipulation of nanoparticles of different shapes inside a scanning electron microscope. *Beilstein Journal of Nanotechnology* 5, 2014: 133–140.

38. E. Buhr, N. Senftleben, T. Klein et al. Characterization of nanoparticles by scanning electron microscopy in transmission mode. *Measurement Science and Technology* 20, 2009: 084025–084034.

39. F. Fanelli, A.M. Mastrangelo, F. Fracassi. Aerosol-assisted atmospheric cold plasma deposition and characterization of superhydrophobic organic–inorganic nanocomposite thin films. *Langmuir* 30, 2014: 857–865.

40. M. Rossi, H. Ye, Y. Gogotsi et al. Environmental scanning electron microscopy study of water in carbon nanopipes. *Nano Letters* 4, 2004: 989–993.

41. T. Ji. Preliminary study on the water permeability and microstructure of concrete incorporating nano-$SiO_2$. *Cement and Concrete Research* 35, 2005: 1943–1947.

42. D.B. Williams, C.B. Carter. *The Transmission Electron Microscope*. Springer, Spain, 1996.

43. A. Bogner, P.-H. Jouneau, G. Thollet et al. A history of scanning electron microscopy developments: Towards "wet-STEM" imaging. *Micron* 38, 2007: 390–401.

44. N. de Jonge, F.M. Ross. Electron microscopy of specimens in liquid. *Nature Nanotechnology* 6, 2011: 695–704.

45. A. Bogner, G. Thollet, D. Basset et al. Wet STEM: A new development in environmental SEM for imaging nano-objects included in a liquid phase. *Ultramicroscopy* 104, 2005: 290–301.

46. E.R. White, M. Mecklenburg, B. Shevitski et al. Charged nanoparticle dynamics in water induced by scanning transmission electron microscopy. *Langmuir* 28, 2012: 3695–3698.

47. J.E. Evans, K.L. Jungjohann, N.D. Browning et al. Controlled growth of nanoparticles from solution with *in situ* liquid transmission electron microscopy. *Nano Letter* 11, 2011: 2809–2813.

48. S. Disch, E. Wetterskog, R.P. Hermann et al. Shape induced symmetry in self-assembled mesocrystals of iron oxide nanocubes. *Nano Letter* 11, 2011: 1651–1656.

49. A.J. Bard, M.V. Mirkin. *Scanning Electrochemical Microscopy*. CRC Press, Boca Raton, FL, 2nd ed., 2012, 660.

50. M.V. Mirkin, W. Nogala, J. Velmurugan et al. Scanning electrochemical microscopy in the 21st century. Update 1: Five years after. *Physical Chemistry Chemical Physics* 13(48), 2011: 21196–21212.

51. G. Wittstock, M. Burchardt, S.E. Pust et al. Scanning electrochemical microscopy for direct imaging of reaction rates. *Angewandte Chemie-International Edition* 46(10), 2007: 1584–1617.

52. P.G. Nicholson, S. Zhou, G. Hinds, A.J. Wain, A. Turnbull. Electrocatalytic activity mapping of model fuel cell catalyst films using scanning electrochemical microscopy. *Electrochimica Acta* 54(19), 2009: 4525–4533.

53. Y. Shen, M. Traeuble G. Wittstock. Electrodeposited noble metal particles in polyelectrolyte multi-layer matrix as electrocatalyst for oxygen reduction studied using SECM. *Physical Chemistry Chemical Physics*, 10(25), 2008: 3635–3644.

54. X. Chen, K. Eckhard, M. Zhou et al. Electrocatalytic activity of spots of electrodeposited noble-metal catalysts on carbon nanotubes modified glassy carbon. *Analytical Chemistry* 81(18), 2009: 7597–7603.

55. T.C. Nagaiah, A. Maljusch, X. Chen et al. Visualization of the local catalytic activity of electrodeposited Pt–Ag catalysts for oxygen reduction by means of SECM. *Chem Phys Chem* 10(15), 2009: 2711–2718.

56. K. Szot, W. Nogala, J. Niedziolka-Jönsson et al. Hydrophilic carbon nanoparticle-laccase thin film electrode for mediatorless dioxygen reduction SECM activity mapping and application in zinc-dioxygen battery. *Electrochimica Acta* 54(20), 2009: 4620–4625.

57. A. Maljusch, T.C. Nagaiah, S. Schwamborn et al. Pt–Ag catalysts as cathode material for oxygen-depolarized electrodes in hydrochloric acid electrolysis. *Analytical Chemistry* 82(5), 2010: 1890–1896.

58. W. Nogala, A. Celebanska, K. Szot et al. Bioelectrocatalytic mediatorless dioxygen reduction at carbon ceramic electrodes modified with bilirubin oxidase. *Electrochimica Acta* 55(20), 2010: 5719–5724.

59. W. Nogala, A. Celebanska, G. Wittstock et al. Bioelectrocatalytic carbon ceramic gas electrode for reduction of dioxygen and its application in a zinc-dioxygen cell. *Fuel Cells* 10(6), 2010: 1157–1163.

60. C.M. Sánchez-Sánchez, J. Solla-Gullón, F.J. Vidal-Iglesias et al. Imaging structure sensitive catalysis on different shape-controlled platinum nanoparticles. *Journal of the American Chemical Society* 132(16), 2010: 5622–5624.

61. C.M. Sánchez-Sánchez, F.J. Vidal-Iglesias, J. Solla-Gullón et al. Scanning electrochemical microscopy for studying electrocatalysis on shape-controlled gold nanoparticles and nanorods. *Electrochimica Acta* 55(27), 2010: 8252–8257.

62. X. Xu, J. Jia, X. Yang et al. A templateless, surfactantless, simple electrochemical route to a dendritic gold nanostructure and its application to oxygen reduction. *Langmuir* 26(10), 2010: 7627–7631.

63. C.M. Sánchez-Sánchez, J. Souza-Garcia, A. Sáez et al. Imaging decorated platinum single crystal electrodes by scanning electrochemical microscopy. *Electrochimica Acta* 56(28), 2011: 10708–10712.

64. W. Li, F.-R.F. Fan, A.J. Bard. The application of scanning electrochemical microscopy to the discovery of Pd–W electrocatalysts for the oxygen reduction reaction that demonstrate high activity, stability, and methanol tolerance. *Journal of Solid State Electrochemistry* 16(7), 2012: 2563–2568.

65. T.C. Nagaiah, D. Schäfer, W. Schuhmann et al. Electrochemically deposited Pd–Pt and Pd–Au code-posits on graphite electrodes for electrocatalytic $H_2O_2$ reduction. *Analytical Chemistry* 85(16), 2013: 7897–7903.

66. A.J. Wain. Imaging size effects on the electrocatalytic activity of gold nanoparticles using scanning electrochemical microscopy. *Electrochimica Acta* 92, 2013: 383–391.

67. A.R Kucernaka, P.B Chowdhurya, C.P Wilde et al. Scanning electrochemical microscopy of a fuel-cell electrocatalyst deposited onto highly oriented pyrolytic graphite. *Electrochimica Acta* 45(27), 2000: 4483–4491.

68. F. Li, P. Bertoncello, I. Ciani et al. Incorporation of functionalized palladium nanoparticles within ultrathin Nafion films: A nanostructured composite for electrolytic and redox-mediated hydrogen evolution. *Advanced Functional Materials* 18(11), 2008: 1685–1693.

69. F. Li, I. Ciani, P. Bertoncello et al. Scanning electrochemical microscopy of redox-mediated hydrogen evolution catalyzed by two-dimensional assemblies of palladium nanoparticles. *Journal of Physical Chemistry C* 112(26), 2008: 9686–9694.

70. S.J. Kwon, A.J. Bard. Analysis of diffusion-controlled stochastic events of iridium oxide single nanoparticle collisions by scanning electrochemical microscopy. *Journal of the American Chemical Society* 134(16), 2012: 7102–7108.

71. S.K. Cho, H.S. Park, He.C. Lee et al. Metal doping of $BiVO_4$ by composite electrodeposition with improved photoelectrochemical water oxidation. *Journal of Physical Chemistry C* 117(44), 2013: 23048–23056.

72. L.A. Näslund, C.M. Sánchez-Sánchez, A.S. Ingason et al. The role of $TiO_2$ doping on $RuO_2$-coated electrodes for the water oxidation reaction. *Journal of Physical Chemistry C* 117(12), 2013: 6126–6135.

73. P. Liljeroth, D. Vanmaekelbergh, V. Ruiz et al. Electron transport in two-dimensional arrays of gold nanocrystals investigated by scanning electrochemical microscopy. *Journal of the American Chemical Society* 126(22), 2004: 7126–7132.

74. P.G. Nicholson, V. Ruiz, J.V. Macphersona et al. Effect of composition on the conductivity and morphology of poly(3-hexylthiophene)/gold nanoparticle composite Langmuir–Schaeffer films. *Physical Chemistry Chemical Physics* 8(43), 2006: 5096–5105.

75. B.M. Quinn, I. Prieto, S.K. Haram et al. Electrochemical observation of a metal/insulator transition by scanning electrochemical microscopy. *Journal of Physical Chemistry B* 105(31), 2001: 7474–7476.

76. F. Li, M. Edwards, J. Guo et al. Silver particle nucleation and growth at liquid/liquid interfaces: A scanning electrochemical microscopy approach. *Journal of Physical Chemistry C* 113(9), 2009: 3553–3565.

77. D.G. Georganopoulou, M.V. Mirkin, R.W. Murray. SECM measurement of the fast electron transfer dynamics between Au-38(1+) nanoparticles and aqueous redox species at a liquid/liquid interface. *Nano Letters* 4(9), 2004: 1763–1767.

78. B.M. Quinn, P. Liljeroth, K. Kontturi. Interfacial reactivity of monolayer-protected clusters studied by scanning electrochemical microscopy. *Journal of the American Chemical Society* 124(43), 2002: 12915–12921.

79. R.R. Peterson, D.E. Cliffel. Scanning electrochemical microscopy determination of organic soluble MPC electron-transfer rates. *Langmuir* 22(25), 2006: 10307–10314.

80. A. Taleba, X. Yanpenga, S. Munteanu et al. Self-assembled thiolate functionalized gold nanoparticles template toward tailoring the morphology of electrochemically deposited silver nanostructure. *Electrochimica Acta* 88, 2013: 621–631.

81. M. Zhang, A. Becue, M. Prudent et al. SECM imaging of MMD-enhanced latent fingermarks. *Chemical Communications* 38, 2007: 3948–3950.

82. M. Zhang, H.H. Girault. SECM for imaging and detection of latent fingerprints. *Analyst* 134(1), 2009: 25–30.

83. S. Lee, Y. Zhang, H.S. White et al. Electrophoretic capture and detection of nanoparticles at the opening of a membrane pore using scanning electrochemical microscopy. *Analytical Chemistry* 76(20), 2004: 6108–6115.

84. R. Tel-Vered, A.J. Bard. Generation and detection of single metal nanoparticles using scanning electrochemical microscopy techniques. *Journal of Physical Chemistry B* 110(50), 2006: 25279–25287.

85. E. Malel, R. Ludwig, L. Gorton et al. Localized deposition of Au nanoparticles by direct electron transfer through cellobiose dehydrogenase. *Chemistry—a European Journal* 16(38), 2010: 11697–11706.

86. J. Wang, F.Y. Song, F.M. Zhou. Silver-enhanced imaging of DNA hybridization at DNA microarrays with scanning electrochemical microscopy. *Langmuir* 18(17), 2002: 6653–6658.

87. H. Fan, F. Jiao, H. Chen et al. Qualitative and quantitative detection of DNA amplified with HRP-modified SiO$_2$ nanoparticles using scanning electrochemical microscopy. *Biosensors & Bioelectronics* 47, 2013: 373–378.

88. H. Fan, X. Wang, F. Jiao et al. Scanning electrochemical microscopy of DNA hybridization on DNA microarrays enhanced by HRP-modified SiO$_2$ nanoparticles. *Analytical Chemistry* 85(13), 2013: 6511–6517.

89. M. Carano, N. Lion, J.P. Abid et al. Detection of proteins on poly(vinylidene difluoride) membranes by scanning electrochemical microscopy. *Electrochemistry Communications* 6(12), 2004: 1217–1221.

90. Z. Chen, S. Xie, L. Shen et al. Investigation of the interactions between silver nanoparticles and Hela cells by scanning electrochemical microscopy. *Analyst* 133(9), 2008: 1221–1228.

91. X. Yang, S.E. Skrabalak, Z.Y. Li et al. Photoacoustic tomography of a rat cerebral cortex *in vivo* with au nanocages as an optical contrast agent. *Nano Letters* 7(12), 2007: 3798–3802.

92. T. Danieli, N. Gaponik, A. Eychmüller et al. Studying the reactions of CdTe nanostructures and thin CdTe films with Ag+ and AuCl$_4$–. *Journal of Physical Chemistry C* 112(24), 2008: 8881–8889.

93. E. Malel, D. Mandler. Localized electroless deposition of gold nanoparticles using scanning electrochemical microscopy. *Journal of the Electrochemical Society* 155(6), 2008: D459–D467.

94. A.P. O'Mullane, S.J. Ippolito, A.M. Bond et al. A study of localised galvanic replacement of copper and silver films with gold using scanning electrochemical microscopy. *Electrochemistry Communications* 12(5), 2010: 611–615.

95. T. Danieli, J. Colleran, D. Mandler, Deposition of Au and Ag nanoparticles on PEDOT. *Physical Chemistry Chemical Physics* 13(45), 2011: 20345–20353.

96. E. Malel, J. Colleran, D. Mandler. Studying the localized deposition of Ag nanoparticles on self-assembled monolayers by scanning electrochemical microscopy (SECM). *Electrochimica Acta* 56(20), 2011: 6954–6961.

97. R.G. Fedorov, D. Mandler. Local deposition of anisotropic nanoparticles using scanning electrochemical microscopy (SECM). *Physical Chemistry Chemical Physics* 15(8), 2013: 2725–2732.

98. M. Yamada, H. Nishihara. Electrodeposition of biferrocene derivative-attached gold nanoparticles: Solvent effects and lithographic assembly. *Langmuir* 19(19), 2003: 8050–8056.

99. T. Danieli, D. Mandler. Local surface patterning by chitosan-stabilized gold nanoparticles using the direct mode of scanning electrochemical microscopy (SECM). *Journal of Solid State Electrochemistry* 17(12), 2013: 2989–2997.

100. J.M. Noël, D. Zigah, J. Simonet et al. Synthesis and immobilization of Ag0 nanoparticles on diazonium modified electrodes: SECM and cyclic voltammetry studies of the modified interfaces. *Langmuir* 26(10), 2010: 7638–7643.

101. V. Ruiz, P. Liljeroth, B.M. Quinn et al. Probing conductivity of polyelectrolyte/nanoparticle composite films by scanning electrochemical microscopy. *Nano Letters* 3(10), 2003: 1459–1462.

102. P. Ahonen, V. Ruiz, K. Kontturi et al. Electrochemical gating in scanning electrochemical microscopy. *Journal of Physical Chemistry C* 112(7), 2008: 2724–2728.

103. K. Huang, A. Anne, M. Ali Bahri et al. Probing individual redox PEGylated gold nanoparticles by electrochemical-atomic force microscopy. *ACS Nano* 7(5), 2013: 4151–4163.

104. A. Celebanska, A. Lesniewski, J. Niedziolka-Jonsson et al. Carbon nanoparticulate film electrode prepared by electrophoretic deposition. Electrochemical oxidation of thiocholine and topography imaging with SECM equipment in dry conditions. *Electrochimica Acta* 144, 2014: 136–140.

105. C.D. O'Connell, M.J. Higgins, R.P. Sullivan et al. Nanoscale platinum printing on insulating substrates. *Nanotechnology* 24(50), 2013: 1–11, doi: 10.1088/0957-4484/24/50/505301.

106. J.R. Cook, W. Frey, S. Emelianov. Quantitative photoacoustic imaging of nanoparticles in cells and tissues. *ACS Nano* 7(2), 2013: 1272–1280.

107. A. Madhankumar, N. Rajendran, T. Nishimura. Influence of Si nanoparticles on the electrochemical behavior of organic coatings on carbon steel in chloride environment. *Journal of Coatings Technology and Research* 9(5), 2012: 609–620.

108. A. Saez, C.M. Sanchez-Sanchez, J. Solla-Gullon et al. Electrosynthesis of L-cysteine on a dispersed Pb/carbon black electrode. *Journal of the Electrochemical Society* 156(11), 2009: E154–E160.

109. N. Njomo, T. Waryoa, M. Masikinia et al. Graphenated tantalum(IV) oxide and poly(4-styrene sulphonic acid)-doped polyaniline nanocomposite as cathode material in an electrochemical capacitor. *Electrochimica Acta* 128, 2014: 226–237.

110. A. Srivatsan, S.V. Jenkins, M. Jeon et al. Gold nanocage-photosensitizer conjugates for dual-modal image-guided enhanced photodynamic therapy. *Theranostics* 4(2), 2014: 163.

111. L.V. Wang, H. Wu. *Biomedical Optics: Principles and Imaging.* Wiley, USA, 2007.

112. B. Sitharaman, P.K. Avti, K. Schaefer et al. A novel nanoparticle-enhanced photoacoustic stimulus for bone tissue engineering. *Tissue Engineering Part A* 17(13–14), 2011: 1851–1858.

113. M. Xu, L.V. Wang. Photoacoustic imaging in biomedicine. *Review of Scientific Instruments* 77(4), 2006: 041101, doi: 10.1063/1.2195024.

114. X. Cai, B.S. Paratala, S. Hu et al. Multiscale photoacoustic microscopy of single-walled carbon nanotube-incorporated tissue engineering scaffolds. *Tissue Engineering Part C: Methods* 18(4), 2011: 310–317.

115. X.D. Wang, Y.J. Pang, G. Ku et al. Noninvasive laser-induced photoacoustic tomography for structural and functional *in vivo* imaging of the brain. *Nature Biotechnology* 21(7), 2003: 803–806.

116. J.T. Oh, G. Stoica, L.V. Wang et al. Three-dimensional imaging of skin melanoma *in vivo* by dual-wavelength photoacoustic microscopy. *Journal of Biomedical Optics* 11(3), 2006: 034032-034032-4.

117. Z. Xu, Q. Zhu, L.V. Wang. In vivo photoacoustic tomography of mouse cerebral edema induced by cold injury. *Journal of Biomedical Optics* 16(6), 2011: 066020-066020-4.

118. Z. Xu, C. Li, L.V. Wang. Photoacoustic tomography of water in phantoms and tissue. *Journal of Biomedical Optics* 15(3), 2010: 036019-036019-6.

119. B. Wang, L.S. Jimmy, J. Amirian et al. Detection of lipid in atherosclerotic vessels using ultrasound-guided spectroscopic intravascular photoacoustic imaging. *Optic Express* 18, 2010: 4889–4897.

120. J. Yao, K. Maslov, S. Hu et al. Evans blue dye-enhanced capillary-resolution photoacoustic microscopy in vivo. *Journal of Biomedical Optics* 14(5), 2009: 054049–054056.

121. Z. Xu, J. Bai. Analysis of finite-element-based methods for reducing the ill-posedness in the reconstruction of fluorescence molecular tomography. *Progress in Natural Science* 19(4), 2009: 501–509.

122. Z. Guo, Z. Xu, L.V. Wang. Dependence of photoacoustic speckles on boundary roughness. *Journal of Biomedical Optics* 17(4), 2012: 0460091–0460096.

123. M.L. Li, J.T. Oh, X.Y. Xie et al. Simultaneous molecular and hypoxia imaging of brain tumors *in vivo* using spectroscopic photoacoustic tomography. *Proceedings of IEEE* 96(3), 2008: 481–489.

124. A. De La Zerda, C. Zavaleta, S. Keren et al. Carbon nanotubes as photoacoustic molecular imaging agents in living mice. *Nature Nanotechnology* 3(9), 2008: 557–562.

125. M. Pramanik, K.H. Song, M. Swierczewska et al. In vivo carbon nanotube-enhanced non-invasive photoacoustic mapping of the sentinel lymph node. *Physics in Medicine and Biology* 54(11), 2009: 3291.

126. M.A. El-Sayed. Some interesting properties of metals confined in time and nanometer space of different shapes. *Accounts of Chemical Research* 34(4), 2001: 257–264.

127. P.K. Jain, X. Huang, I.H. El-Sayed et al. Noble metals on the nanoscale: Optical and photothermal properties and some applications in imaging, sensing, biology, and medicine. *Accounts of Chemical Research* 41(12), 2008: 1578–1586.

128. M.L. Li, J.C. Wang, J.A. Schwartz et al. In-vivo photoacoustic microscopy of nanoshell extravasation from solid tumor vasculature. *Journal of Biomedical Optics* 14(1), 2009: 010507-010507-3.

129. J. Chen, F. Saeki, B.J. Wiley et al. Gold nanocages: Bioconjugation and their potential use as optical imaging contrast agents. *Nano Letters* 5(3), 2005: 473–477.

130. Q. Zhang, N. Iwakuma, P. Sharma et al. Gold nanoparticles as a contrast agent for *in vivo* tumor imaging with photoacoustic tomography. *Nanotechnology* 20(39), 2009: 395102.

131. V.P. Zharov, J.W. Kim, D.T. Curiel et al. Self-assembling nanoclusters in living systems: Application for integrated photothermal nanodiagnostics and nanotherapy. *Nanomedicine: Nanotechnology, Biology and Medicine* 1(4), 2005: 326–345.

132. V.P. Zharov. Ultrasharp nonlinear photothermal and photoacoustic resonances and holes beyond the spectral limit. *Nature Photonics* 5(2), 2011: 110–116.

133. L. Christopher, A. Lowery, N. Halas et al. Immunotargeted nanoshells for integrated cancer imaging and therapy. *Nano Letters* 5(4), 2005: 709–711.

134. L. Wu, X. Cai, K. Nelson et al. A green synthesis of carbon nanoparticles from honey and their use in real-time photoacoustic imaging. *Nano Research* 6(5), 2013: 312–325.

135. X. Liu, C. Lee, W.C. Law et al. Au–$Cu_2$–x Se heterodimer nanoparticles with broad localized surface plasmon resonance as contrast agents for deep tissue imaging. *Nano Letters* 13(9), 2013: 4333–4339.

136. Y. Zhang, M. Jeon, L.J. Rich et al. Non-invasive multimodal functional imaging of the intestine with frozen micellar naphthalocyanines. *Nature Nanotechnology* 9, 2014: 631–638.

137. A. Curry, L.H. William, A. Wax. Epi-illumination through the microscope objective applied to dark-field imaging and microspectroscopy of nanoparticle interaction with cells in culture. *Optics Express* 14, 2006: 6535–6542.

138. K. Seekell, J.C. Matthew, M. Stella et al. Hyperspectral molecular imaging of multiple receptors using immunolabeled plasmonic nanoparticles. *Journal of Biomedical Optics* 16(11), 2011: 116003-1-7.

139. A.C. Curry, M. Crow, A. Wax. Molecular imaging of epidermal growth factor receptor in live cells with refractive index sensitivity using dark-field microspectroscopy and immunotargeted nanoparticles. *Journal of Biomedical Optics* 13, 2008: 014022-1-8.

140. S. Konstantin, M. Follen, J. Aaron et al. Real-time vital optical imaging of precancer using anti-epidermal growth factor receptor antibodies conjugated to gold nanoparticles. *Cancer Research* 63.9, 2003: 1999–2004.

141. L. Christopher, L. Hirsch, M.H. Lee et al. Gold nanoshell bioconjugates for molecular imaging in living cells. *Optics Letters* 30(9), 2005: 1012–1014.

142. A. Wax, K. Sokolov. Molecular imaging and darkfield microspectroscopy of live cells using gold plasmonic nanoparticles. *Laser & Photonics Review* 3(1–2), 2009: 146–158.

143. M.J. Crow, G. Grant, J.M. Provenzale et al. Molecular imaging and quantitative measurement of epidermal growth factor receptor expression in live cancer cells using immunolabeled gold nanoparticles. *American Journal of Roentgenology* 192(4), 2009: 1021–1028.

144. M.J. Crow, K. Seekell, J.H. Ostrander et al. Monitoring of receptor dimerization using plasmonic coupling of gold nanoparticles. *ACS Nano* 5.11, 2011: 8532–8540.

145. M.J. Crow, K. Seekell, A. Wax. Polarization mapping of nanoparticle plasmonic coupling. *Optics Letters* 36.5, 2011: 757–759.

146. S. Eustis, M.A. El-Sayed. Why gold nanoparticles are more precious than pretty gold: Noble metal surface plasmon resonance and its enhancement of the radiative and nonradiative properties of nanocrystals of different shapes. *Chemical Society Reviews* 35(3), 2006: 209–217.

147. K. Seekell, P. Hillel, S. Marinakos et al. Optimization of immunolabeled plasmonic nanoparticles for cell surface receptor analysis. *Methods* 56.2, 2012: 310–316.

148. K. Seekell, S. Lewis, C. Wilson et al. Feasibility study of brain tumor delineation using immunolabeled gold nanorods. *Biomedical Optics Express* 4(11), 2013: 2284–2295.

# 6 Mechanical Properties of Nanomaterials

*Shirin Shokoohi, Ghasem Naderi, and Aliasghar Davoodi*

## CONTENTS

## 6.1 INTRODUCTION

Nanomaterials are an attractive class of modern materials that have created a high interest in recent years by virtue of their noble properties superior to that of their constitutive counterparts. This is mainly due to small dimensions consequently leading to large fractions of surface atoms, high surface energy, spatial confinement, and reduced imperfections in nanoscale materials, which do not exist in the corresponding bulk materials since most microstructured materials have properties similar to the corresponding bulk material, while the properties of materials with nanometer dimensions are in between those of atoms and the bulk material.

Indirect investigations have demonstrated that mechanical characteristics of nanoparticles are different from the corresponding bulk materials. Crystals are composed of alternative molecules as repeating units with quantized electronic structures. Continuous electronic band structures of crystals are, in fact, the result of a combination and an overlap of repeating molecules orbitals and that is why isolated molecules exhibit quantum mechanical properties unlike the properties of bulk crystals that undergo the classical mechanical laws. When the crystal size falls under the nanoscale dimensions, the crystal electronic bands start to be quantized and the individual nanocrystals moderately behave between molecules and crystals [1]. In another phrase, crystals present at the topmost layers of nanomaterials are not buried in the same charge density encountered by those of the bulk. According to the Hellmann–Feynman theorem remarked in quantum mechanics, once the spatial distribution of the electrons has been fixed, electrostatic interactions govern the forces on the ionic cores. According to this principle, an oscillatory-type relaxation behavior inwardly oriented into the region of higher charge density is performed by the surface layer. This phenomenon is intensified in more open surfaces [2,3] such as nanostructured materials in which the high surface/volume ratio would also lead to the strain relief mechanisms arisen from the epitaxial growth of inorganic crystal layers that is absent in large-crystal structures. Accordingly, since crystal layers present on the surface have the capacity to relive stress more easily, miniaturization of crystals to nanodimension leads to the development of functional nanomaterials with large surface/volume ratios facilitating stress and strain relaxations [4,5]. As mechanical property inherently depends on the correlations between stresses and strains encountered within the material bulk, the mechanical properties of nanomaterials would be different compared to the macro counterparts [6]. Accordingly, the present chapter is assigned to survey the mechanical properties of nanomaterials as it is crucial for obtaining deep insights to the physical properties of nanocomposites.

Mechanical properties are generally referred to as strength, hardness, creep, fatigue, fracture toughness, and so forth. Unlike optical, electrical or microhardness characterizations, which do not require any special sample preparation, measuring mechanical properties of nanoparticles encounters practical problems arisen from the nanoscale dimensions. Recently, miniaturized high-precision deformation instruments were technically progressed facilitating applying and controlling loads of mN order and nanometric displacements in the case of nanoparticles characterization but restricted in certain respects [7,8]. Regarding the fact that experimental investigations on the mechanical properties of individual nanoparticles are still in their early stages due to the practical difficulties, the following sections represent the information available on different mechanical properties of nanostructured materials. Mechanical properties are, in principle, described on the basis of the material structure; therefore, exploring mechanical properties of nanoparticles is commonly performed using transmission electron microscopy (TEM) for obtaining required crystallographic knowledge of the nanoparticle [7].

## 6.2 MECHANICAL CHARACTERIZATION OF NANOPARTICLES

### 6.2.1 Strength

Intensive deformation of nanomaterials provides the opportunity for the traditional measurement of tensile properties. Uniaxial tensile measurements have been utilized to evaluate the strength of materials [9]. At low stresses, material deforms reversibly illustrating a linear elastic stress–strain trend ending at the onset of plastic deformation defined as yield stress. Young's modulus or elasticity modulus in the elastic region is determined by Hook's law as below:

$$E = \frac{\sigma}{\varepsilon} \tag{6.1}$$

However, elastic properties of nanostructured materials basically depend on the structure [10]. MaacKenzie suggested Taylor series development for porosity ($p$) dependence of Young's modulus:

$$E = E_0(1 + \alpha_1 p + \alpha_2 p^2 + \cdots) \tag{6.2}$$

In which $\alpha_n$ ($n = 1, 2, \ldots$) are fitting parameters. For materials with low porosity, the linear term is enough. Young's modulus of porous materials is also estimated using

$$E = E_0 \exp\left(\frac{-\beta \Delta \rho_m}{\rho_m}\right) \tag{6.3}$$

where $\Delta \rho_m$, $E$, and $E_0$ refer to porosity, apparent elastic modulus, and reference elastic modulus (Young's modulus of fully dense material) and $\beta$ is 3–4.5 [11]. Porosity is defined as

$$p = 1 - \frac{\rho_{\text{sample}}}{\rho_{\text{theory}}} \tag{6.4}$$

where $\rho_{sample}$ and $\rho_{theory}$, respectively, refer to the experimentally measured density and estimated density of fully dense material [10]. Anyhow, Young's modulus of carbon nanotubes ($3.7 \times 10^{12}$ Pa) is slightly below theoretical predictions (5 TPa). It has been reported that when a tensile loading is applied to a carbon nanotube perpendicular to C–C bonds, ductile behavior is observed whereas in the case of longitudinal parallel tension force brittle rupture occurs. This is ascribed to the stress relaxation via C–C transverse rotational motions called Stone–Wales transformation [12,13].

Recently, a novel method has been suggested to study the tensile behavior of one-dimensional nano-structures, i.e., nanobelt/nanowire/nanotube according to the image profiled obtained utilizing atomic force microscopy. Ding et al. [14] investigated the tensile behavior of boron nanowires using atomic force microscopy cantilever as the force sensor and scanning electron microscopy to track the deformations in the nanoparticle simultaneously monitored by scanning electron microscopy and atomic force microscopy. Tensile loading with a continuously increasing trend was applied until the nanowire fracture. Deflection angle ($\theta$) and deflection ($\delta$) under loading the cantilever tip could be obtained:

$$\delta = \frac{PL^3}{3EI} \tag{6.5}$$

$$\theta = \frac{PL^2}{2EI} \tag{6.6}$$

where $P$, $L$, $E$, and $I$ are the applied load, cantilever length, elastic modulus, and moment inertia of the cantilever, respectively. The following equation was then obtained for the cantilever deflection:

$$\delta = \frac{2}{3}\theta L \tag{6.7}$$

To prevent misalignments of nanowire axis with the forced loaded, tensile load $F$ is decomposed into three terms as below:

$$F_y = F_x \tan\alpha$$
$$F_z = F_{xy}\tan\beta = F_x \tan\beta\sqrt{1 + \tan^2\alpha} \tag{6.8}$$

Since $F_x$ is the major component of the force applied to the nanowire specimen, $F_y$ and $F_z$ are written in terms of $F_x$.

$$k\Delta x = F_x - F_y \frac{H}{L} \tag{6.9}$$

In which $k$, $\Delta x$, $H$, and $L$, respectively, refer to cantilever force constant, cantilever deflection, the distance from the clamped end of nanowire to the cantilever central plane, and cantilever length. Ignoring $F_y$ due to its dispensible effect, the relationship between tensile load and cantilever deflection is simplified to:

$$F = \sqrt{F_x^2 + F_y^2 + F_z^2} = F_x\sqrt{1 + \tan^2\alpha + \tan^2\beta + \tan^2\alpha\tan^2\beta}$$
$$= \frac{k\Delta x}{1 - H/L\tan\alpha}\sqrt{1 + \tan^2\alpha + \tan^2\beta + \tan^2\alpha\tan^2\beta} \tag{6.10}$$

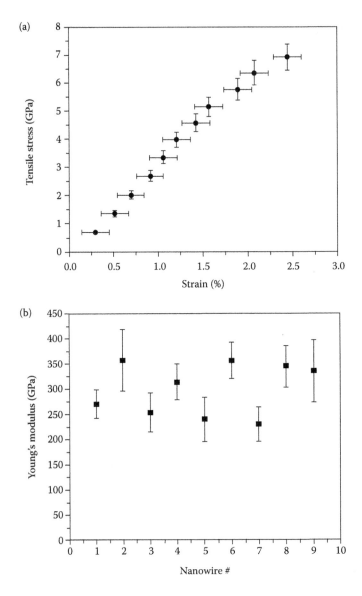

**FIGURE 6.1**  Boron nanowires under tension. (a) Stress–strain behavior and (b) tensile modulus. (Adapted from Ding, W. et al., *Composites Science and Technology*, 66(9), 1112–1124, 2006.)

Figure 6.1 shows the stress–strain curves and tensile modulus data for the nine boron nanowires (diameter = 3 nm) investigated under the tension according to the experimental data and theoritical calculations.

Reversed loading leading to the buckling of the nanoparticle sample was also examined. The deformations undergone by the sample was digitally analyzed and simulated applying the elastic theory according to which displacement coordinates (Figure 6.2) could be described at any point was calculated upon nanowire stiffness $EI$ (where $E$ and $I$, respectively, refer to elastic modulus and geometrical moment of inertia) under longitudinal force P is obtained as below [15]:

$$y = \sqrt{\frac{EI}{4P}} \int_0^\alpha \frac{\sin\theta \, d\theta}{\sqrt{\sin^2 \alpha/2 - \sin^2 \theta/2}} \qquad (6.11)$$

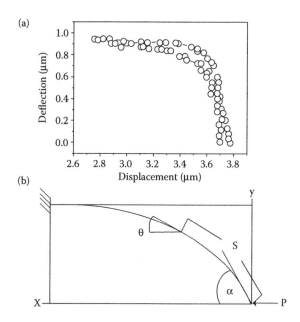

**FIGURE 6.2** Schematic illustration of parameters involved in buckling test used for the elastic theory equations. (Adapted from Kaplan-Ashiri, I. et al., *Proceedings of the National Academy of Sciences of the United States of America*, 103(3), 523–528, 2006.)

Young's modulus at the edges of the nanotube, where $\alpha$ is identical to $\theta$, was obtained equal to 150 GPa.

The three-point bend test has also been utilized to investigate the tensile properties of two-dimensional nanoparticles such as nanowires or nanotubes [9]. Young's modulus of the nanofiber positioned in the way that the middle section of the fiber is suspended over a hole (see Figure 6.3) is obtained using [16]:

$$E = \frac{PL^3}{192vI} \tag{6.12}$$

In which $P$ is the maximum force applied, $L$ is the fiber length suspended on the groove, $v$ is the deflection of the rod-like fiber at midspan, and $I$ is the second moment of area of the beam where:

$$I = \frac{\pi D^4}{64} \tag{6.13}$$

Assuming $D$ as the beam diameter. The maximum force is

$$P = k_l D_{max} \tag{6.14}$$

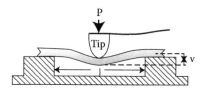

**FIGURE 6.3** Schematic of three-point bend test utilized to characterize mechanical properties of nanofibers. (Adapted from Tan, E. and Lim, C., *Composites Science and Technology*, 66(9), 1102–1111, 2006.)

In which $k_l$ and $D_{max}$ are defined as the spring constant of the cantilever and maximum deflection undergone by the cantilever. It is worth noting that usually there exists enough adhesion between the sample and substrate upon deposition in order to avoid slippage during conducting the bend test. Otherwise, the friction force between the nanorod and the substrate should be considered as well as the loading force.

Conventional strength, i.e., ambient temperature plasticity, of materials depends on average crystallite (grain) size expressed by the well-known Hall–Petch relation as below:

$$\sigma_y = \sigma_0 + \frac{k}{\sqrt{D}} \qquad (6.15)$$

In which $\sigma$, $D$, and $\sigma_0$, respectively, refer to yield strength, average grain diameter, and friction stress representing the strating stress for dislocation motion of crystal lattice. $k$ is a material parameter called the Hall–Petch coefficient [11].

Figure 6.4 shows strength-grain size dependance for alumina particles illustrating yield stress increases with decreasing grain diameter. This phenomenon is attributed to the pile-up of dislocations in grain boundries and the consequent back stress on the dislocation source leading to and enhanced resistance against plastic deformation. In materials with smaller grains, the mean free path to the grain boundries is shorter. Although a higher content of grain boundary is present in the nanometer range, e.g., 14%–27% of whole atoms lie in regions 0.5–1.0 nm abut grain boundary, it is expected that grain boundaries play a key role in the overall properties of the material. Grain boundaries as sources and sinks for dislocation phenomenon facilitate stress-relaxation mechanisms such as grain boundary sliding [17]. However, in materials with grain sizes smaller than 10 nm ($3.9 \times 10^{-7}$ in.) mechanisms are altered and the Hall–Petch relation is violated since material strength would either

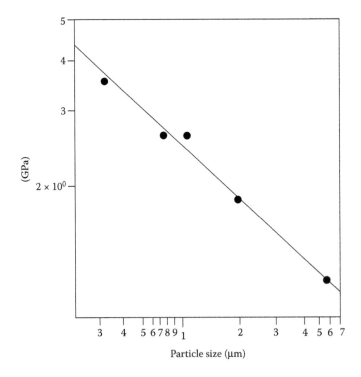

**FIGURE 6.4** Strength of alumina particles in different grain sizes. (Reprinted from Nogi, K. et al., *Nanoparticle Technology Handbook*, Copyright 2012, with permission from Elsevier.)

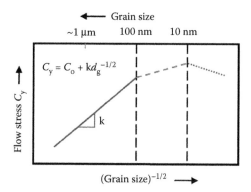

**FIGURE 6.5** Hall–Petch breakdown schematic. (Adapted from Kumar, K., H. Van Swygenhoven, and S. Suresh, *Acta Materialia*, 51(19), 5743–5774, 2003.)

decrease or remain constant with further decreasing in grain size [18,19]. The unusual behavior has been expressed as an inverse Hall–Petch relation justified by deficiency of mobile dislocations able to cross grains down to nanoscale grain sizes. It is believed that plasticity, below a few nanometers, is derived by the small-amplitude sliding motions at grain boundaries and no longer by the dislocation motions [20]. Figure 6.5 illustrates the Hall–Petch relation trend in grain size ranges.

Despite the growing quantity of experimental investigations observing similar unusual deformation responses in nanocrystalline materials, at present a verdict is still difficult to reach since reliable measurement of mechanical properties for particles in the order of a few nanometers is stringently restricted [21]. Furthermore, influence of subsidiary factors such as dislocations, material porosity, different processing roots, and distinct internal stresses should not be ignored [11]. Hence, according to the knowledge available, nanostructured material strength exhibits its maximum value at a crucial grain size averagely occurring in the range of 10–20 nm.

### 6.2.2 Hardness

Hardness of a material is defined as the ratio of load applied to a hard indenter ($P$) to the deformation left after removing the load:

$$H = \frac{P}{A} \tag{6.16}$$

where $H$ and $A$, respectively, refer to material hardness and residual indent area. The contact area ($A$) is obtained using area functions depending on the indenter geometry.

Famed with the Moh's scale, hardness, in principle, indicates the material resistance to plastic deformation under a hard indenter and therefore is inherently related to the material strength. A simple empirical relationship between hardness and material strength is as follows [9,21,22]:

$$H = 3\sigma_y \tag{6.17}$$

In fact, experimental data confirming the validity of the Hall–Petch relationship for nanostructured materials have been obtained measuring hardness according to Equation 6.17 [11].

### 6.2.2.1 Microindentation Analysis

Regarding the sample size restrictions in measuring mechanical properties of individual nanoparticles using conventional tensile and/or compression tests, quasi-static microindentation (loads in the range

0.1–10 N) has been suggested by Oliver and Pharr [23] as a quick inexpensive method alternatively used to evaluate strength of nanomaterials since it requires no special sample preparation [20,21].

### 6.2.2.2 Nanoindentation Analysis

Microindentation analysis cannot be used to measure the mechanical properties of very thin films or surface treated materials since it would measure the substrate property piercing the top layer. Nanoindentation analysis is equipped with a genuine mechanical microprobe applying loads in the order of mN and utilized to mechanical characterization of very thin films (see Figure 6.6) [20]. Available commercial nanoindenters respectively offer load and displacement resolution about 0.1 Å and 1 nano-Newton [22]. Spontaneous load-displacement recording in nanoindentation techniques provides the opportunity to detect the material elastic part relaxation, which is ignored in conventional indentation methods measuring the residual indented depth after withdrawing the indenter [20]. A mathematical solution to the elastic stress beneath an indenter was calculated by Hertz as below:

$$P = \frac{16}{9} \frac{1}{\left(1 - v_i^2/E_i + 1 - v_s^2/E_s\right)^2} \frac{R_i R_s}{R_i + R_s} \delta^{3/2} \tag{6.18}$$

In which $P$ is the force applied on the indenter probe. While subscripts $i$ and $s$ refer to the indenter and sample, $\delta$, $R$, and $E$ respectively describe indented depth, curvature radius, and modulus. According to Equation 6.18, a sample modulus is obtained calculating the line slope drawing force applied on the indenter tip ($P$) versus $\delta^{3/2}$ [22].

In Figure 6.7, load is applied on the 20 nm grain sized nanocrystalline nickel sample with a standard loading rate until the specific maximum load is reached. The load then is reduced to zero with the same rate. Besides the hardness value calculated according to the maximal load of the loading cycle and contact area between the indenter and material, Young's modulus is obtainable analyzing the unloading cycle. This is not unexpected since unlike the twofold plastic-elastic nature of the loading cycle, the unloading curve reflects the pure elastic response [20].

Figure 6.8 illustrates the defined parameters in the schematic of an elastoplastic material surface under indentation process. According to this schematic:

$$h = h_s + h_c \tag{6.19}$$

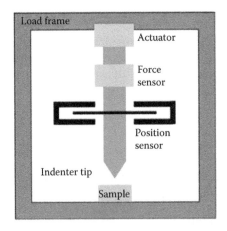

**FIGURE 6.6** Schematic of nanoindenter. (With kind permission from Springer Science+Business Media: *Nanomaterials: Mechanics and Mechanisms*, 2009, Ramesh, K.T.)

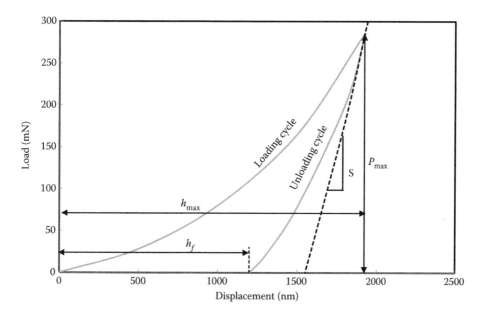

**FIGURE 6.7** Loading/unloading nanoindentation curve for nanocrystalline Ni (20 nm grain size). Data provided by Brian Schuster (With kind permission from Springer Science+Business Media: *Nanomaterials: Mechanics and Mechanisms*, 2009, Ramesh, K.T.)

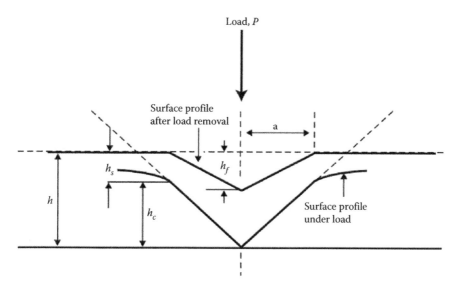

**FIGURE 6.8** Indentation surface profile schematic. (With kind permission from Springer Science+Business Media: *Nanomaterials: Mechanics and Mechanisms*, 2009, Ramesh, K.T., Oliver, W.C., and Pharr, G.M., *Journal of Materials Research*, 19(01), 3–20, 2004.)

where $h_c$, $h_s$, and $h_f$, respectively, refer to the penetration depth during indentation, circumferential displacement due to elastic depression, and the residual surface displacement after unloading.

Tabor [24] considered nonrigidity of indenter defining $E_r$ as reduced Young's modulus:

$$\frac{1}{E_r} = \frac{1-\vartheta^2}{E} + \frac{1-\vartheta_i^2}{E_i} \qquad (6.20)$$

In which $\vartheta$ is the Poisson's ratio and subscript $i$ refers to the indenter. This reduced Young's modulus could be related to rigidity ($S$) as below [25–27]:

$$S = \frac{dP}{dh} = \beta \frac{2}{\sqrt{\pi}} E_r \sqrt{A(h_c)} \qquad (6.21)$$

In which $A = 24.5\,h_c^2$ for conventional axially symmetric nanoindentation probe geometries and $\beta$ is between 1 and 1.034 depending on the indenter geometry. It is worth noting that rigidity is determined calculating the initial slope of unloading curve at maximal load ($S = dP/dh$) [20].

### 6.2.3 FRACTURE TOUGHNESS

High stresses existing within the nanoparticle prior to fracture are responsible for improved fracture resistance of these materials with decreasing grain size. Fracture on a nanoparticle is most likely to initiate concentrating the stress through a crack on the oxide film topping the nanoparticle surface. Critical fracture toughness of nanoparticles evaluated by critical stress intensity factor, $K_{Ic}$, shall be obtained using

$$K_{Ic} = 1.1\sigma\sqrt{\pi t_{ox}} \qquad (6.22)$$

where $t_{ox}$ refers to the flaw length initiated by the fracture [7].

Estimating elastic strain energy release rate, $G_{Ic}$, holding the elastic modulus ($E$) could also be conducted averagely considering the work per unit fracture area as

$$G_{IC} = \frac{W}{A_f} = \frac{\bar{\sigma}^2}{2E} \times \frac{V}{A_f} = \frac{2\sigma^2 r}{3E} \qquad (6.23)$$

In which $\bar{\sigma}^2/2E$ represents the elastic strain energy density and $V$ is the nanoparticle volume. The fracture toughness is then calculated using

$$K_{Ic} = \sqrt{\frac{EG_{IC}}{1-\nu}} \qquad (6.24)$$

where $\nu$ is the Poisson's ratio.

### 6.2.4 FATIGUE BEHAVIOR

Fatigue life is defined as the material endurance under controlled cyclic loading dominated by crack initiation and propagation generally nucleated at the free surface. Resistance to fatigue is more pronounced in coarser grained particles [21]. Tensile measurements data might be used to estimate fatigue behavior of materials. Accordingly, total strain, $\Delta\varepsilon$, could be written as the sum of elastic and plastic strain ranges

$$\frac{\Delta\varepsilon}{2} = \left(\frac{\sigma_f}{E}\right)(2N_f)^b + \varepsilon_f(2N_f)^c \qquad (6.25)$$

In which $\sigma_f$, $E$, $N_f$, and $\varepsilon_f$, respectively, refer to fracture strength, Young's modulus, cycles to failure, and fracture strain. Assuming the dominancy of elastic behavior, we have

$$\frac{\Delta\varepsilon}{2} = \left(\frac{\sigma_f}{E}\right)(2N_f)^b \tag{6.26}$$

$$\Delta\sigma = \sigma_f(2N_f)^b \tag{6.27}$$

$\Delta\sigma$ is the stress amplitude, $\sigma_f$ fatigue strength coefficient, and $b$ fatigue strength exponent. Experimental results have shown that most fatigue life values predicted on the basis of tensile data are reliable compared to the fatigue data [9].

### 6.2.5 CREEP

Creep or cold flow is the tendency of solid materials to deform permanently with time upon application of mechanical stresses. Such deformations occur easier in materials with smaller grain sizes due to the large fraction of grain boundary atoms and their unusual diffusion rates [21]. In such conditions, diffusional creep mechanism at intermediate temperatures of $0.4T_m < T < 0.5T_m$ is observed to be dominant. At temperatures higher than $0.5T_m$ applying low strain rates ($10^{-4} - 10^{-3}$ s$^{-1}$) to the nanocrystalline particles leads to the superplastic behavior defined as the ability of the material to deform several thousand percent [8]. Conventional semiempirical constitutive models suggested for predicting creep behavior of materials are of the type

$$\dot{\varepsilon} = \frac{d\varepsilon}{dt} = \frac{AGb}{k_BT}\left(\frac{b}{d}\right)^p\left(\frac{\sigma-\sigma_0}{G}\right)^n \quad D = \alpha\frac{\sigma^n}{d^p} \tag{6.28}$$

In which $\dot{\varepsilon}$, $\sigma$, $\sigma_0$, $G$, $n$, $p$, $D$, and $d$ respectively represent plastic strain rate, threshold stress, creep stress, shear modulus, stress exponent, grain size exponent, diffusion coefficient, and grain size [28]. Assuming $Q$ and $k$ as activation energy and Boltzmann constant, diffusion coefficient is calculated using Boltzmann's theory:

$$D = D_0\exp\left(-\frac{Q}{k_BT}\right) \tag{6.29}$$

In optimal superplasticity, $p = 2$ and $n = 2$. At high stresses, vacancy diffusion controls the dislocation creep ($p = 0$ and $n > 3$) but at lower stress levels Coble creep ($p = 3$ and $n = 1$) or Nabarro–Herring takes place ($p = 2$ and $n = 1$) [8].

The dominant mechanism giving rise to plastic flow localization in nanometrials bulk is grain boundary sliding accommodated by grain boundary diffusion or slippage. High concentrations of segregants in grain boundaries conduct the formation of a liquid phase at temperatures higher than $0.5T_m$ at the interfaces facilitating the grain boundary sliding [8]. Hahn [29] suggested a model, schematically illustrated as Figure 6.9, indicating that grain boundary migration leads to the formation of a zone in which all grain boundary faces lie parallel to each other. This would facilitate the sliding motions of the grains yielding superplastic behavior. At very small grain sizes, Nabarro–Herring or Coble diffusion mechanisms respectively representing diffusion across grain boundaries and diffusion along grain boundaries dominate the creep behavior violating the Hall–Petch relationship [20].

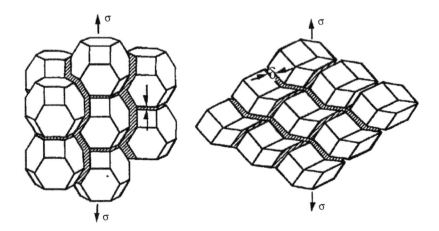

**FIGURE 6.9** Schematic of the superplastic deformation mechanism (a) intact grains and (b) grain boundary migration leading to plastic shear through grain boundary. (Adapted from Hahn, H., Mondal, P., and Padmanabhan, K., *Nanostructured Materials*, 9(1), 603–606, 1997.)

Coble assumes that the deviation from the Hall–Petch equation is attributed to the intrinsic behavior of nanostructured crystals. Coble creep is dominated by grain boundary diffusion mechanism and could occur rapidly at ambient temperature. The strain rate rapidly increases with decreasing grain size as shown below [11]:

$$\frac{d\varepsilon}{dt} = \frac{148\delta D_b V_a \sigma}{\pi k T d^3} \tag{6.30}$$

where $\sigma$, $D_b$, $V_a$, and $\delta$, respectively, refer to applied stress, grain boundary diffusion coefficient, atomic volume, and grain boundary width.

Nabarro–Herring assumes that the deformation is originated by the homogeneous diffusion of atoms through the equiaxial geometry of grain volume from the compressioned regions to the localities under tension. Evidentially, cavities diffuse in the opposite direction [20]. Cavity concentration gradient under stress could be written as

$$C = C_0 \exp\left(\frac{-\Delta G}{kT}\right) \tag{6.31}$$

where $\Delta G$ represents the free enthalpy of cavity formation calculated using

$$\Delta G = \pm \sigma \Omega \tag{6.32}$$

in which $\Omega$ refers to the cavity volume. The cavity flow induced is

$$J = -D\nabla C \tag{6.33}$$

The transportation rate of cavities through the area $d^2$ could be written as

$$\phi = -J \, d^2 \tag{6.34}$$

Combination of the equations developed above results in

$$\phi = -2DdC \sinh\left(\frac{\sigma\Omega}{kT}\right) \qquad (6.35)$$

Displacement amplitude ($u$) of a cavity by the diffusion mechanism is equal to $u = -\Omega/d^2$. The strain is then calculated as $\varepsilon = u/d = -\Omega/d^3$ assuming $d$ as the distance cavity shifts through the grain volume. The strain rate would finally be obtained as below:

$$\frac{d\varepsilon}{dt} = 2DC\Omega \frac{\sinh(\sigma\Omega/kT)}{d^2} \qquad (6.36)$$

Substituting $C = v/\Omega$ in the equation below expressed by the atomic fraction of cavities ($v$) and approximating $\sinh(\sigma\Omega/kT) \approx \sigma\Omega/kT$ as $\sigma\Omega \ll kT$, Nabarro–Herring reduces to

$$\frac{d\varepsilon}{dt} = 2Dv \frac{\sigma\Omega}{kTd^2} \qquad (6.37)$$

It would be interesting if the diffusions are restricted to grain boundaries as that of Coble, with $\delta$ as the boundary thickness, a $\delta/d$ fraction of the atoms involved in the material could diffuse through the grain boundaries, and the equation obtained above would reduce to the relationship suggested by Coble. The grain size dependence of strain rate is illustrated in Figure 6.10.

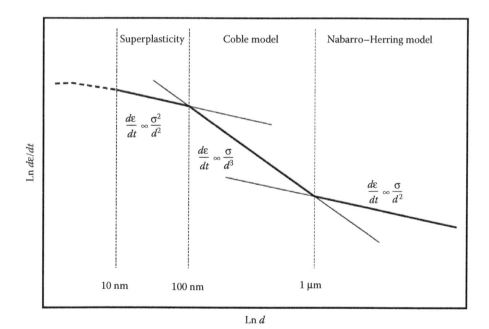

**FIGURE 6.10** Schematic creep behavior of nanostructured materials with different grain sizes. (With kind permission from Springer Science+Business Media: *Nanomaterials and Nanochemistry*, 2008, Bréchignac, C. and Houdy, P.)

## 6.3 NUMERICAL MODELING

Regarding the fact that experimental measurement of mechanical properties for nanoparticles is often difficult, if not impossible, numerical methods are conventionally utilized to estimate the properties. Nowadays, availability of powerful computers has provided the opportunity for the numerical simulation of finite samples; accordingly several computational methods have been developed to model and estimate properties of nanomaterials [21,31]. On the other hand, interpretation of experimental data obtained conducting experimental measurement techniques such as solid-state nuclear magnetic resonance or neutron diffraction studies is required to be directly assisted by the application of computer simulations [32].

However, missestimations would be encountered if the clusters were too large to be modeled by quantum-mechanical methods such as Hartree–Fock model and too small for applying density functional theory, which have been traditionally utilized to physically interpret and understand many properties of small molecules and macroscopic solid-state systems [21,31]. This approach to research can be worth it if good agreements are observed between the computer simulation results with the experimentally obtained data. Computational modeling tools such as molecular dynamics (MD), ab initio, Mont Carlo, finite element modeling methods, or continuum mechanics models have been successfully applied to estimate the properties of carbon nanotubes based on given input variables such as temperature, geometry, and defects [33]. Anyhow, successful application of classical simulation methods to the nanoparticles requires the use of reliable interatomic potentials, i.e., force field as the basis of calculations representing the interactions between atoms (see Figure 6.11). The atomic interaction parameters involved in the force fields are conventionally calculated on the basis of chemical quantum chemical points of

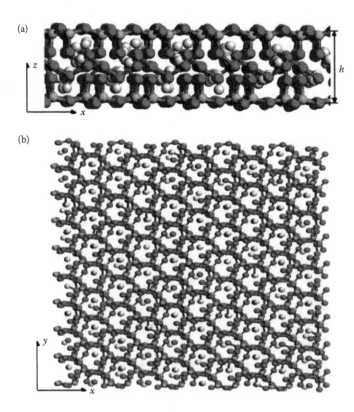

**FIGURE 6.11** (**See color insert.**) Structure of a pyrophyllite single layer computational cell. (Adapted from Mazo, M.A. et al., *The Journal of Physical Chemistry B*, 112(10), 2964–2969, 2008.)

view or a few reference compounds assuming the nanoparticles (clay, etc.) as a constrained lattice. It should be noted that a rigid lattice with immobile atoms ignores the degrees of freedom for atomic motions and hence provides a faster computer proceeding. Some harmonic valence (bonded) interactions have been successfully applied as the basis of simulation of kaolinite, smectite ... clay structures. But in this approach, each bond should be individually evaluated and this is not possible except for small-scale simulations of a few reference structures as the large number of parameters to be evaluated for a harmonic valence force field are not easily obtained for complex systems [34].

Properties of clay nanoparticles have been numerically modeled using molecular dynamics and Monte Carlo methods, respectively, on the basis of interatomic interactions and coarse-grained models. Results have shown that applying density functional theory (DFT) for small units of nanomaterials in order to estimate the nanoscale elastic properties leads to better agreements with laboratory measurements in comparison with classic force field methods [36]. Oleg [37] presented calculated 250–260 N/m for $E_1h$ and $E_2h$, and 166 N/m for the in-plane shear response, $G_3h$, where $h$ is the thickness as elastic properties of a single lamella of montmorillonite clay as an infinite "nanoplate" on the basis of force fields theory assuming both bonded and nonbonded interatomic contributions with periodic boundary conditions in two dimensions. Simulating flexibility of smectite clay minerals using molecular dynamics method, the single layer was found to be more flexible along one direction than along the perpendicular directions which was ascribed to the changes occurred in Si–O–Si angles in the silicate tetrahedral sheets as the main origin of flexibility rather than the changes in bond lengths [38].

## 6.4 CONCLUSION

A large fraction of surface atoms, high surface energy, spatial confinement, and reduced imperfections make nanoscale materials attractive compared to the microstructured counterparts. In fact, epitaxial growth of inorganic crystal layers in nanocrystal structures facilitates stress and strain relaxations presenting different mechanical behavior. Since measuring mechanical properties of nanoparticles, including strength, hardness, fracture toughness fatigue and creep, encounters practical difficulties arisen from the nanoscale dimensions, there is little information on the mechanical properties of individual nanoparticles.

Tensile strength of particulate nanostructures would be predicted according to Hook's law coupled with MaacKenzie relation of porosity oriented on Taylor series. Atomic force microscopy has also been used to investigate strength in one-dimensional nanostructures lately. Nanoindentation analysis recording load-displacement is a conventional method characterizing nanomaterials in view of the hardness properties following Hertz relation. Critical fracture toughness of nanoparticles is evaluated by critical stress intensity factor calculated measuring elastic strain energy density and $V$ is the nanoparticle volume. Tensile measurements data, i.e., fracture strength, Young's modulus, and cycles to failure could be reliably used to estimate fatigue behavior of materials. Creep deformation has been proved to be dominated by diffusional mechanisms at intermediate temperatures. Superplastic behavior is observed at higher temperatures modeled by semi-empirical constitutive models. Numerical methods are conventionally available to estimate the properties where experimental measurement of mechanical properties are difficult or impossible.

## REFERENCES

1. Rotello, V.M., *Nanoparticles: Building Blocks for Nanotechnology*. 2004, New York: Kluwer Academic/Plenum Publishers.
2. Wandelt, K., *Surface and Interface Science: Concepts and Methods*. Vol. 1. 2012, Weinheim: John Wiley & Sons.
3. Kolasinski, K.K., *Surface Science: Foundations of Catalysis and Nanoscience*. 2012, Chichester: John Wiley & Sons.

4.  Al-Kaysi, R.O., A.M. Müller, and C.J. Bardeen, Photochemically driven shape changes of crystalline organic nanorods. *Journal of the American Chemical Society*, 2006, 128(50):15938–15939.

5.  Snyder, C. et al., Effect of strain on surface morphology in highly strained InGaAs films. *Physical Review Letters*, 1991, 66(23):3032–3035.

6.  Reddy, C.M., G.R. Krishna, and S. Ghosh, Mechanical properties of molecular crystals—Applications to crystal engineering. *CrystEngComm.*, 2010, 12(8):2296–2314.

7.  Nowak, J.D. and University of Minnesota, *Characterization of Surfaces and Interfaces in Nanoparticles Using Transmission Electron Microscopy*. 2007, University of Minnesota.

8.  Nogi, K. et al., *Nanoparticle Technology Handbook*. 2012, Amsterdam: Elsevier.

9.  Yang, F. and J.C.M. Li, *Micro and Nano Mechanical Testing of Materials and Devices*. 2009, New York: Springer.

10. Vollath, D., *Nanomaterials: An Introduction to Synthesis, Properties and Applications*. 2013, Weinheim: Wiley.

11. Goddard, W.A. et al., *Handbook of Nanoscience, Engineering, and Technology*. 2002, Boca Raton: Taylor & Francis.

12. Stone, A. and D. Wales, Theoretical studies of icosahedral $C_{60}$ and some related species. *Chemical Physics Letters*, 1986, 128(5):501–503.

13. Schaefer, H.E., *Nanoscience: The Science of the Small in Physics, Engineering, Chemistry, Biology and Medicine*. 2010, Heidelberg: Springer.

14. Ding, W. et al., Mechanics of crystalline boron nanowires. *Composites Science and Technology*, 2006, 66(9):1112–1124.

15. Kaplan-Ashiri, I. et al., On the mechanical behavior of WS2 nanotubes under axial tension and compression. *Proceedings of the National Academy of Sciences of the United States of America*, 2006, 103(3):523–528.

16. Tan, E. and C. Lim, Mechanical characterization of nanofibers—A review. *Composites Science and Technology*, 2006, 66(9):1102–1111.

17. Kumar, K., H. Van Swygenhoven, and S. Suresh, Mechanical behavior of nanocrystalline metals and alloys. *Acta Materialia*, 2003, 51(19):5743–5774.

18. Funk, M.F., *Microstructural Stability of Nanostructured Fcc Metals During Cyclic Deformation and Fatigue*. 2014, Karlsruhe: KIT Scientific Publication.

19. Conrad, H. and J. Narayan, On the grain size softening in nanocrystalline materials. *Scripta Materialia*, 2000, 42(11):1025–1030.

20. Bréchignac, C. and P. Houdy, *Nanomaterials and Nanochemistry*. 2008, Heidelberg: Springer.

21. Gogotsi, Y., *Nanomaterials Handbook*. 2006, Boca Raton: Taylor & Francis.

22. Ramesh, K.T., *Nanomaterials: Mechanics and Mechanisms*. 2009, New York: Springer.

23. Oliver, W.C. and G.M. Pharr, Measurement of hardness and elastic modulus by instrumented indentation: Advances in understanding and refinements to methodology. *Journal of Materials Research*, 2004, 19(01):3–20.

24. Tabor, D., The hardness of solids. *Review of Physics in Technology*, 1970, 1(3):145–179.

25. Bulychev, S. et al., Determining Young's modulus from the indentor penetration diagram. *Ind. Lab.*, 1975, 41(9):1409–1412.

26. Shorshorov, M.K., S. Bulychev, and V. Alekhin. Work of plastic and elastic deformation during indenter indentation. *Soviet Physics Doklady*, 1981, 26: 769–771.

27. Sneddon, I.N., The relation between load and penetration in the axisymmetric Boussinesq problem for a punch of arbitrary profile. *International Journal of Engineering Science*, 1965, 3(1):47–57.

28. Mukherjee, A., J. Bird, and J. Dorn, Cones and Vietoris-Begle type theorems. *Transactions of the American Society of Metals*, 1969, 62:155–174.

29. Hahn, H. and K. Padmanabhan, A model for the deformation of nanocrystalline materials. *Philosophical Magazine B*, 1997, 76(4):559–571.

30. Hahn, H., P. Mondal, and K. Padmanabhan, Plastic deformation of nanocrystalline materials. *Nanostructured Materials*, 1997, 9(1):603–606.

31. Balbuena, P. and J.M. Seminario, *Nanomaterials: Design and Simulation*. 2006, Amsterdam: Elsevier Science.

32. Greenwell, H.C. et al., On the application of computer simulation techniques to anionic and cationic clays: A materials chemistry perspective. *Journal of Materials Chemistry*, 2006, 16(8):708–723.

33. Vijayaraghavan, V. et al., Estimation of mechanical properties of nanomaterials using artificial intelligence methods. *Applied Physics A*, 2014, 116(3): 1099–1107.

34. Cygan, R.T., J.-J. Liang, and A.G. Kalinichev, Molecular models of hydroxide, oxyhydroxide, and clay phases and the development of a general force field. *The Journal of Physical Chemistry B*, 2004, 108(4):1255–1266.

35. Mazo, M.A. et al., Molecular dynamics simulation of thermomechanical properties of montmorillonite crystal. 1. Isolated clay nanoplate. *The Journal of Physical Chemistry B*, 2008, 112(10):2964–2969.

36. Galimberti, M., *Rubber-Clay Nanocomposites: Science, Technology, and Applications*. 2011, New Jersey: Wiley.

37. Manevitch, O.L. and G.C. Rutledge, Elastic properties of a single lamella of montmorillonite by molecular dynamics simulation. *The Journal of Physical Chemistry B*, 2004, 108(4):1428–1435.

38. Sato, H., A. Yamagishi, and K. Kawamura, Molecular simulation for flexibility of a single clay layer. *The Journal of Physical Chemistry B*, 2001, 105(33):7990–7997.

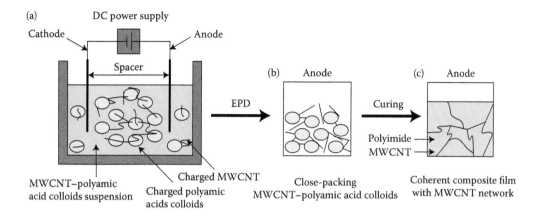

**FIGURE 2.2** Schematic description of the fabrication of MWCNT/polyimide composite films through EPD. (Reprinted from *Polymer*, 51, Wu, D. C. et al., Multi-walled carbon nanotube/polyimide composite film fabricated through electrophoretic deposition, 2155–60, Copyright 2010, with permission from Elsevier.)

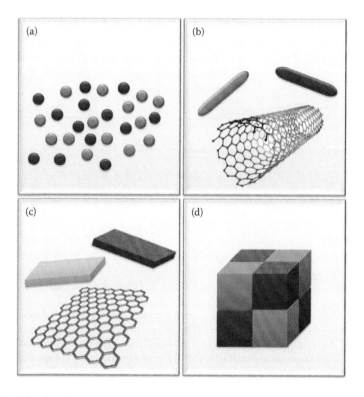

**FIGURE 3.1** Classification of nanomaterials (a) 0D spheres and clusters, (b) 1D nanofibers, wires, and rods, (c) 2D films, plates, and networks, and (d) 3D nanomaterials.

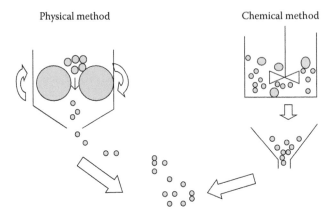

**FIGURE 3.2** Physical and chemical approach to produce nanomaterials. (Adapted from I. Rahman, V. Padavettan, *Journal of Nanomaterials*, 1, 2012, 1–15.)

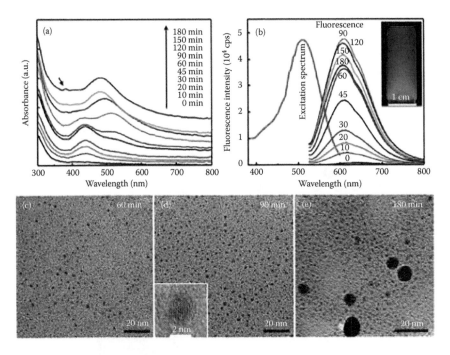

**FIGURE 3.3** (a) UV–vis spectra and (b) fluorescence emission spectra of the solution containing PMAA and AgNO₃ in different length of sonication time; TEM images of as prepared Ag nanoclusters from different lengths of sonication: (c) 60 min, (d) 90 min, and (e) 180 min. (Reprinted with permission from Xu Hangxun, S. Kenneth Suslick, Sonochemical synthesis of highly fluorescent nanoclusters, *American Chemical Society*, 4, 3209–3214. Copyright 2010 American Chemical Society.)

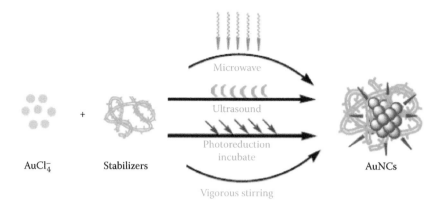

**FIGURE 3.4** Synthesis of gold nanoclusters (AuNCs). (Adapted from M. Cui, Y. Zhao, Q. Song, *Trends in Analytical Chemistry*, 57, 2014, 73–82.)

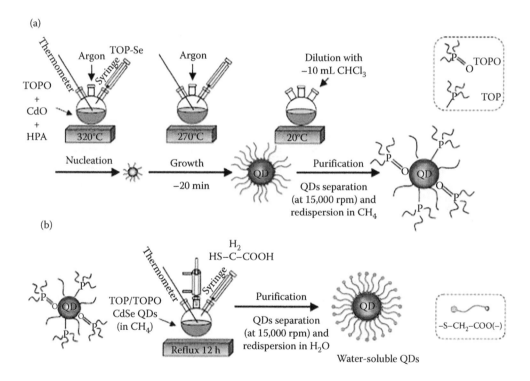

**FIGURE 3.5** Schematic illustration of a typical synthesis process and surface-modification of a luminescent QD based on the use of CdO as precursor. (a) Nanoparticle synthesis and (b) surface modification. (Adapted from M. Costa-Fernandez, R. Pereiro, A. Sanz-Medel, *Trends in Analytical Chemistry*, 25, 2006, 207–218.)

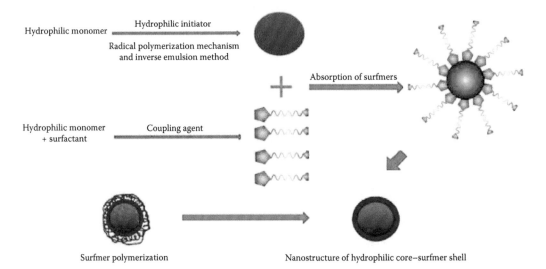

**FIGURE 3.10** The schematic diagram for inverse emulsion polymerization of hydrophilic polymer–surfmer core–shell nanostructure. (Adapted from Y. Tamsilian, A. Ramazani S.A., *The International Conference on Nanotechnology: Fundamentals and Applications (ICNFA 2013)*, Canada, 2013.)

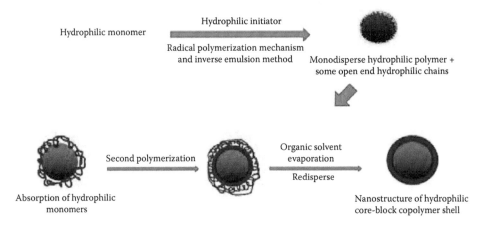

**FIGURE 3.11** The schematic diagram for inverse emulsion polymerization of hydrophilic polymer-block copolymer core–shell nanostructure. (Adapted from Y. Tamsilian, A. Ramazani S.A., *The International Conference on Nanotechnology: Fundamentals and Applications (ICNFA 2013)*, Canada, 2013.)

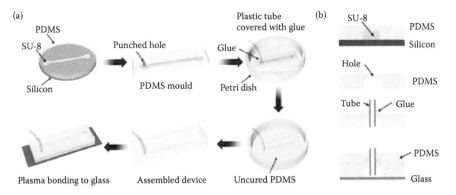

**FIGURE 3.14** (a) Schematics of fabricating a tubing interconnection for PDMS microfluidic devices. (b) A cross-sectional view of the tubing interconnection fabrication. (Adapted from J. Wang et al., *Lab on Chip*, 14, 2014, 1673–1677.)

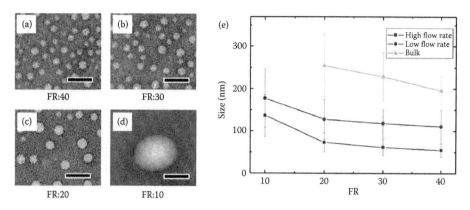

**FIGURE 3.15** TEM images of PLGA nanoparticles with (a) FR: 40, (b) FR: 30, (c) FR: 20, (d) FR: 10, and (e) size distribution of PLGA nanoparticles versus FR. (Adapted from J. Wang et al., *Lab on Chip*, 14, 2014, 1673–1677.)

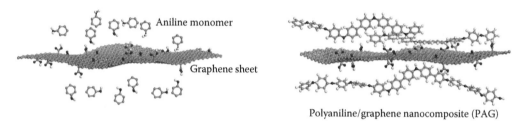

**FIGURE 3.25** Schematic process of preparing PANI/graphene nanocomposites. (Adpted from I. Harada, Y. Furukawa, F. Ueda, *Synthetic Metals*, 29, 1989, 303–312.)

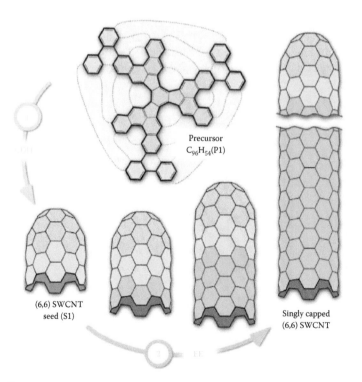

**FIGURE 3.26** Two step bottom-up synthesis of SWCNTs: (1) singly capped ultrashort (6,6) SWCNT seed formation via cyclodehydrogenation (CDH) as polycyclic hydrocarbon precursor $C_{96}H_{54}$ (P1) and (2) nanotube growth via epitaxial elongation (EE). Orange and blue: short CNT segment of the seed S1; red dashed lines: new C–C bonds; green: epitaxial elongation. (Adapted from J.R. Sanchez-Valencia et al., *Nature*, 512, 2014, 61–64.)

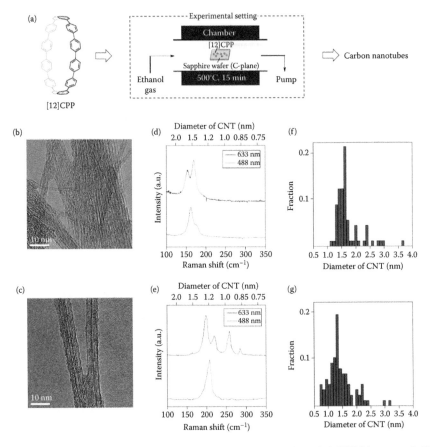

**FIGURE 3.28** (a) Schematic presentation of growth experiments, (b and c) TEM images of CNTs synthesized from CPP, (d and e) Raman spectra of CNTs, and (f and g) diameter distribution histograms. Data for CNTs grown from CPP are given in (d and f) and data for CNTs grown from CPP are given in (e and g). (Adapted from H. Omachi et al., *Nature Chemistry*, 5, 2013, 572–576.)

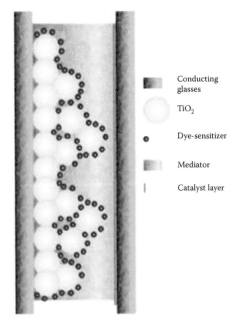

**FIGURE 4.11** Schematic arrangement of a dye-sensitized solar cell. (With kind permission from Springer Science+Business Media: *Nanoenergy: Nanotechnology Applied for Energy Production*, 2012, de Souza, F. L. and Leite, E. R.)

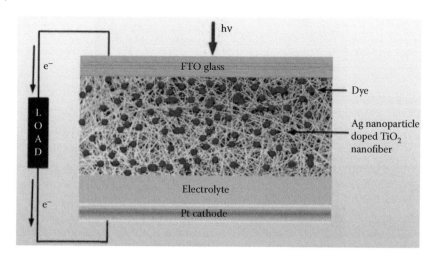

**FIGURE 4.14** Scheme of a solar cell combining quantum dots and dye-sensitizers. (Adapted from J. Li et al., *Chem. Phys. Lett.,* 514(1–3), 141–145, Sep. 2011.)

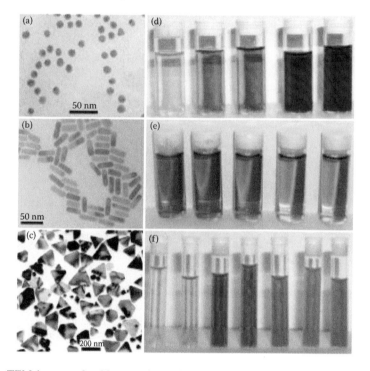

**FIGURE 4.18** TEM images of gold nanospheres (a), gold nanorods (b), and silver nanoprisms (c); and visual appearance of colloidal dispersions of AuAg alloy nanoparticles with increasing Au concentration (d), Au nanorods of increasing aspect ratio (e), and Ag nanoprisms with increasing lateral size (f). (Adapted from M. Liz-Marzán, *Mater. Today,* 7(2), 26–31, Feb. 2004.)

**FIGURE 4.19** Variation of color of silver nanoparticles with different dilutions. (Adapted from A. Chhatre et al., *Colloids Surf. Physicochem. Eng. Asp.,* 404, 83–92, June 2012.)

**FIGURE 4.20** Nanoclay-based pigments. (Adapted from M. Pomares, *Diario Informacion*, 2, Apr. 2009.)

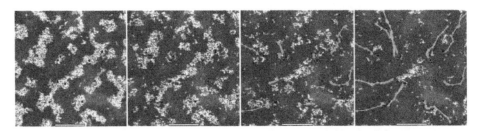

**FIGURE 5.5** Time series of STEM images taken with 350 nm × 350 nm from the center of the field of view. The images are provided ~7 s apart, by the passage of time to the right. The trajectories of 10 particles are presented by the green tracks, with starting points assigned by red dots and the shade of green increasing among the frames. (Reprinted with permission from E.R. White et al. Charged nanoparticle dynamics in water induced by scanning transmission electron microscopy. *Langmuir* 28, 3695–3698. Copyright 2012 American Chemical Society.)

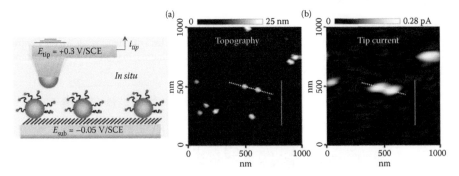

**FIGURE 5.9** AFM-SECM tapping mode imaging of a gold surface bearing a random array of 20 nm Fc-PEGylated AuNP. Simultaneously acquired topography (a) and tip current images (b). (Reprinted with permission from A. Srivatsan et al. Gold nanocage-photosensitizer conjugates for dual-modal image-guided enhanced photodynamic therapy. *Theranostics* 4(2), 163. Copyright 2014 American Chemical Society.)

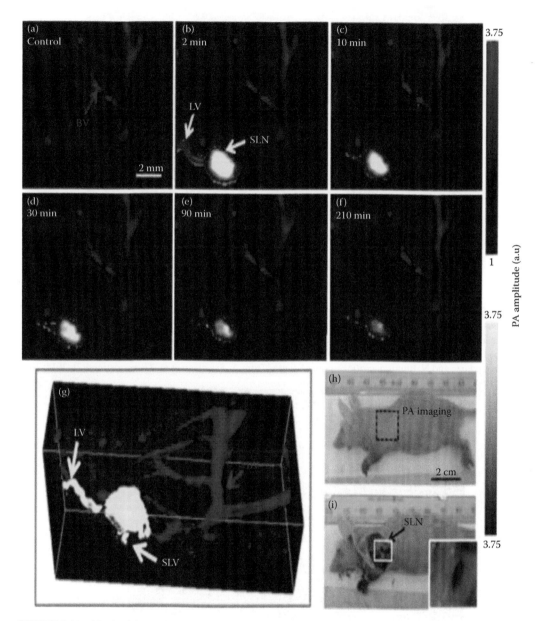

**FIGURE 5.11** Noninvasive real-time *in vivo* PA imaging of SLN in nude mouse: For all PA images, the laser was tuned to 650 nm wavelength. (a) Control PA image acquired before OCN injection. Red parts represent optical absorption from blood vessels (BV); (b) PA image acquired immediately (2 min) after the OCN injection, blood vessel (BV), lymph vessel (LV), and sentinel lymph node are marked with arrows, and the SLN is visible in (b–e), however, the contrast is much weaker after 210 min postinjection in (f). (g) 3D depiction of the SLN and BVs immediately after OCN particles injection, (h) photograph of the nude mouse before taking the PA images. The scanning region is marked with a black dotted square. (i) Photograph of the mouse with the skin removed after PA imaging, accumulation of dark-colored OCN particles are visible in the lymph node. (Adapted from L. Wu, X. Cai, K. Nelson et al. *Nano Research* 6(5), 2013: 312–325.)

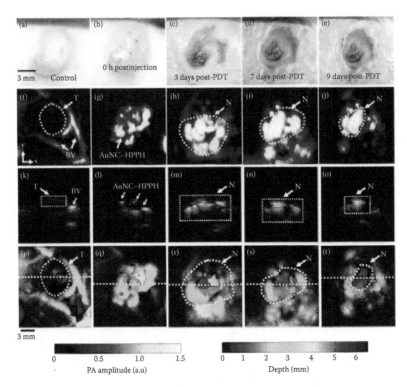

**FIGURE 5.12** (a–e) Photograph of the tumor taken before and after PDT with an injection of AuNC–HPPH at various time points up to 9 days. PA images acquired (f) before intratumoral injection of AuNC–HPPH, (g) after injection, and (h–j) 3, 7, and 9 days posttreatment. (k–o) Depth-resolved PA B-scan images cut along the dotted lines in (f–j), respectively. (p–t) Depth-encoded PA images of (f–j), respectively. BV, blood vessels; T, tumor boundary; and N, tumor necrotic region. (Adapted from A. Srivatsan et al. *Theranostics* 4(2), 2014: 163.)

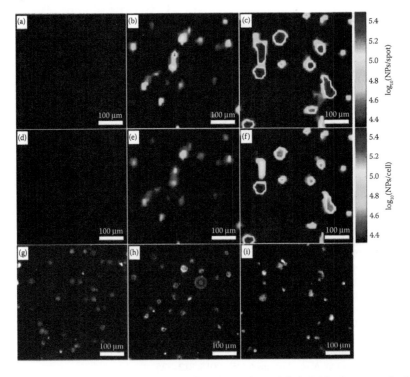

**FIGURE 5.13** Quantitative PA imaging results (a–f) and coregistered dark-field microscopy (g–i) of J774A.1 cells incubated with and without NPs. Quantitative PA images scaled to the number of NPs per focal spot (a–c) and number of NPs per cell (d–f) are shown. Images of cells incubated without NPs (a, d, and g), with $1.9 \times 10^{12}$ NPs per milliliter of cell culture media (b, e, and h), and with $3.7 \times 10^{12}$ NPs per milliliter of cell culture media (c, f, and i) are shown. (Adapted from J.R. Cook, W. Frey, S. Emelianov. *ACS Nano* 7(2), 2013: 1272–1280.)

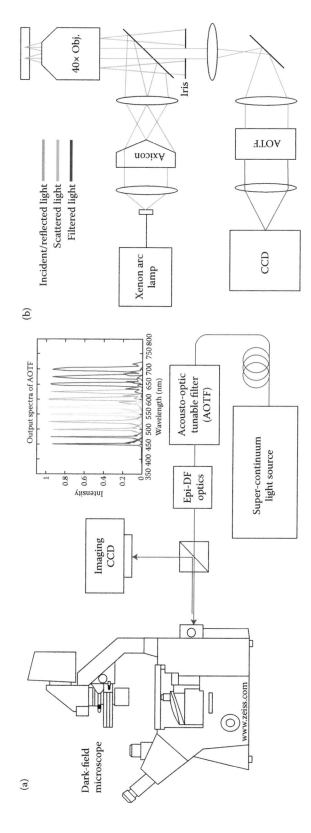

**FIGURE 5.14** (a) Block diagram of the hyperspectral dark-field system. Inset: Example output spectra of white light as filtered by the AOTF. (Adapted from K. Seekell et al. *Methods* 56.2, 2012: 310–316; K. Seekell et al. *Biomedical Optics Express* 4(11), 2013: 2284–2295.) (b) Schematic of the epi-illumination and detection trains of the dark-field microspectroscopy system showing the optical paths of the incident/reflected illuminating light (gray), the backscattered light (green), and the filtered light (red). Note that the AOTF can be placed in either the illumination (a) or detection (b) path.

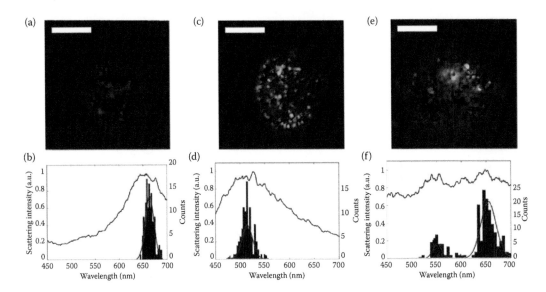

**FIGURE 5.15** (a) Dark-field image of MDA-MB-468 cell tagged with anti-EGFR functionalized GNRs. (b) Example scattering spectrum and distribution of peak scattering wavelengths (Gaussian fit with peak of $664.0 \pm 9.3$ nm) of MDA-MB-468 cells tagged with anti-EGFR GNRs. (c) A549 cell tagged with anti-IGF-1R silver nanospheres. (d) Example scattering spectrum and distribution of peak scattering wavelengths (Gaussian fit with peak $520.8 \pm 11.3$ nm) of A549 cells tagged with anti-IGF-1R silver nanospheres. (e) A549 cell tagged with both anti-EGFR GNRs and anti-HER-2 gold nanospheres. (f). Example scattering profile with peak scattering distribution showing two peaks due to the two nanoparticle species employed in (e).

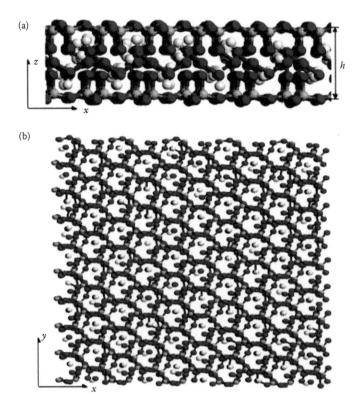

**FIGURE 6.11** Structure of a pyrophyllite single layer computational cell. (Adapted from Mazo, M.A. et al., *The Journal of Physical Chemistry B*, 112(10), 2964–2969, 2008.)

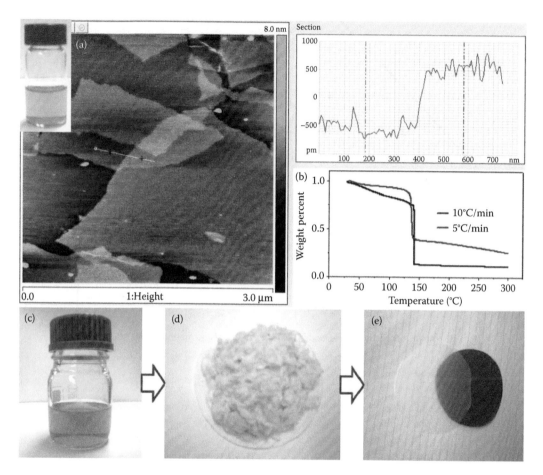

**FIGURE 7.4** (a) Atomic force microscopic image of GO sheets deposited from an aqueous dispersion (inset) onto mica. The thickness scan across GO sheets (central dash line) gives a value of about 1 nm. (b) TGA curves of GO powder showing an effective thermal reduction of GO at 150°C and above. (c) Suspension of GO (1 mg mL⁻¹) and PVDF in DMF. (d) GO–PVDF composite mixture obtained via coagulation of the suspension in distilled water and vacuum filtration. (e) Hot-pressed PVDF circular sheet with semitransparent feature (left) and 0.12 vol% TRG/PVDF composite sheet (right) with dark feature. (He, L. and S. C. Tjong. A graphene oxide–polyvinylidene fluoride mixture as a precursor for fabricating thermally reduced graphene oxide-polyvinylidene composites. *RSC Advances* 3, 2013:22981–87. Reproduced by permission of The Royal Society of Chemistry.)

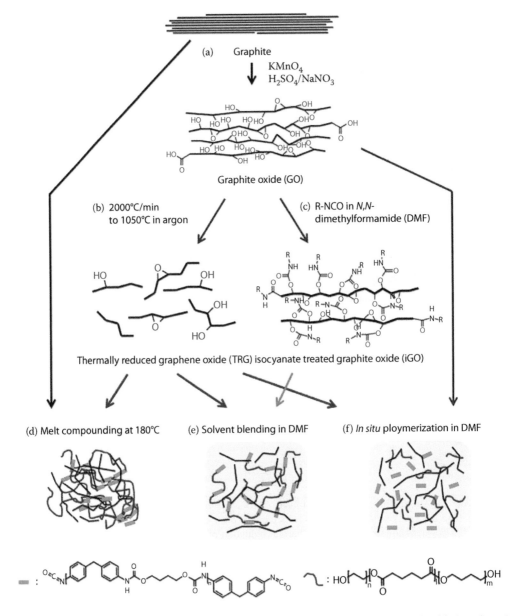

**FIGURE 7.5** Preparation routes of TRG/PU and iGO/TPU nanocomposites. (a) Chemical oxidation of graphite flakes to graphite oxide, (b) thermal reduction of graphite oxide to TRG, (c) isocyanate treatment of graphite oxide in DMF to generate iGO, (d) melt compounding of TRG and TPU in forming TRG/TPU nanocomposites, (e) solvent blending of either TRG or iGO with TPU in DMF, (f) *in situ* polymerization of TRG with TPU in forming TPU-based nanocomposites. (Reprinted with permission from Kim, H., Y. Miura, and C. W. Macosko. Graphene/polyurethane nanocomposites for improved gas barrier and electrical conductivity. *Chemistry of Materials* 22, 3441–50. Copyright 2010 American Chemical Society.)

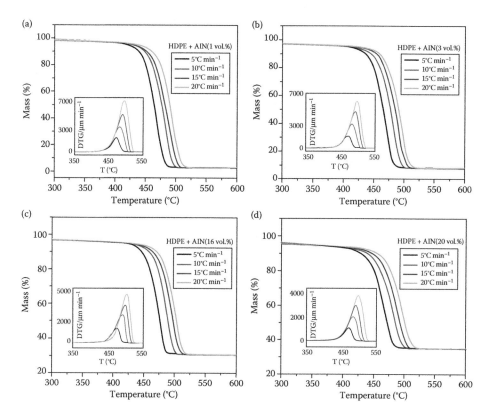

**FIGURE 8.8** TGA mass loss (%) and inset images are derivative mass loss (DTG) versus temperature curves, at four different heating rates: 5°C/min, 10°C/min, 15°C/min, and 20°C/min, for different volume fraction loadings of nanocomposites: (a) HDPE/1 vol.% n-AlN, (b) HDPE/3 vol.% n-AlN, (c) HDPE/16 vol.% n-AlN, and (d) HDPE/20 vol.% n-AlN. (With kind permission from Springer Science+Business Media: *J. Therm. Anal. Calorim.*, Structural and thermal properties of HDPE/n-AlN polymer nanocomposites, 118, 2014, 1513–1530, Rajeshwari, P., Dey, T.K.)

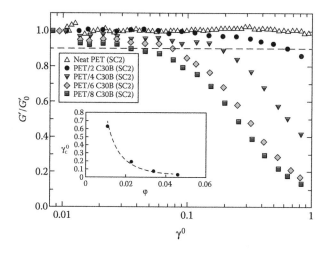

**FIGURE 10.3** Normalized storage modulus versus strain amplitude for neat PET and PET/C30B nanocomposites at 6.28 rad/s. The inset shows the maximum strain amplitude for the linear viscoelastic behavior as a function of the clay volume fraction. (With kind permission from Springer Science + Business Media: *Rheologica Acta,* Morphological and rheological properties of PET/clay nanocomposites, 52, 2013, 59–74, Ghanbari, A. et al.)

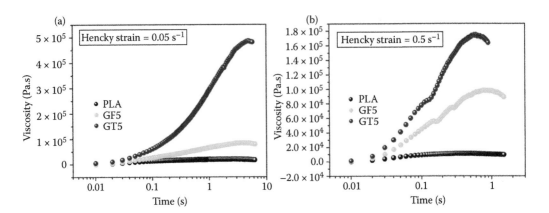

**FIGURE 10.23** Time variation of elongational viscosity $\eta_E(\dot{\varepsilon},t)$ for molten PLA and its silica nanocomposites at 158.5°C. (a) $\dot{\varepsilon} = 0.05\,\text{s}^{-1}$ and (b) $\dot{\varepsilon} = 0.5\,\text{s}^{-1}$. GF5 contains 5 wt% PLA-treated silica prepared by the "grafting from" method, and GT5 contains 5 wt% PLA-treated silica prepared by the "grafting to" method. (Reprinted from *Polymer*, 55, Wu, F. et al., Inorganic silica functionalized with PLLA chains via grafting methods to enhance the melt strength of PLLA/silica nanocomposites, 5760–72, Copyright 2014, with permission from Elsevier.)

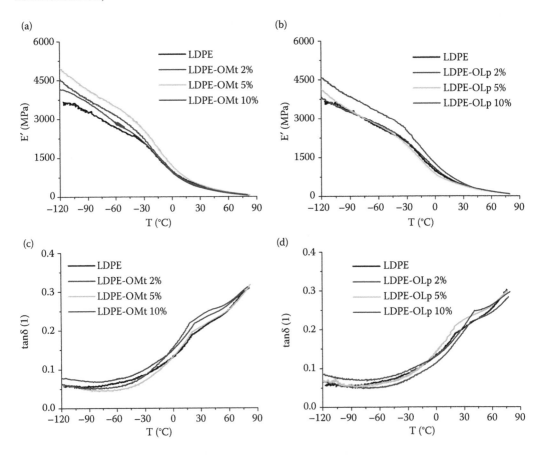

**FIGURE 11.5** Storage modulus E′ (a and b) and loss modulus E″ (c and d) as a function of temperature of LDPE modified nanocomposites with various loadings of (a) and (c) OMt and (b) and (d) OLp. (Adapted from Grigoriadi, K. et al., *Polymer Engineering and Science*, 53(2), 301–308, 2013. Copyright © 2012, Society of Plastics Engineers, with permission.)

# 7 Preparation of Nanocomposites

*Linxiang He and Sie Chin Tjong*

## CONTENTS

## 7.1 INTRODUCTION

Polymers find attractive applications in industrial sectors due to their light weight, ease of fabrication, good processability, and relatively low cost. However, polymers exhibit relatively poor electrical, mechanical, and thermal properties when compared with metals. By adding fillers of micrometer sizes to polymers used as matrices, new materials that exhibit greatly enhanced performances have been obtained. Some inorganic microparticles are originally introduced into polymers as additives. Those fillers are recognized to be very effective for reinforcing polymers. The new obtained materials can exhibit higher mechanical stiffness and strength, excellent dimensional and thermal stability, and can be obtained with lower production cost in comparison with neat polymers. In general, high filler loadings (ca. 20–40 wt%) are incorporated into polymers to achieve desired mechanical properties. Such high filler contents often lead to poor processability and heavy weight, thus minimizing the expected property enhancements of the resulting composites.[1–3]

Recent progress in nanotechnology enables chemists and material scientists to synthesize a wide variety of nanomaterials with unique chemical, mechanical, and physical properties. Among them, one-dimensional carbon nanotubes (CNTs) and two-dimensional graphene nanosheets have drawn special interest of global researchers and industrial sectors due to their exceptional high mechanical strength and elastic modulus as well as superior electrical and thermal properties. The introduction of low loading levels of carbon nanofillers into polymers leads to the formation of novel nanocomposites with functional and structural properties.[4] In particular, the natural abundance, low cost, and multifunctionality of graphene make it an attractive filler material for fabricating electrically conductive polymer nanocomposites designed for sensor, electromagnetic shielding interference, and biomedical applications. By achieving homogeneous dispersion of carbonaceous fillers in the polymer matrix, nanocomposites can exhibit improved mechanical, thermal, electrical, and gas barrier properties. Homogeneous distribution of nanofillers depends on several factors such as the type of surface modification or functionalization, the composite preparation process, the presence of polar groups in the polymer matrix, etc.

CNTs and graphene have poor dispersibility in the polymer matrix due to their chemical inertness. Carbonaceous materials with large surface area and aspect ratio tend to agglomerate into clusters. The van der Waals interactions between individual carbon tubes often lead to the formation of bundles. The filler bundles can impair mechanical and physical properties of the resulting composites. In general, strong interfacial interactions between the fillers and polymer matrix are required for the polymer nanocomposites designed for structural engineering applications. This is because strong interfacial filler-matrix bonding can ensure efficient "stress-transfer" effect during tensile testing. In this respect, functionalization of fillers is needed in order to enhance their compatibility and interaction with the polymer matrix. Surface modification of carbonaceous nanomaterials can be realized by covalent and noncovalent functionalization. Covalent functionalization can induce defects on CNTs and graphene sheets by disrupting the $sp^2$-hybridization required for electron conduction. Thus, covalent functionalization of carbonaceous fillers is detrimental for making electrically conductive polymer composites, but beneficial for forming structural polymer nanocomposites due to enhanced interfacial bonding. In contrast, noncovalent functionalization of CNTs and graphene with surfactants or small aromatic molecules can prevent formation of surface defects or disruption of $\pi-\pi$ conjugation.

Thus far, several processing techniques have been employed for the preparation of polymer-based nanocomposites. These include solution mixing, melt mixing and *in situ* polymerization. The main goal of these techniques is to achieve homogeneous dispersion of fillers in the polymer matrix. Overall, solution mixing and *in situ* polymerization give better dispersion of nanofillers than melt mixing. However, melt compounding offers several advantages including cost effectiveness, mass production, free from organic solvent, and versatility. This processing route is well adopted for manufacturing commercial products in the plastic industries using existing facilities such as extruders and injection molders. This chapter gives the state-of-the-art review on the preparation of polymer nanocomposites reinforced with CNTs, graphene or its derivatives by different processing techniques. Particular attention is paid to polymer nanocomposites with graphene-based fillers.

## 7.2  FUNCTIONALIZATION

### 7.2.1  CARBON NANOTUBES

CNTs are long cylinders of covalently bonded carbon atoms. A single-wall nanotube (SWNT) is formed by rolling a single graphene sheet into a seamless cylinder, while a multiwall nanotube (MWNT) consists of nested graphene cylinders coaxially arranged around a central hollow core with interlayer separations of ~0.34 nm. Functionalization exfoliates CNT bundles prior to composite preparation, thus allowing their uniform dispersion in the polymer matrix. A considerable number of articles have been published reviewing the functionalization of CNTs.[5-9] Therefore, a brief description is given on the nanotube functionalization herein. Oxidation of CNTs in strong acids introduces carboxyl and hydroxyl groups on the sidewalls as well as on the end caps. These functional groups act as anchoring sites for further derivatization reactions for improving their solubility or dispersibility in organic solvents. Moreover, carboxylic acid groups of CNTs can be converted to highly reactive acyl chloride by reacting with thionyl chloride ($SOCl_2$). In this respect, polymer chains terminated with amino ($RNH_2$) or hydroxyl (ROH) can react with acyl chloride groups through amidation or esterification reactions. Covalent linkages can also be established by zwitterionic interactions between the carboxyl groups of CNTs and amine groups of the polymer (Figure 7.1a and b). Under certain conditions, for example, cycloaddition, alkylation, and fluorination, prior wet oxidation of CNTs is unnecessary for covalent sidewall functionalization. The cycloaddition of azomethine ylides onto CNT gives rise to a pyrrolidine ring on its surface.[5,7] Alkylation takes place by reacting CNTs with alkyl groups of long chain molecules. The long polymer chains facilitate the dissolution of bundled nanotubes in organic solvents.

Noncovalent functionalization involves the adsorption of weakly bonded functional groups on CNTs by wrapping with surfactants or polymers through physical interactions.[10–13] Consequently, structural integrity and intrinsic properties of CNTs are preserved. Physical adsorption of surfactants on nanotube surfaces reduces their surface tension markedly, preventing formation of the agglomerates. Surfactants are surface active agents having hydrophilic head groups and hydrophobic hydrocarbon chain molecules. The effective interactions between the surfactants and CNTs depend greatly on the nature of surfactants such as headgroup size and alkyl chain length. Ionic surfactants like sodium dodecyl sulfate (SDS), sodium dodecylbenzene sulfonate (SDBS), and cetyltrimethylammonium bromide (CTAB), and nonionic surfactant such as octyl phenol ethoxylate (Triton X-100) have been reported to be effective dispersants for CNTs in water.[10,11] The order of dispersibility is SDBS > Triton X-100 > SDS.

### 7.2.2 GRAPHENE AND ITS DERIVATIVES

Pristine graphene obtained by the so-called "Scotch-tape" technique is ineffective to produce a sufficient amount of filler materials for reinforcing polymers. In this respect, wet chemical oxidation of graphite flakes with $KMnO_4$ and $NaNO_3$ in concentrated $H_2SO_4$ (Hummer's process) is frequently used to yield graphene oxide (GO) in large quantities.[14] GO sheets are hydrophilic in nature, comprising carboxylic acid groups at the edges with epoxide and hydroxyl groups attached on the basal plane.[15] Those functional groups weaken the van der Waals forces between graphene interlayers, facilitating its exfoliation into thin sheets. Furthermore, oxygenated groups also render GO an electrically nonconductive material. To resume its electrical conductivity, hydrazine or other reducing agents are frequently employed for eliminating oxygen functionalities, producing reduced graphene oxide (rGO).[16,17] The oxygen functionalities of GO can also be removed by either rapid thermal annealing through the release of water and gas molecules (e.g., $CO_2$ and CO), producing thermally reduced graphene oxide (TRG), or via solvothermal reduction of GO in the polymer solution.[18,19] It is noteworthy that rGO and TRG sheets still contain some remnant oxygen groups despite chemical reducing and thermal annealing treatments. Alternatively, pure graphene sheets can be obtained by liquid-phase exfoliation of graphite flakes in N-methyl-pyrrolidone (NMP) solvent using a high shear mixer followed by filtering. The obtained graphene powder is then dried in a vacuum oven, forming high yield and high purity graphene sheets.[20]

Graphene oxide is incompatible with hydrophobic polymers, and thus does not form homogeneous nanocomposites. GO sheets are amphiphilic and can be dispersed in polar solvents such as water, ethanol, and N,N-dimethylformamide (DMF).[21] However, hydrophobic and π–π interactions in the reduced graphene favor restacking of the sheets. Covalent functionalization with isocyanates or polyamines has been reported to be effective for dispersing rGO in several organic solvents, facilitating the fabrication of graphene/polymer nanocomposites.[22,23] In the former case, GO reacts with isocyanate to form isocyanate-treated GO (iGO) during which the carboxyl and hydroxyl groups of GO react with isocyanate to generate amides and carbamate esters, respectively.[22] TRGs generally disperse well in organic solvents such as DMF, 1,2-dichlorobenzene, nitromethane and tetrahydrofuran under sonication.[20] The presence of remained oxygenenated groups also enables TRGs

(a)

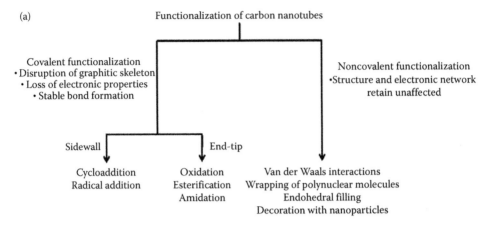

(b)

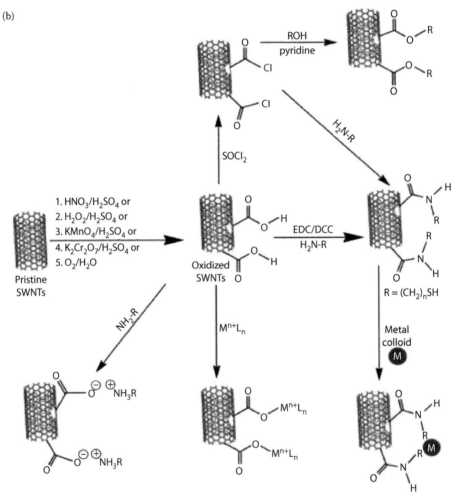

**FIGURE 7.1** (a) Scheme of the functionalization of carbon nanotubes. (Reprinted with permission from Nikolaos, K. and T. Nikos. Current progress on the chemical modification of carbon nanotubes. *Chemical Reviews* 110, 5366–97. Copyright 2010 American Chemical Society.) (b) Scheme of the covalent functionalization of SWNTs at ends and defect sites. (Banerjee, S., T. Hemraj-Benny, and S. S. Wong: Covalent surface chemistry of single-walled carbon nanotubes. *Advanced Materials.* 2005. 17. 17–29. Copyright Wiley-VCH Verlag GmbH & Co. KGaA. Reproduced with permission.)

to react with water-soluble polymer, that is, polyethylene oxide (PEO), and polar polymers such as poly(methyl methacrylate) (PMMA), poly(acrylonitrile), and poly(acrylic acid).[24,25]

In terms of noncovalent functionalization, Stankovich et al. reported the first example of modifying rGO with poly(sodium 4-styrenesulphonate).[26] Choi et al. obtained stable dispersions of rGO in various organic solvents by reacting it with amine-terminated polystyrene (PS-NH$_2$).[27] GO sheets can also be functionalized noncovalently with a water-soluble pyrene derivative (1-pyrenebutyrate) by means of $\pi$–$\pi$ stacking interactions.[28,29] The pyrene derivative tends to react with the basal plane of graphene to form stable $\pi$–$\pi$ stacking bonds.

## 7.3  SOLUTION MIXING

### 7.3.1  CNT/POLYMER NANOCOMPOSITES

Solution mixing is a widely employed processing route to prepare CNT/polymer nanocomposites, especially in the academic sector. The process involves the dissolution of polymer in a solvent followed by mixing the polymer solution with CNT suspension under mechanical stirring and/or sonication, which facilitates disentanglement of nanotubes in the solution. The CNT/polymer composites are formed by removing solvent via evaporation or precipitation, or casting the suspension into a film. Sonication can be carried out by mild vibration in a bath or high-power ultrasonication. The limitations of this route include the use of organic solvents that are expensive and environmentally hazardous, and high-power sonication can reduce the length or aspect ratio of CNTs. This processing route has been used to blend CNTs with a wide variety of polymers including polystyrene (PS), epoxy, polyvinylidene fluoride (PVDF), poly($\varepsilon$-caprolactone), polyurethane (PU), etc.[30–36] For example, Qian et al. prepared MWNT/PS composite film by mixing pristine MWNT with PS in toluene under sonication and solvent evaporation.[30] They reported that the elastic modulus and break stress of the film increase significantly by adding 1 wt% MWNT only. Recently, He and Tjong synthesized MWNT/PVDF nanocomposites using solution mixing followed by coagulation.[33] Both pristine and carboxylated MWNTs were employed as the filler materials. The obtained composite powders were finally hot-compressed into bulky materials with low electrical percolation thresholds. Some pristine MWNTs were dispersed into a few agglomerates and independent tubes, while carboxylated MWNTs were all distributed uniformly in the PVDF matrix.

### 7.3.2  GRAPHENE/POLYMER NANOCOMPOSITES

For solution mixing, GO or graphene-like (rGO and TRG) fillers are dispersed in a solvent followed by mixing it with a polymer solution under mechanical/magnetic stirring or high-energy sonication. The composite material is then obtained by vaporizing the solvent. GO, rGO, and TRG exhibit typical flat sheet morphology, so entangled bundles are not a critical issue like in the case of CNTs. However, restacking of graphene sheets often occurs in rGO after chemical reduction treatment. In this respect, hydrazine is typically added to a mixed polymer–GO dispersion rather than directly added to GO itself. The polymer chains can thus insert into the galleries between the layers to yield intercalated or exfoliated nanocomposites during solution mixing. This strategy is generally used to create nanocomposites filled with layered silicates.[36] Exfoliated nanofillers usually give the best property enhancements due to their large interfacial area and homogeneous dispersion in the polymer matrix.

Graphite oxide exhibits a larger interlayer spacing compared to graphite, so hydrophilic polymers such as polyethylene oxide (PEO), polyvinyl alcohol (PVA), and poly(furfuryl alcohol) can be readily inserted into its galleries during solution mixing to produce intercalated nanocomposites.[37–39] Using this approach, several workers also fabricated rGO/PVA and GO/PVA nanocomposites recently.[40–43] They achieved enhanced tensile strength and modulus in the resulting composites due to the strong hydrogen bonding and homogeneous dispersion of GO/rGO fillers in the polymer

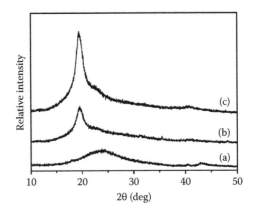

**FIGURE 7.2**  XRD patterns of (a) rGO, (b) PVA, and (c) solution-mixed 0.6 vol% rGO/PVA nanocomposite. (Reprinted with permission from Zhao, X., Q. Zhang, and D. Chen. Enhanced mechanical properties of graphene-based poly(vinyl alcohol) composites. *Macromolecules* 43, 2357–63. Copyright 2010 American Chemical Society.)

matrix, ensuring effective stress-transfer effect across the matrix-filler interface. Figure 7.2 shows typical x-ray diffraction (XRD) patters of rGO, PVA, and 0.6 vol% rGO/PVA nanocomposite. Apparently, rGO exhibits a broad peak; the characteristic peak of PVA appears at $2\theta = 19.5°$. By dispersing rGO into PVA, the broad peak of rGO disappears, demonstrating the loss of structure regularity of rGO sheets due to their exfoliation into single graphene sheets. Figure 7.3 is a SEM image showing that most rGO sheets are fully exfoliated and well dispersed in the PVA matrix.

It is noteworthy that TRG with remnant oxygen functionalities can be mixed with polar PMMA, leading to homogeneous dispersion of TRG fillers in the matrix and large electrical conductivity of the resulting composites.[44] He and Tjong prepared TRG/PVDF nanocomposites by first suspending GO and PVDF in DMF.[45] The suspension of GO and PVDF was then coagulated in distilled water. The composite mixture precipitated out immediately due to its insolubility in water. The mixture was dried and further subjected to hot pressing at 200°C to reduce GO to TRG. A very low electrical

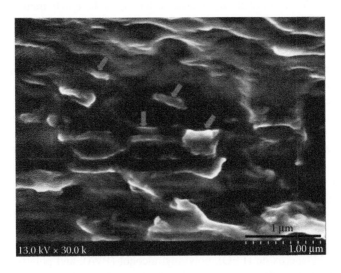

**FIGURE 7.3**  SEM image showing homogeneous dispersion of rGO in solution-mixed rGO/PVA nanocomposite. rGO fillers are indicated with arrows. (Reprinted with permission from Zhao, X., Q. Zhang, and D. Chen. Enhanced mechanical properties of graphene-based poly(vinyl alcohol) composites. *Macromolecules* 43, 2357–63. Copyright 2010 American Chemical Society.)

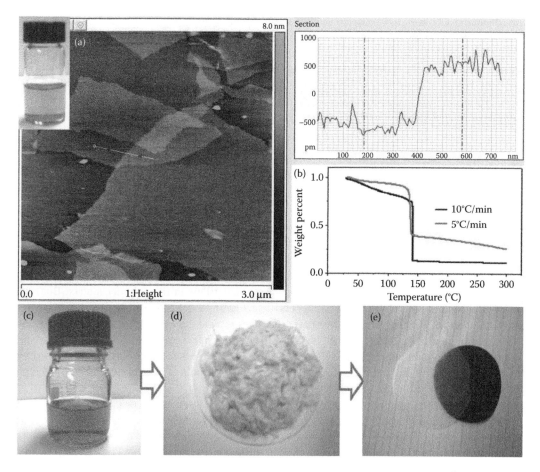

**FIGURE 7.4** **(See color insert.)** (a) Atomic force microscopic image of GO sheets deposited from an aqueous dispersion (inset) onto mica. The thickness scan across GO sheets (central dash line) gives a value of about 1 nm. (b) TGA curves of GO powder showing an effective thermal reduction of GO at 150°C and above. (c) Suspension of GO (1 mg mL⁻¹) and PVDF in DMF. (d) GO–PVDF composite mixture obtained via coagulation of the suspension in distilled water and vacuum filtration. (e) Hot-pressed PVDF circular sheet with semitransparent feature (left) and 0.12 vol% TRG/PVDF composite sheet (right) with dark feature. (He, L. and S. C. Tjong. A graphene oxide–polyvinylidene fluoride mixture as a precursor for fabricating thermally reduced graphene oxide–polyvinylidene composites. *RSC Advances* 3, 2013:22981–87. Reproduced by permission of The Royal Society of Chemistry.)

percolation threshold of 0.12 vol% was obtained due to homogeneous dispersion of the TRGs in the polymer matrix. Figure 7.4 illustrates typical processing steps for forming 0.12 vol% TRG/PVDF nanocomposite.

For polymer nanocomposites designed for structural engineering applications, both homogeneous dispersion of nanofillers and strong filler-matrix interactions are important factors for achieving efficient stress-transfer effect. In this respect, chemical functionalization of GO is needed to improve its solubility and interactions with organic solvents and polymers. Stankovich et al. synthesized rGO/PS nanocomposite by mixing iGO with PS in DMF followed by reduction with dimethylhydrazine.[46] The composite material was precipitated by adding methanol into the DMF solution. Methanol is widely employed for precipitating rGO/polymer composite from its solution medium. Kim at al. fabricated graphene/thermoplastic polyurethane (TPU) by mixing iGO with TPU in DMF.[47] Up to a tenfold increase in tensile stiffness was achieved in TPU by adding 1.6 vol% (3 wt%) iGO. Kuila et al. functionalized rGO with docecyl amine and then solution-mixed with linear low-density polyethylene (LLDPE). They reported that the tensile strength and modulus of LLDPE increase with increasing filler content.[48] The improvement in tensile strength is attributed to the formation of strong interfacial bonding between the polymer matrix and the filler. More recently, Nawaz et al. covalently functionalized GO with octadecylamine (ODA) to form GO–ODA.[49] Such modified GO and TPU elastomer were then dispersed in THF to give graphene/TPU nanocomposites. Composite films were formed by casting the dispersions into Teflon trays. The ultimate tensile strength increased linearly with graphene content up to 3 wt%, and then decreased considerably with increasing filler content.

## 7.4   MELT COMPOUNDING

### 7.4.1   CNT/Polymer Nanocomposites

Melt compounding is a versatile and cost-effective route for industrial sectors to fabricate commercial plastic products in large quantities. Melt processing involves the melting of polymer pellets to form a viscous liquid at high temperatures and the application of high shear force to disperse nanofillers using extruders and/or injection molders.[50] It is environmentally friendly due to the absence of organic solvents during the processing. In general, a high shear force on molten mixture resulting from the high screw speed is beneficial for dispersing nanofillers in the polymer matrix. But high shear force can damage the aspect ratio of CNTs, deteriorating mechanical and electrical properties of the nanocomposites considerably. Moreover, polymer viscosity and the nature of CNTs (pristine or functionalized) also influence the state of dispersion of nanotubes.[51–53] Comparing with solution mixing, melt compounding is less effective at dispersing CNTs more uniformly in the polymer matrix and is confined to lower filler contents. At high filler loadings, the mixture melts with high viscosities impairing dispersion of the nanotubes.

McNally et al. fabricated MWNT-filled polyethylene (PE) composites via melt compounding using pristine MWNTs in a Haake twin screw extruder.[52] The nanotubes distributed mainly as small aggregates in the PE matrix. Therefore, the yield strength and breaking stress of the nanocomposites decreased with increasing filler content, resulting from poor interfacial filler–matrix interactions. To improve the dispersion of nanotubes and interfacial filler–matrix bonding, functionalized nanotubes are needed. Zhang et al. compounded acylated MWNTs with PA6 in a Brabender twin-screw mixer. The mechanical properties of resulting composites improved considerably by using functionalized nanotubes.[53] Very recently, Cohen et al. prepared MWNT/poly(ethylene-co-(methacrylic acid)) (PEMAA) composites by mixing composite constituents and 4-(aminomethyl)pyridine-modified PEMAA, acting as a noncovalent compatibilizer.[54] They reported that the compatibilizer improves the dispersion of nanotubes due to the pyridine ring reacting with MWNT surface via $\pi$–$\pi$ interactions. Melt compounding has been extensively used to fabricate CNT/polymer nanocomposites with various polymeric materials.[50,55–58]

### 7.4.2   Graphene/Polymer Nanocomposites

Thermally reduced graphene is typically selected as the filler material to compound with polymers to form nanocomposites.[47,59–63] In the process, polymer chains can be inserted between the layers, forming nanocomposites with an intercalated structure.[47,59,61] Occasionally, an exfoliated structure can also be produced.[60] Kim et al. investigated systematically the preparation, microstructure, and properties of melt compounded polymer composites with TRG fillers. The polymer matrix materials included TPU, polyester, polycarbonate, and PE.[47,59–61] They also compared the dispersion state of TRGs in the matrix of nanocomposites prepared by melt compounding, solution mixing, and *in situ* polymerization techniques.[47,61] In addition, conventional composite filled with graphite flakes was also melt compounded. Figure 7.5 shows schematic diagrams of TRG/TPU nanocomposites prepared by different processing routes. Transmission electron microscopy (TEM) image of melt compounded graphite/TPU composite reveals graphite flakes comprised of aggregates of tactoids (Figure 7.6a). Thus, polymer chains cannot intercalate into unmodified graphite galleries during

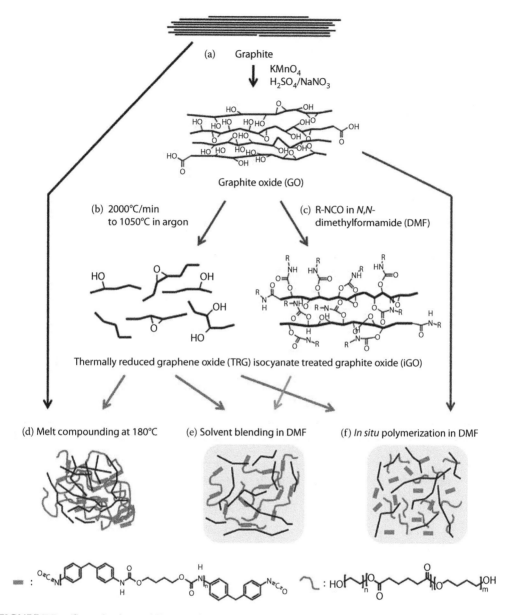

**FIGURE 7.5** **(See color insert.)** Preparation routes of TRG/PU and iGO/TPU nanocomposites. (a) Chemical oxidation of graphite flakes to graphite oxide, (b) thermal reduction of graphite oxide to TRG, (c) isocyanate treatment of graphite oxide in DMF to generate iGO, (d) melt compounding of TRG and TPU in forming TRG/ TPU nanocomposites, (e) solvent blending of either TRG or iGO with TPU in DMF, (f) *in situ* polymerization of TRG with TPU in forming TPU-based nanocomposites. (Reprinted with permission from Kim, H., Y. Miura, and C. W. Macosko. Graphene/polyurethane nanocomposites for improved gas barrier and electrical conductivity. *Chemistry of Materials* 22, 3441–50. Copyright 2010 American Chemical Society.)

compounding as expected. By employing TRG fillers, a stacked sheet feature is observed, revealing the formation of intercalated nanocomposite (Figure 7.6b and c). Their studies also demonstrated that the TRG sheets disperse more uniformly and independently in the matrix of nanocomposites prepared from solution mixing and *in situ* polymerization (Figure 7.6d and f). Such exfoliated graphene sheets improve electrical conductivity of solution-mixed and *in situ* polymerized TRG/ TPU nanocomposites (Figure 7.7). In a recent study, Istrate et al. fabricated graphene/polyethylene

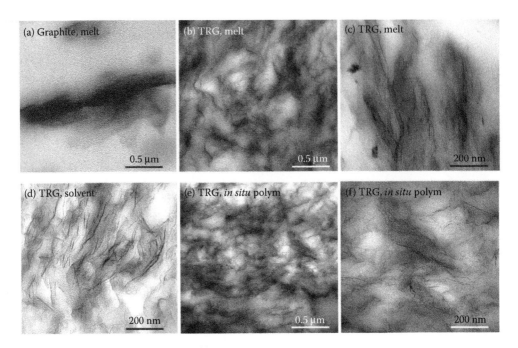

**FIGURE 7.6** TEM images of TPU with (a) 5 wt% (2.7 vol%) graphite, (b and c) melt compounded, (d) solvent-mixed, (e and f) *in situ* polymerized 3 wt% (1.6 vol%) TRG. (Reprinted with permission from Kim, H., Y. Miura, and C. W. Macosko. Graphene/polyurethane nanocomposites for improved gas barrier and electrical conductivity. *Chemistry of Materials* 22, 3441–50. Copyright 2010 American Chemical Society.)

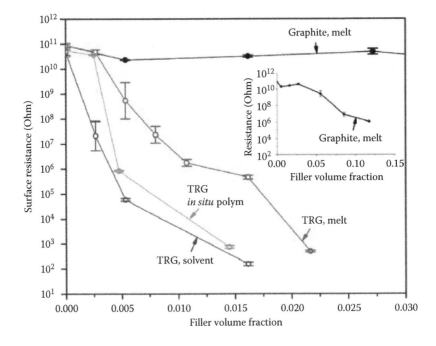

**FIGURE 7.7** DC surface resistance of melt compounded graphite/TPU composites (closed symbols, also in inset), melt compounded and solution-mixed, as well as *in situ* polymerized TRG/TPU composites (open symbols). (Reprinted with permission from Kim, H., Y. Miura, and C. W. Macosko. Graphene/polyurethane nanocomposites for improved gas barrier and electrical conductivity. *Chemistry of Materials* 22 3441–50. Copyright 2010 American Chemical Society.)

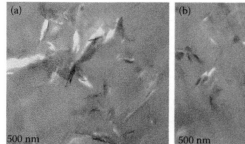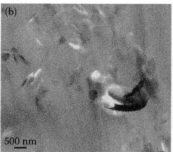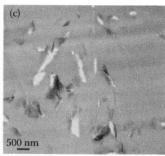

**FIGURE 7.8** TEM images of composites containing 3 wt% (1.2 vol%) TRG with the matrix materials of (a) low viscosity LLDPE-g-MA and (b) high viscosity LLDPE-g-MA showing a mixture of TRG aggregates and nanoplatelets; (c) low viscosity LLDPE-g-pyridine containing individually dispersed platelets. (Reprinted with permission from Vasileiou, A. A., M. Kontopoulou, and A. Docoslis. A noncovalent compatibilization approach to improve the filler dispersion and properties of polyethylene/graphene composites. *ACS Applied Materials and Interfaces* 6, 1916–25. Copyright 2014 American Chemical Society.)

terephthalate (PET) nanocomposites by melt blending PET with pure graphene obtained from liquid-phase exfoliation.[64] The addition of only 0.07 wt% exfoliated graphene to PET led to the resulting nanocomposites with excellent mechanical properties.

Vasileiou et al. incorporated TRGs into two kinds of LLDPE with high and low viscosity and into their respective aminopyridine derivatives by melt compounding.[65] Large graphene aggregates formed when both types of maleated LLDPE were used. However, noncovalent functionalization of maleated LLDPE with aminopyridine improved the dispersion of TRGs considerably (Figure 7.8). This was due to the interactions between the aromatic moieties on the pyridine grafted matrix through π–π stacking with the surface of TRG.

## 7.5  *IN SITU* POLYMERIZATION

*In situ* polymerization techniques such as bulk, solution, ring opening, and emulsion polymerization have been employed for preparing different polymer resins. Each technique has its own advantages and limitations. Bulk polymerization involves mixing monomer and initiator in a heated reactor under stirring in a protective atmosphere. For solution polymerization, the monomer is first dissolved in a solvent containing a catalyst. The heat released by the reaction is absorbed by the solvent. The main disadvantage of this technique is associated with the removal of solvent.[66] Ring-opening polymerization (ROP) involves the use of cyclic monomers and initiators (anionic, cationic, or radical). Cyclic monomers such as alkanes, alkenes, and lactams are typically selected for synthesizing desired polymer resins. Emulsion polymerization is used to produce latex particles with a wide variety of desired properties. Emulsion polymerization involves emulsification of a monomer with poor solubility in a continuous phase (e.g., water), forming an oil-in-water emulsifier in the presence of a surfactant. The surfactant facilitates formation of the micelles in water, where the monomer diffuses inside for polymerization. However, the surfactant molecules that trap in the synthesized polymers are difficult to remove. Under certain conditions, surfactant-free composites are desirable because the surfactant may affect their transparency, and mechanical and thermal properties. The polymerization reaction is initiated by a water-soluble initiator (e.g., potassium persulfate [$K_2S_2O_8$; KPS] or an oil-soluble initiator such as 2,2′-azobisisobutyronitrile [AIBN]). Generally, KPS is often used and it decomposes in water to form radicals, which also diffuses inside the micelles to initiate or terminate the polymerization. Water is a good heat conductor, facilitating homogeneous viscosity control of the polymerization reaction and promoting fast transfer of heat from the polymerizing medium. Emulsion polymerization is effective in increasing polymer molecular weight without sacrificing the polymerization rate. In the miniemulsion process, a system containing water, oil, a surfactant, and

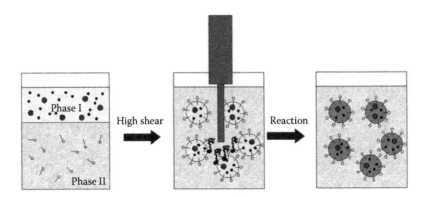

**FIGURE 7.9** Miniemulsion polymerization. Phase I (oil), Phase II (aqueous phase with surfactant molecules. (Landfester, K: Miniemulsion polymerization and the structure of polymer and hybrid nanoparticles. *Angewandte Chemie International Edition*. 2009. 48. 4488–07. Copyright Wiley-VCH Verlag GmbH & Co. KGaA. Reproduced with permission.)

an osmotic pressure agent (hydrophobe) insoluble in the continuous phase is subjected to high shear force or high-power sonication energy.[67] This results in the formation of small, homogeneous, and narrowly distributed droplets with the size ranging from 50 to 500 nm (Figure 7.9). Comparing with conventional emulsions, the surfactant concentration in miniemulsions is much smaller.

To fabricate polymer nanocomposites, pristine or functionalized nanofillers are first prepared, followed by adding a monomer of interest to undergo subsequent polymerization via free radical, ring opening, or emulsion reactions under mechanical stirring or sonication. *In situ* polymerization facilitates stronger interactions between the nanofillers and polymeric phase, thus achieving homogeneous dispersion of nanofillers in the polymer matrix. This method is attractive for preparing polymer nanocomposites that cannot be processed by solution mixing and melt blending.

### 7.5.1  CNT–Polymer Nanocomposites

#### 7.5.1.1  Bulk Polymerization

Bulk polymerization has been used to fabricate MWNT/PMMA and MWNT/PS nanocomposites.[68–72] For example, Jia et al. prepared MWNT/PMMA nanocomposites using both pristine and carboxylated nanotubes, and methyl methacrylate (MMA) monomer with the aid of AIBN initiator to induce radicals. This initiator opened the π-bonds of nanotubes, facilitating MMA polymerization.[68] Meng et al. used benzoyl peroxide (BP) initiator for preparing MWNT/PMMA composites.[70] In the process, carboxylated MWNT and BP were added into MMA monomer under sonication. The mixture was heated to 80–85°C for decomposing BP into radicals. Then the C=C bonds of the MMA monomer and the π-bonds of MWNTs opened up accordingly, forming strong filler–matrix interface. Very recently, Shrivastava et al. modified the synthesized process of MWNT/PMMA nanocomposites that involved an initial bulk polymerization of MMA monomer with BP in the presence of pristine MWNTs at 85°C.[71] Commercial PMMA beads were then added into a reactor during the polymerization process (Figure 7.10). As a result, MWNTs were selectively dispersed and formed a continuous network structure in *in situ* bulk polymerized PMMA phase, outside the PMMA beads, leaving the beads free from the nanotubes. The beads in which MWNTs failed to penetrate were considered the "excluded volume" sites (Figure 7.11). The formation of continuous MWNT network was responsible for the very low electrical percolation threshold (0.12 wt%) in the nanocomposites (Figure 7.12).

#### 7.5.1.2  Ring-Opening Polymerization

In general, the so-called "grafting-from" method, in which monomers are first attached to the nanotubes, can give effective activators for subsequent *in situ* polymerization. Qu et al. functionalized SWNT

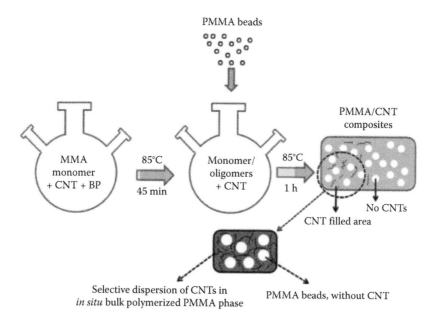

**FIGURE 7.10** Schematic representation of the composite fabrication process. (Shrivastava, N. K. et al.: A facile route to develop electrical conductivity with minimum possible multiwall carbon nanotube (MWCNT) loading in poly(methyl methacrylate)/MWCNT nanocomposites. *Polymer International.* 2012. 61. 1683–92. Copyright Wiley-VCH Verlag GmbH & Co. KGaA. Reproduced with permission.)

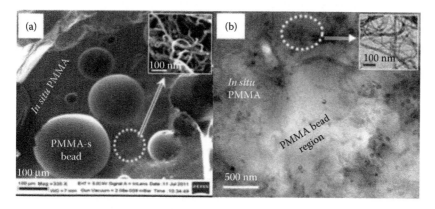

**FIGURE 7.11** (a) SEM micrograph of 0.2 wt% MWNT/PMMA composite containing 25 wt% PMMA beads. (b) TEM micrograph of 0.2 wt% MWNT/PMMA composite containing 70 wt% PMMA beads. (Shrivastava, N. K. et al.: A facile route to develop electrical conductivity with minimum possible multi-wall carbon nanotube (MWCNT) loading in poly(methyl methacrylate)/MWCNT nanocomposites. *Polymer International.* 2012. 61. 1683–92. Copyright Wiley-VCH Verlag GmbH & Co. KGaA. Reproduced with permission.)

with PA6 via covalent attachment of ε-caprolactam molecules to acyl modified nanotubes followed by the anionic ring-opening polymerization of these bound ε-caprolactam species.[73] The synthesis scheme is shown in Figure 7.13. Yang et al. also employed the "grafting-from" strategy by initially attached acyl caprolactam to MWNTs followed by *in situ* polymerization to form PA6-functionalized MWNTs.[74] Very recently, Yan and Yang prepared MWNT/PA6 nanocomposites by reacting MWNT–OH with toluene 2, 4-diisocyanate (TDI) through esterification to produce TDI functionalized MWNT (MWNT-NCO).[75] The resultant nanotubes were used as *in situ* activators for anionic ring-opening polymerization

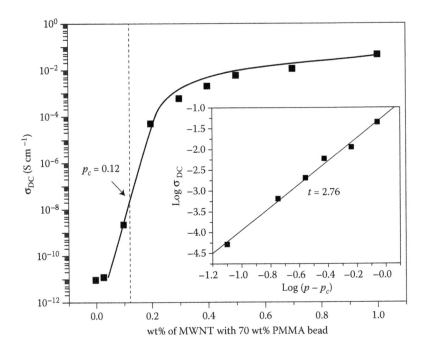

**FIGURE 7.12** DC electrical conductivity versus MWNT content of *in situ* polymerized MWNT/PMMA nanocomposites. $p_c$ denotes percolation threshold and $t$ is the exponential exponent. (Shrivastava, N. K. et al.: A facile route to develop electrical conductivity with minimum possible multi-wall carbon nanotube (MWCNT) loading in poly(methyl methacrylate)/MWCNT nanocomposites. *Polymer International*. 2012. 61. 1683–92. Copyright Wiley-VCH Verlag GmbH & Co. KGaA. Reproduced with permission.)

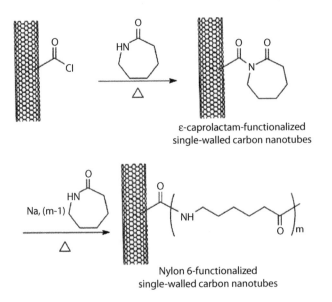

ε-caprolactam-functionalized
single-walled carbon nanotubes

Nylon 6-functionalized
single-walled carbon nanotubes

**FIGURE 7.13** Synthesis of PA6-functionalized SWNT. (Reprinted with permission from Qu, L. et al. Soluble nylon-functionalized carbon nanotubes from anionic ring-opening polymerization from nanotube surface. *Macromolecules* 38, 10328–31. Copyright 2005 American Chemical Society.)

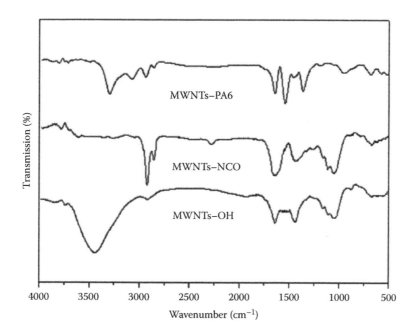

**FIGURE 7.14** Fourier transform infrared spectra of MWNTs–OH, MWNTs–NCO, and MWNTs–PA6. (Yan, D. and G. Yang: Synthesis and properties of homogeneously dispersed polyamide 6/MWNTs nanocomposites via simultaneous *in situ* anionic ring-opening polymerization and compatibilization. *Journal of Applied Polymer Science*. 2009. 12. 3620–26. Copyright Wiley-VCH Verlag GmbH & Co. KGaA. Reproduced with permission.)

of ε-caprolactam to produce MWNT/PA6 nanocomposites. Figure 7.14 shows the Fourier transform infrared (FTIR) spectra of MWNT–OH, MWNT–NCO, and MWNT–PA6 samples. For MWNT–OH, the peak near 3400 cm$^{-1}$ is associated with the O–H stretching of the hydroxyl groups of MWNT–OH. This peak disappears in the spectrum of MWNT–NCO. However, a small peak occurs at 2283 cm$^{-1}$, corresponding to asymmetric stretching of isocyanate groups. The peaks at 2900 and 2825 cm$^{-1}$ are due to the C–H stretching of methyl groups of TDI. The peak 1245 cm$^{-1}$ is related to the C–N stretching of carbamate groups, resulting from the esterification of hydroxyl and isocyanate groups. For the MWNT–PA6 material, the absence of 2283 cm$^{-1}$ peak clearly implies that the isocyanate groups are consumed during the ROP process. Moreover, characteristic bands for PA6 can be readily observed including the peaks at 3297 (N–H stretching), 3060 and 1637 (amide I band), as well as 1540 cm$^{-1}$ (amide II band).

### 7.5.1.3 Miniemulsion Polymerization

Miniemulsion polymerization is a useful tool to synthesize latex nanoparticles containing fillers. In the process, the oil phase (monomer) and filler is dispersed in an aqueous phase in the presence of a surfactant/hydrophobe to stabilize the emulsion under sonication. Surfactant molecule prevents the droplets from coalescing while hydrophobe prevents Ostwald ripening. Coalescence arises from the collision of droplets and Ostwald ripening is due to degradation of the droplets through diffusion. Droplet nucleation occurs when radicals formed in the aqueous phase enter monomer droplets. The stabilized monomer/filler droplets with large interfacial areas are then polymerized to form composite particles through an emulsion polymerization step. One monomer droplet generates one polymer/composite particle accordingly. The oil-soluble or water-soluble radical initiators under mild temperature conditions promote the polymerization, forming composite nanoparticles with different morphologies, that is, polymer encapsulated filler, polymer core with attached inorganic fillers, or a mixture of both.

Barraza et al. fabricated SWNT/PS and SWNT/styrene–isoprene composites by miniemulsion polymerization using CTAB surfactant and AIBN initiator.[76] Under these conditions, the resulting composites appear homogeneous without any considerable aggregation of nanotubes, thereby

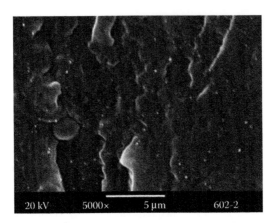

**FIGURE 7.15** SEM micrograph showing cryogenically fractured surface of 1% PBA–encapsulated MWNT/ PA6 nanocomposite. (Reprinted with permission from Xia, H., Q. Wang, and G. Qiu. Polymer-encapsulated carbon nanotubes prepared through ultrasonically initiated *in situ* emulsion polymerization. *Chemistry of Materials* 15, 3879–86. Copyright 2003 American Chemical Society.)

achieving some improvements in electrical conductivity. Xia et al. encapsulated MWNTs with PMMA and polybutyl acrylate (PBA) using *in situ* emulsion polymerization under intense ultrasonic irradiation.[77,78] The multiple effects of ultrasound, that is, dispersion, pulverizing, activation, and initiation, can prevent the agglomeration and entanglement of MWNTs in aqueous solutions. Furthermore, ultrasonic cavitation generates local high temperature and high pressure, producing a very rigorous environment for chemical reactions. Under these conditions, radicals can be induced due to the decomposition of water, monomer, and surfactant. As a result, polymerization of monomer *n*-butyl acrylate (BA) or MMA on the nanotube surface proceeds without addition of chemical initiator, producing polymer encapsulated nanotubes. The PBA-encapsulated nanotubes were then melt-blended with PA6 to form nanocomposites. Figure 7.15 is a SEM micrograph showing homogenously dispersed MWNTs in the polymeric matrix.

### 7.5.2 Graphene–Polymer Nanocomposites

#### 7.5.2.1 Bulk Polymerization

Several researchers have utilized free radical bulk polymerization to prepare rGO/PMMA, GO/PS, and GO/PU nanocomposites.[79–83] Graphene or its derivative is initially added to the monomer, followed by dispersing desired initiator to the suspension. Polymerization reaction is initiated by either heat or electromagnetic radiation. The monomers prevent the restacking of the graphene sheets, thereby producing uniformly dispersed graphene in the polymer matrix. In certain conditions, a solvent is needed, but solvent removal becomes the main issue after polymerization.[81] Recently, Jang et al. fabricated GO/PMMA composites using AIBN and a special macroazoinitiator (MAI) with a PEO segment.[82] For the composites prepared with AIBN only, an intercalated structure is produced after polymerization. In contrast, an exfoliated structure is formed in the composites prepared with both AIBN and MAI due to the MMA monomers, which can intercalate easily into the interlayer galleries. As a result, individual GO sheets are delaminated and randomly dispersed in the polymer matrix during the polymerization. The uniform dispersion of graphene sheets and strong interfacial adhesion between the polymer matrix and graphene lead to significant improvement in both physical and mechanical properties of exfoliated nanocomposites.

The bulk polymerization approach can also be used to prepare polymer composites with hybrid nanofillers. As an example, Liu et al. prepared magnetic $Fe_3O_4$–GO/PS nanocomposite by *in situ* radical bulk polymerization of styrene in the presence of the $Fe_3O_4$ modified graphene oxide

(Fe$_3$O$_4$–GO) and oleic acid (OA) as the surface modifier.[84] In general, GO with large surface area provides a good support for loading magnetic Fe$_3$O$_4$ nanoparticles. The Fe$_3$O$_4$–GO shows attractive use for bioimaging, magnetic energy storage, catalysis, etc.

### 7.5.2.2 Ring-Opening Polymerization

This approach is often adopted by the researchers to prepare graphene/PA6 nanocomposites.[85–87] For example, Zhang et al. prepared PA6 grafted GO by first reacting iGO with ε-caprolactam (CL) to form GO–NCL followed by anionic ring-opening polymerization in the presence of caprolactam magnesium bromide (C1) initiator and hexamethylene-1,6-dicarbamoylcaprolactam (C20) activator (Figure 7.16). The PA6 grafted graphene oxide (denoted as g-GO) was then examined with FTIR, nuclear magnetic resonance, and x-ray photoelectron spectroscopic (XPS) techniques. Figure 7.17 shows typical FTIR spectra of GO, GO–NCL, and g-GO. Apparently, GO exhibits characteristic peaks at 3310, 1718, 1604, and 1062 cm$^{-1}$, resulting from the OH stretching, C=O in carboxylic acid and carbonyl moieties, C=C in skeletal vibrations of GO, and C–O stretching, respectively (Trace a). By reacting GO with CL coupling by 4,4′-methylenebis(phenyl isocyanate) to yield GO–NCL, the

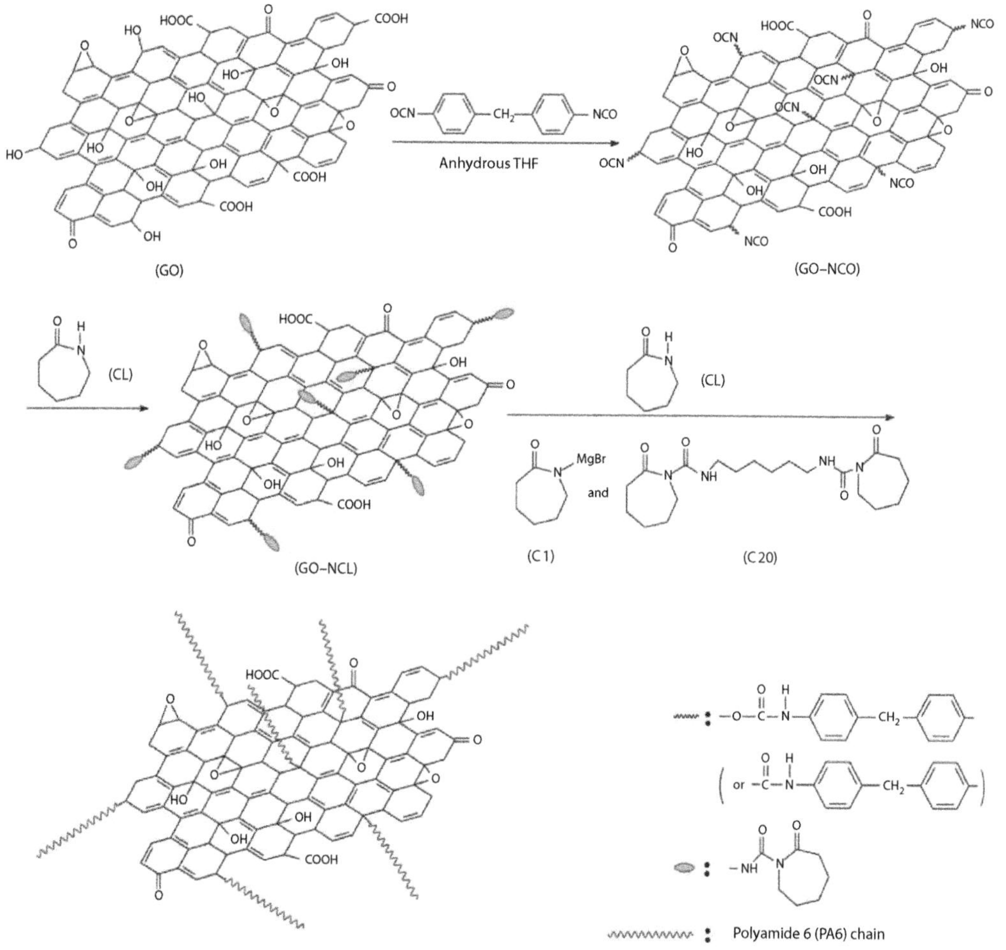

**FIGURE 7.16** Schematic scheme for the synthesis of PA6 grafted GO via ring-opening polymerization. (Zhang, X. et al. Facile preparation routes for graphene oxide reinforced polyamide 6 composites via anionic ring-opening polymerization. *Journal of Materials Chemistry* 22, 2012:24081–91. Reproduced by permission of The Royal Society of Chemistry.)

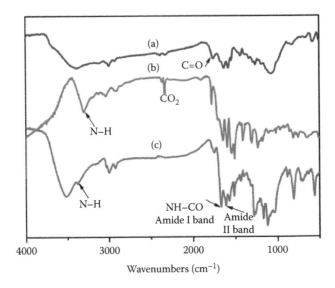

**FIGURE 7.17**  FTIR spectra of (a) GO–GO, (b) GO–NCL and (c) PA6 grafted GO. (Zhang, X. et al. Facile preparation routes for graphene oxide reinforced polyamide 6 composites via anionic ring-opening polymerization. *Journal of Materials Chemistry* 22, 2012:24081–91. Reproduced by permission of The Royal Society of Chemistry.)

peaks at 2960 and 2845 cm$^{-1}$ are due to the methylene (–CH$_2$–) stretching vibration, while the N–H stretching peaks at 3217 and 1611 cm$^{-1}$ are attributed to the presence of amide (–CO–NH–) or carbamate esters (Trace b). For the g-GO (Trace c), the peaks at 1640 and 1583 cm$^{-1}$ are assigned to amide I and amide II, resulting from the C=O stretching vibration, the N–H bending mode, and the C–N stretching vibration of amide groups, respectively. Figure 7.18 shows the XPS C$_{1s}$ spectra of GO and g-GO. The C$_{1s}$ spectrum of GO can be deconvoluted into three peaks located at 284.6 eV (C–C/C–H), 286.6 eV (C–O/C–OH), and 288.2 eV (C=O/O=C–OH). For the g-GO, two new peaks at 285.4 eV (C–N) and 287.8 eV (O=C–NH) appear in the deconvoluted C$_{1s}$ spectrum in addition to the existing three peaks mentioned earlier.

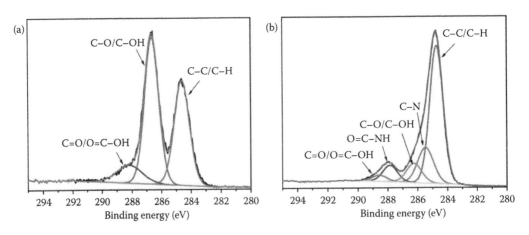

**FIGURE 7.18**  XPS C$_{1s}$ spectra of (a) GO and (b) PA6 grafted GO. (Zhang, X. et al. Facile preparation routes for graphene oxide reinforced polyamide 6 composites via anionic ring-opening polymerization. *Journal of Materials Chemistry* 22, 2012:24081–91. Reproduced by permission of The Royal Society of Chemistry.)

### 7.5.2.3 Miniemulsion Polymerization

This synthesis route can be used to prepare graphene/PMMA, graphene/PS, and graphene/poly(styrene-*co*-butyl acrylate) (poly(St-*co*-BA) nanocomposite latices.[88–93] Etmimi and Mallon synthesized poly(St-*co*-BA) nanocomposite latices containing GO very recently.[91–91] They dispersed an aqueous solution of GO in a monomer phase (styrene and BA), followed by emulsification in the presence of a surfactant (SDS) and a hydrophobe (hexadecane) into miniemulsions (Figure 7.19). XRD patterns revealed the formation of an exfoliated structure in latex particles containing 2 and 4 wt% GO. At 6 wt% GO content, an intercalated structure was produced. In another study, GO was first modified with a reactive surfactant, 2-acrylamido-2-methyl-1-propanesulfonic acid (AMPS) for widening the gap between the graphene interlayers, thus promoting monomer intercalation into the GO galleries.[93] At 1 wt% AMPS-modified GO, TEM micrograph reveals that the modified GO nanosheets were encapsulated in the copolymer particles (Figure 7.20a). Moreover, TEM micrograph of microtomed latex film shows that the AMPS-modified GO sheets are mostly exfoliated in the polymer matrix (Figure 7.20b). In this case, the GO sheets are smaller in size due to exfoliation, thus they can readily enter the polymer latex particles to form polymer shell–filler core structure. Without AMPS modification, many GO sheets exist outside the polymer particles, that is, GO sheets are not encapsulated by the copolymer shell (Figure 7.21a). Unmodified GO contains many stacked graphene with small interlayer spacing. Such stacked nanosheets are relatively large compared to the polymer particles, thus stacked sheets are unable to enter the polymer particles. The TEM micrograph of the microtomed latex film shows that the GO sheets exhibit a typical

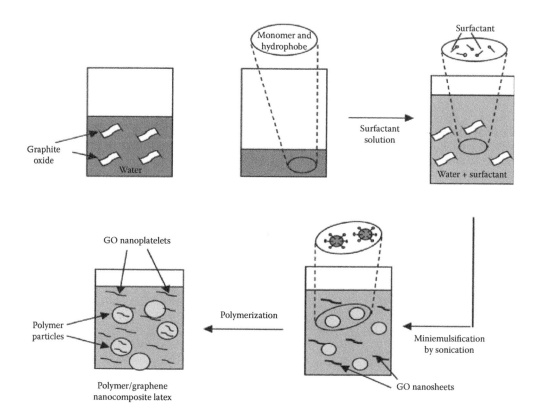

**FIGURE 7.19** Synthesis of polymer nanocomposite latices with GO using miniemulsion polymerization. (Reprinted from *Polymer*, 54, Etmimi, H. M., and P. E. Mallon, *In situ* exfoliation of graphite oxide nanosheets in polymer nanocomposites using miniemulsion polymerization, 6078–88, Copyright 2013, with permission from Elsevier.)

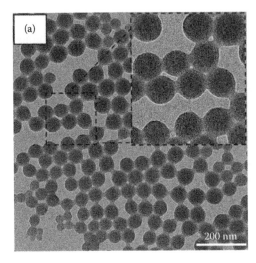
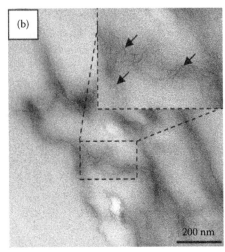

**FIGURE 7.20** TEM micrographs of poly(St-BA)/GO nanocomposite synthesized from 1 wt% AMPS-modified GO: (a) latex particles and (b) a microtomed film cast from the same latex. (Reprinted with permission from Etmimi, H. M. and P. E. Mallon. New approach to the synthesis of exfoliated polymer/graphite nanocomposites by miniemulsion polymerization using functionalized graphene. *Macromolecules* 44, 8504–15. Copyright 2011 American Chemical Society.)

intercalated morphology (Figure 7.21b). In a very recent study, Park et al. also modified rGO with AMPS to facilitate intercalation of monomers (ST and BA) into the rGO nanogalleries to form poly(St-*co*-BA)/rGO composite latex nanoparticles.[88]

As recognized, an emulsion stabilized by solid particles rather than surfactant molecules is termed the "Pickering emulsion." Solid particles with special characteristics can self-assemble at immiscible liquid–liquid interface to lower its interfacial tension.[94] Particulate emulsifiers offer some advantages over conventional emulsifying agents including the ease of fabrication, low cost, and less pollution. It is noteworthy that GO can be employed as a stabilizer of Pickering emulsion due to its amphiphilic nature with the unique combination of hydrophilic edges and hydrophobic carbon basal plane.[95] This implies that GO sheets can be assembled and preferentially adsorbed at

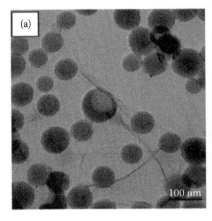
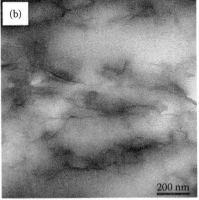

**FIGURE 7.21** TEM micrographs of poly(St-BA)/GO nanocomposite synthesized from 1 wt% GO relative to monomer: (a) latex particles and (b) a microtomed film cast from the same latex. (Reprinted with permission from Etmimi, H. M. and P. E. Mallon. New approach to the synthesis of exfoliated polymer/graphite nanocomposites by miniemulsion polymerization using functionalized graphene. *Macromolecules* 44, 8504–15. Copyright 2011 American Chemical Society.)

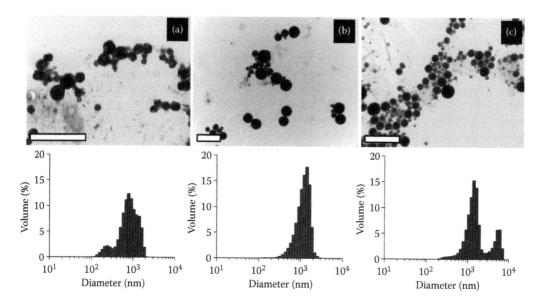

**FIGURE 7.22** TEM micrographs and the corresponding volume-based particle size distribution of polymer particles synthesized at (a) 19, (b) 47, and (c) 90% conversion in AIBN initiated miniemulsion polymerization of styrene at 70°C using GO as a sole surfactant at 7.5 wt% solids content. All scale bars: 2 μm. (Man, S. H., S. C. Thickett, P. B. Zetterlund et al.: Synthesis of polystyrene nanoparticles armoured with nanodimensional graphene oxide sheets by miniemulsion polymerization. *Journal of Polymer Science Part A: Polymer Chemistry*. 2013. 51. 47–58. Copyright Wiley-VCH Verlag GmbH & Co. KGaA. Reproduced with permission.)

the oil–water interface, occupying a substantial interfacial region to prevent droplet coalescence. Therefore, the use of GO for forming composite latex nanoparticles is an environmentally friendly process because it is an aqueous-based and soap-free emulsifying agent. In general, several factors affect the properties of Pickering emulsion including particle size, wettability, initial location, and interparticle interactions.[96] From the zeta potential measurements, GO is highly negatively charged when dispersed in water, resulting from the ionization of carboxylic groups and hydroxyl groups. Stable GO colloids are resulted from the electrostatic repulsion among adjacent GO.[97] Furthermore, the presence of salt is critical for the stability of emulsions. Yoon et al. reported that the addition of NaCl to GO dispersions decreases the zeta potential and interfacial tension, leading to the adsorption of GO to the oil/water interface.[98]

To the present, little information is available in the literature relating the synthesis and properties of GO/polymer composites with GO as a sole surfactant.[99–104] Song et al. synthesized GO/PS colloidal particles using GO as a stabilizer in which stabilized emulsion droplets act as nanoreactors for polymerization.[102] Man et al. produced GO/PS nanoparticles via Pickering emulsion polymerization at extremely low GO loadings (<0.1 wt% relative to styrene).[103] At larger GO loadings, a highly aggregated, fibrous morphology was formed. They then conducted in-depth studies on the synthesis of GO/PS colloidal particles.[104–106] Polymerization proceeded to high conversion with minor coagulation, with an average particle diameter of ~500 nm, but relatively broad particle size distributions (Figure 7.22).[104] In another research, they investigated the effect of pH suspension and ionic strength on the formation of stable GO/PS colloidal particles.[106] The presence of controlled NaCl concentrations led to enhanced colloidal stability and narrow particle size distribution.

## 7.6 CONCLUSIONS

The preparation of nanocomposites of different polymer matrices having one-dimensional CNTs and two-dimensional graphene sheets was briefly reviewed. The preparation techniques include solution mixing, melt compounding and *in situ* polymerization. The main difficulty of using CNTs for preparing nanocomposites is their poor dispersibility in the polymer matrix due to their high tendency for agglomeration. This results in poor mechanical and physical properties of the resulting composites. The dispersion of CNTs can be enhanced greatly by means of covalent and noncovalent functionalization. Graphene derivatives exhibit typical flat sheet morphology, thus entangled bundles are not a critical issue. However, restacking of graphene sheets often occurs especially after chemical reduction treatment and it can be prevented by functionalization with selected organic moieties. Functionalization of carbon nanofillers also enhances their interaction with the polymer matrix, facilitating the stress-transfer effect during tensile testing. This factor is crucial for making polymer nanocomposites for structural engineering applications. Graphene and its derivatives are more attractive fillers than CNTs due to their facile synthesis, low production cost, and the absence of purification for removing catalyst nanoparticles.

Generally, solution-mixing and *in situ* polymerization routes offer better dispersion of carbon nanofillers in the polymer matrix than the melt-blending process. For graphene and its derivatives, either polymer chains or monomers can intercalate into the interlayer galleries during wet processing and melt-mixing routes, forming intercalative nanocomposites. However, solution mixing process requires the use of solvents that are costly and environmental unfriendly. *In situ* polymerization involves the use of monomer, initiator, surfactant, and solvent (in certain cases). Miniemulsion polymerization is effective for synthesizing composite particles with controlled morphologies. In particular, Pickering emulsion polymerization using GO in aqueous medium is a green synthesis process because it excludes the use of organic surfactants. Finally, melt blending offers advantages of manufacturing polymer nanocomposites in large quantities employing commercially available facilities such as extruders and injection molders.

## REFERENCES

1. Tjong, S. C., S. L. Liu, and R. K. Y. Li. Mechanical properties of injection molded blends of polypropylene with thermotropic liquid crystalline polymer. *Journal of Materials Science* 31, 1996:479–84.
2. Meng, Y. Z. and S. C. Tjong. Rheology and morphology of compatibilized polyamide 6 blends containing liquid crystalline polyesters. *Polymer* 39, 1998:99–107.
3. Li, X. J., S. C. Tjong, Y. Z. Meng et al. Fabrication and properties of polypropylene carbonate)/calcium carbonate composites. *Journal of Polymer Science Part B: Polymer Physics* 41, 2003:1806–13.
4. Tjong, S. C. *Polymer Composites with Carbonaceous Nanofillers: Properties and Applications.* Weinhem: Wiley-VCH, 2012.
5. Nikos, T. and P. Maurizio. Functionalization of carbon nanotubes via 1,3-dipolar cycloadditions. *Journal of Materials Chemistry* 14, 2004:437–39.
6. Banerjee, S., T. Hemraj-Benny, and S. S. Wong. Covalent surface chemistry of single-walled carbon nanotubes. *Advanced Materials* 17, 2005:17–29.
7. Dimitrios, T., T. Nikos, B. Alberto et al. Chemistry of carbon nanotubes. *Chemical Reviews* 106, 2006:1105–36.
8. Nikolaos, K. and T. Nikos. Current progress on the chemical modification of carbon nanotubes. *Chemical Reviews* 110, 2010:5366–97.
9. Byrne, M. T. and K. Gun'ko Yurii. Recent advances in research of carbon nanotube–polymer composites. *Advanced Materials* 22, 2012:1672–88.
10. Islam, M. F., E. Rojas, D. M. Bergey et al. High weight fraction surfactant solubilization of single-wall carbon nanotubes in water. *Nano Letters* 3, 2003:269–73.
11. Rastogi, R., R. Kaushal, S. K. Tripathi et al. Comparative study of carbon nanotube dispersion using surfactants. *Journal of Colloid and Interface Science* 328, 2008:421–28.
12. O'Connell, M. J., P. Boul, L. M. Ericson et al. Reversible water-solubilization of single-walled carbon nanotubes by polymer wrapping. *Chemical Physics Letters* 342, 2001:265–71.
13. Curran, S. A., P. M. Ajayan, W. J. Blau et al. A composite from poly(m-phenylenevinylene-*co*-2,5-dioctoxy-p-phenylenevinylene) and carbon nanotubes: A novel material for molecular optoelectronics. *Advanced Materials* 10, 1998:1091–93.
14. Hummers, W. S. and R. E. Offeman. Preparation of graphitic oxide. *Journal of the American Chemical Society* 80, 1958:1339–41.

15. Lerf, A., H. Y. He, M. Forster et al. Structure of graphite oxide revisited. *Journal of Physical Chemistry B* 102, 1998:4477–82.

16. Park, S. and R. S. Ruoff. Chemical methods for the production of graphenes. *Nature Nanotechnology* 4, 2009:217–24.

17. Stankovich, S., D. A. Dikin, R. D. Piner et al. Synthesis of graphene-based nanosheets via chemical reduction of exfoliated graphite oxide. *Carbon* 45, 2007:1558–65.

18. McAllister, M. J., J. L. Li, D. H. Adamson et al. Single sheet functionalized graphene by oxidation and thermal expansion of graphite. *Chemistry of Materials* 19, 2007:4396–04.

19. He, L. and S. C. Tjong. Low percolation threshold of graphene/polymer composites prepared by solvothermal reduction of graphene oxide in the polymer solution. *Nanoscale Research Letters* 8, 2013:132.

20. Hernandez, Y., V. Nicolosi, M. Lotya et al. High-yield production of graphene by liquid-phase exfoliation of graphite. *Nature Nanotechnology* 3, 2008:563–68.

21. Paredes, J. I., S. Villar-Rodil, A. Martinez-Alonso et al. Graphene oxide dispersions in organic solvents. *Langmuir* 24, 2008:10560–64.

22. Stankovich, S., R. D. Piner, S. B. Nguyen et al. Synthesis and exfoliation of isocyanate-treated grapheme oxide nanoplatelets. *Carbon* 44, 2006:3342–47.

23. Park, S., D. A. Dikin, S. B. Nguyen et al. Grapheme oxide sheets chemically cross-linked by polyallylamine. *Journal of Physical Chemistry C* 113, 2009:15801–04.

24. Mahmoud, W. E. Morphology and physical properties of poly(ethylene oxide) loaded graphene nanocomposites prepared by two different techniques. *European Polymer Journal* 47, 2011:1534–40.

25. Ramanathan, T., A. A. Abdala, S. Stankovich et al. Functionalized graphene sheets for polymer nanocomposites. *Nature Nanotechnology* 3, 2008:327–31.

26. Stankovich, S., R. D. Piner, and X. Chen. Stable aqueous dispersions of graphitic nanoplatelets via the reduction of exfoliated graphite oxide in the presence of poly(sodium 4-styrenesulfonate). *Journal of Materials Chemistry* 16, 2006:155–58.

27. Choi, E. Y., T. H. Han, J. Y. Hong et al. Noncovalent functionalization of graphene with end-functional polymers. *Journal of Materials Chemistry* 20, 2010:1907–12.

28. Xu, Y., H. Bai, G. W. Lu et al. Flexible graphene films via the filtration of water-soluble noncovalent functionalized graphene sheets. *Journal of the American Chemical Society* 130, 2008:5856–57.

29. An, X. H., T. Simmons, R. Shah et al. Stable aqueous dispersions of noncovalently functionalized graphene from graphite and their multifunctional high-performance applications. *Nano Letters* 10, 2010:4295–01.

30. Qian, D., E. C. Dickey, R. Andrews et al. Load transfer and deformation mechanisms in carbon nanotube–polystyrene composites. *Applied Physics Letters* 76, 2000:2868–70.

31. He, L. and S. C. Tjong. Carbon nanotube/epoxy resin nanocomposite: Correlation between state of nanotube dispersion and Zener tunneling parameters. *Synthetic Metals* 162, 2012:2277–81.

32. He, L. and S. C. Tjong. Electrical conductivity of polyvinylidene fluoride nanocomposites with carbon nanotubes and nanofibers. *Journal of Nanoscience and Nanotechnology* 11, 2011:10668–72.

33. He, L. and S. C. Tjong. Effect of temperature on electrical conduction behavior of polyvinylidene fluoride nanocomposites with carbon nanotubes and nanofibers. *Current Nanoscience* 6, 2010:520–24.

34. Dottori, M., I. Armentano, E. Fortunat et al. Production and properties of solvent-cast poly(ε caprolactone) composites with carbon nanostructures. *Journal of Applied Polymer Science* 119, 2011:3544–52.

35. Fernández-d'Arlas, B., U. Khan, L. Rueda et al. Study of the mechanical, electrical and morphological properties of PU/MWCNT composites obtained by two different processing routes. *Composites Science and Technology* 72, 2012:235–42.

36. Tjong, S. C. Structural and mechanical properties of polymer nanocomposites. *Materials Science and Engineering R: Reports* 53, 2006:73–197.

37. Matsuo, Y., K. Tahara, and Y. Sugie. Synthesis of poly(ethylene oxide)-intercalated graphite oxide. *Carbon* 34, 1996:672–74.

38. Matsuo, Y., K. Tahara, and Y. Sugie. Structure and thermal properties of poly(ethylene oxide)-intercalated graphite oxide. *Carbon* 35, 1997:113–20.

39. Kyotani, T., H. Moriyama, and A. Tomita. High temperature treatment of polyfurfuryl alcohol/graphite oxide intercalation compound. *Carbon* 35, 1997:1185–87.

40. Salavagione, H. J., G. Martinez, and M. A. Gomez. Synthesis of poly(vinyl alcohol)/reduced graphite oxide nanocomposites with improved thermal and electrical properties. *Journal of Materials Chemistry* 19, 2009:5027–32.

41. Zhao, X., Q. Zhang, and D. Chen. Enhanced mechanical properties of graphene-based poly(vinyl alcohol) composites. *Macromolecules* 43, 2010:2357–63.
42. Wang, J., X. Wang, and C. Xu. Preparation of graphene/poly(vinyl alcohol) nanocomposites with enhanced mechanical properties and water resistance. *Polymer International* 60, 2011:816–22.
43. Tantis, I., G. C. Psarras, and D. Dasis. Functionalized graphene–poly(vinyl alcohol) nanocomposites: Physical and dielectric properties. *Express Polymer Letters* 6, 2012:283–92.
44. Ramanathan, T., A. A. Abdala, S. Stankovich et al. Functionalized graphene sheets for polymer nanocomposites. *Nature Nanotechnology* 3, 2008:327–31.
45. He, L. and S. C. Tjong. A graphene oxide–polyvinylidene fluoride mixture as a precursor for fabricating thermally reduced graphene oxide–polyvinylidene composites. *RSC Advances* 3, 2013:22981–87.
46. Stankovich, S., D. A. Dikin, G. H. Dommett et al. Graphene-based composite materials. *Nature* 442, 2006:282–86.
47. Kim, H., Y. Miura, and C. W. Macosko. Graphene/polyurethane nanocomposites for improved gas barrier and electrical conductivity. *Chemistry of Materials* 22, 2010:3441–50.
48. Kuila, T., S. Bose, C. E. Hong et al. Preparation of functionalized graphene/linear low density polyethylene composites by a solution mixing method. *Carbon* 49, 2011:1033–51.
49. Khalid, N., K. Umar, U. H. Noaman et al. Observation of mechanical percolation in functionalized graphene oxide/elastomer composites. *Carbon* 50, 2012:4489–94.
50. Tjong, S.C. and S. P. Bao. Fracture toughness of high density polyethylene/SEBS-g-MA/montmorillonite nanocomposites. *Composites Science and Technology* 67, 2007:314–23.
51. Robert, S., B. Krause, M. T. Müller et al. The influence of matrix viscosity on MWCNT dispersion and electrical properties in different thermoplastic nanocomposites. *Polymer* 53, 2012:495–04.
52. McNally, T., P. Pötschke, P. Halley et al. Polyethylene multiwalled carbon nanotube composites. *Polymer* 46, 2005:8222–32.
53. Zhang, W. D., L. Shen, I. Y. Phang et al. Carbon nanotubes reinforced nylon-6 composite prepared by simple melt-compounding. *Macromolecules* 37, 2004:256–59.
54. Cohen, E., A. Ophir, S. Kenig et al. Pyridine-modified polymer as a non-covalent compatibilizer for multiwalled CNT/poly[ethylene-*co*-(methacrylic acid)] composites fabricated by direct melt mixing. *Macromolecular Materials and Engineering* 298, 2013:419–28.
55. Liang, G. D., S. P. Bao, and S. C. Tjong. Microstructure and properties of polypropylene composites filler with silver and carbon nanotube nanoparticles prepared by melt-compounding. *Materials Science and Engineering B* 142, 2007:55–61.
56. Bao, S. P. and S. C. Tjong. Mechanical behavior of polypropylene/carbon nanotube nanocomposites: The effect of loading rate and temperature. *Materials Science and Engineering A* 485, 2008:508–16.
57. Pan, Y., L. Li, S. H. Chan et al. Correlation between dispersion state and electrical conductivity of MWCNTs/PP composites prepared by melt blending. *Composites Part A* 41, 2010:419–26.
58. McClory, C., P. Pötschke, T. Mcnally et al. Influence of screw speed on electrical and rheological percolation of melt-mixed high-impact polystyrene/MWCNT nanocomposites. *Macromolecular Materials and Engineering* 296, 2011:59–69.
59. Kim, H. and C. W. Macosko. Morphology and properties of polyester/exfoliated graphite nanocomposites. *Macromolecules* 41, 2008:3317–27.
60. Kim, H. and C. W. Macosko. Processing-property relationships of polycarbonate–graphene composites. *Polymer* 50, 2009:3797–09.
61. Kim, H., S. Kobayashi, M. A. Abdurrahim et al. Graphene/polyethylene nanocomposites: Effect of polyethylene functionalization and blending methods. *Polymer* 52, 2011: 1837–46.
62. Kim, I. H. and Y. G. Jeong. Polylactide/exfoliated graphite nanocomposites with enhanced thermal stability, mechanical modulus, and electrical conductivity. *Journal of Polymer Science Part B: Polymer Physics* 48, 2010: 850–58.
63. Zhang, H. B., W. G. Zheng, Q. Yan et al. Electrically conductive polyethylene terephthalate/graphene nanocomposites prepared by melt compounding. *Polymer* 51, 2010:1191–96.
64. Istrate, O. M., K. R. Paton, U. Khan et al. Reinforcement in melt-processed polymer–graphene composites at extremely low graphene loading level. *Carbon* 78, 2014:243–49.
65. Vasileiou, A. A., M. Kontopoulou, and A. Docoslis. A noncovalent compatibilization approach to improve the filler dispersion and properties of polyethylene/graphene composites. *ACS Applied Materials and Interfaces* 6, 2014:1916–25.
66. Meng, Y. Z., S. C. Tjong, A. S. Hay et al. Synthesis and proton conductivities of phosphonic acid containing poly-(arylene ether)s. *Journal of Polymer Science Part A: Polymer Chemistry* 39, 2001: 3218–26.

67. Landfester, K. Miniemulsion polymerization and the structure of polymer and hybrid nanoparticles. *Angewandte Chemie International Edition* 48, 2009:4488–07.
68. Jia, Z., Z. Wang, C. Xu et al. Study on poly(methyl methacrylate)/carbon nanotube composites. *Materials Science and Engineering A* 271, 1999:395–400.
69. Park, S. J., M. S. Cho, S. T. Lim et al. Synthesis and dispersion characteristics of multiwalled carbon nanotube composites with poly(methyl methacrylate) prepared by *in situ* bulk polymerization. *Macromolecular Rapid Communications* 24, 2003:1070–73.
70. Meng, Q. J., X. X. Zhang, S. H. Bai et al. Preparation and characterization of poly(methyl metahcrylate)-functionalized carboxyl multiwalled carbon nanotubes. *Chinese Journal of Chemical Physics* 20, 2007:660–64.
71. Shrivastava, N. K., P. Kar, S. Maiti et al. A facile route to develop electrical conductivity with minimum possible multi-wall carbon nanotube (MWCNT) loading in poly(methyl methacrylate)/MWCNT nano-composites. *Polymer International* 61, 2012:1683–92.
72. Choi, J. J., K. Zhang, and J. Y. Lim. Multi-walled carbon nanotube/polystyrene composites prepared by *in situ* bulk sonochemical polymerization. *Journal of Nanoscience and Nanotechnology* 7, 2007:3400–03.
73. Qu, L., L. M. Veca, Y. Lin et al. Soluble nylon-functionalized carbon nanotubes from anionic ring-opening polymerization from nanotube surface. *Macromolecules* 38, 2005:10328–31.
74. Yang, M., Y. Gao, H. Li et al. Functionalization of multiwalled carbon nanotubes with polyamide 6 by anionic ring-opening polymerization. *Carbon* 45, 2007:2327–33.
75. Yan, D. and G. Yang. Synthesis and properties of homogeneously dispersed polyamide 6/MWNTs nanocomposites via simultaneous *in situ* anionic ring-opening polymerization and compatibilization. *Journal of Applied Polymer Science* 112, 2009:3620–26.
76. Barraza, H. J., F. Pompeo, E. A. O'Rear et al. SWNT-filled thermoplastic and elastomeric composites prepared by miniemulsion polymerization. *Nano Letters* 8, 2002:797–02.
77. Xia, H., Q. Wang, and G. Qiu. Polymer-encapsulated carbon nanotubes prepared through ultrasonically initiated *in situ* emulsion polymerization. *Chemistry of Materials* 15, 2003:3879–86.
78. Xia, H., G. Qiu, and Q. Wang. Polymer/carbon nanotube composite prepared through ultrasonically assisted *in situ* emulsion polymerization. *Journal of Applied Polymer Science* 100, 2006:3123–30.
79. Aldosari, M. A., A. A. Othman, and E. H. Alsharaeh. Synthesis and characterization of the *in situ* bulk polymerization of PMMA containing graphene sheets using microwave irradiation. *Macromolecules* 18, 2013:3152–67.
80. Yuan, X. Y., L. L. Zou, C. C. Liao et al. Improved properties of chemically modified graphene/poly(methyl methacrylate) nanocomposites via a facile *in situ* bulk polymerization. *Express Polymer Letters* 6, 2012:847–58.
81. Wang, J., Z. Shi, Y. Ge et al. Solvent exfoliated graphene for reinforcement of PMMA composites prepared by *in situ* polymerization. *Materials Chemistry and Physics* 136, 2012:43–50.
82. Jang, J. Y., M. S. Kim, H. M. Jeong et al. Graphite oxide/poly(methyl methacrylate) nanocomposites prepared by a novel method utilizing macroazoinitiator. *Composites Science and Technology* 69, 2009:186–91.
83. Pokharel, P. and D. S. Lee. High performance polyurethane nanocomposite films prepared from a masterbatch of graphene oxide in polyether polyol. *Chemical Engineering Journal* 253, 2014:356–65.
84. Liu, P., W. Zhong, X. Wu et al. Facile synergetic dispersion approach for magnetic $Fe_3O_4$@graphene oxide/polystyrene tri-component nanocomposite via radical bulk polymerization. *Chemical Engineering Journal* 219, 2013:10–18.
85. Xu, Z. and C. Gao. *In situ* polymerization approach to graphene-reinforced nylon-6 composites. *Macromolecules* 43, 2010:6716–23.
86. Zhang, X., X. Fan, H. Li et al. Facile preparation routes for graphene oxide reinforced polyamide 6 composites via anionic ring-opening polymerization. *Journal of Materials Chemistry* 22, 2012:24081–91.
87. Zheng, D., G. Tang, H. Zhang et al. *In situ* thermal reduction of graphene oxide for high electrical conductivity and low percolation threshold in polyamide 6 nanocomposites. *Composites Science and Technology* 72, 2012:284–89.
88. Park, N., J. Lee, H. Min et al. Preparation of highly conductive reduced graphite oxide/poly(styrene-co-butyl acrylate) composites via miniemulsion polymerization. *Polymer* 55, 2014:5088–94.
89. Kuila, T., S. Bose, P. Khanra et al. Characterization and properties of *in situ* emulsion polymerized poly(methylmethacrylate)/graphene nanocomposites. *Composites Part A* 42, 2011:1856–61.
90. Hassan, M., K. R. Reddy, E. Haque et al. High-yield aqueous phase exfoliation of graphene for facile nanocomposite synthesis via emulsion polymerization. *Journal of Colloid and Interface Science* 410, 2013:43–51.

91. Etmimi, H. M. and P. E. Mallon. *In situ* exfoliation of graphite oxide nanosheets in polymer nanocomposites using miniemulsion polymerization. *Polymer* 54, 2013:6078–88.

92. Etmimi, H. M., P. E. Mallon, and R. D. Sanderson. Polymer/graphite nanocomposites: Effect of reducing the functional groups of graphite oxide on water barrier properties. *European Polymer Journal* 49, 2013:3460–70.

93. Etmimi, H. M. and P. E. Mallon. New approach to the synthesis of exfoliated polymer/graphite nanocomposites by miniemulsion polymerization using functionalized graphene. *Macromolecules* 44, 2011:8504–15.

94. Hu, L., M. Chen, X. Fang et al. Oil–water interfacial self-assembly: A novel strategy for nanofilm and nanodevice fabrication. *Chemical Society Reviews* 41, 2012:1350–62.

95. Kim, J., L. J. Cote, F. Kim et al. Graphene oxide sheets at interfaces. *Journal of the American Chemical Society* 132, 2010:8180–86.

96. Aveyard, R., B. P. Binks, J. H. Clint et al. Emulsions stabilised solely by colloidal particles. *Advances in Colloid and Interface Science* 100–102, 2003:503–46.

97. Cote, L. J., F. Kim, and J. Huang, Langmuir–Blodgett assembly of graphite oxide single layers. *Journal of the American Chemical Society* 131, 2009:1043–49.

98. Yoon, K. Y., S. J. An, Y. Chen et al. Graphene oxide nanoplatelet dispersions in concentrated NaCl and stabilization of oil/water emulsions. *Journal of Colloid and Interface Science* 403, 2013:1–6.

99. Huang, Y., X. Wang, X. Jin et al. Study on the PMMA/GO nanocomposites with good thermal stability prepared by *in situ* Pickering emulsion polymerization. *Journal of Thermal Analysis and Calorimetry* 117, 2014:755–63.

100. Yin, G., Z. Zheng, H. Wang et al. Preparation of graphene oxide coated polystyrene microspheres by Pickering emulsion polymerization. *Journal of Colloid and Interface Science* 394, 2013:192–8.

101. Sun, J. and H. Bi. Pickering emulsion fabrication and enhanced supercapacity of graphene oxide-covered polyaniline nanoparticles. *Materials Letters* 81, 2012:48–51.

102. Song, X., Y. Yang, J. Liu et al. PS colloidal particles stabilized by graphene oxide. *Langmuir* 27, 2011:1186–91.

103. Thickett, S. and P. B. Zetterlund. Preparation of composite materials by using graphene oxide as a surfactant in ab initio emulsion polymerization systems. *ACS Macro Letters* 2, 2013:630–4.

104. Man, S. H., S. C. Thickett, P. B. Zetterlund et al. Synthesis of polystyrene nanoparticles armoured with nanodimensional graphene oxide sheets by miniemulsion polymerization. *Journal of Polymer Science Part A: Polymer Chemistry* 51, 2013:47–58.

105. Man, S. H., Y. N. Y. Mohd, and M. R. Whittaker. Influence of monomer type on miniemulsion polymerizations systems stabilized by graphene oxide as sole surfactant. *Journal of Polymer Science Part A: Polymer Chemistry* 51, 2013:5153–62.

106. Man, S. H., D. Ly, and M. B. Whittaker. Nano-sized graphene oxide as sole surfactant in miniemulsion polymerization for nanocomposite synthesis: Effect of pH and ionic strength. *Polymer* 55, 2014:3490–97.

# 8 Thermal Properties of Novel Polymer Nanocomposites

*Jerzy J. Chruściel*

## CONTENTS

## 8.1 INTRODUCTION

Thermal behavior of polymers and polymer composites and nanocomposites (PNCs) is crucial in many demanding practical applications. In order to develop durable industrial products, it is necessary to study the thermal stability of these nanocomposites.

## 8.2 HEAT RESISTANCE AND THERMAL STABILITY OF POLYMERS

Two fundamental features determine thermal characteristics of polymers: *heat resistance* and *thermal stability*. The heat resistance defines the ability of the polymer to undergo softening with increasing temperature, while the thermal stability means an unchanged chemical structure at high temperatures.

The *heat resistance* is expressed by the temperature value, in which under conditions of defined constant load a deformation of a sample does not exceed a certain fixed boundary. There are many normalized methods of the determination of the heat resistance. These methods differ in shape and dimensions of the samples, a kind of deformation, a rate of temperature increase, etc. Most popular of them are the *Vicat* and *Martens* methods. The simple Vicat method, based on determination of a softening point $T_m$, is applied only for thermoplastic polymers—it is measured at the temperature at which a needle with a cross-sectional area of 1 mm² dips into the appropriate burden of material to a depth of 1 mm. The speed of temperature rise of the measurement device is 50°C/h, and the thickness of the sample is at least 3 mm.[1] The Martens point test (according to DIN 53458) is used to measure the heat resistance at which a bar, attached to the specimen, under constant bending moment, deflects a specified amount of flexure (6 mm). This method is applied for studies of thermo- and chemocurable polymeric materials, and even some thermoplastic materials of a higher modulus.[2]

Finally, the thermal stability of polymers can be detected by a number of other methods, such as DSC, TGA, DTA, DMA, TMA, etc. This will be discussed in more detail in the coming sections.

### 8.2.1 FACTORS AFFECTING THERMAL PROPERTIES OF POLYMERIC MATERIALS

The *thermal stability* of polymers and polymeric materials (plastics, composites, and nanocomposites) depends on many factors. First of all, it depends on a chemical structure of polymer chains and properties and a content of additives used for their fabrication. A presence of aromatic and polyaromatic rings, imide or benzoxazine groups strongly improves melting temperature, glass temperature, and thermal properties of polymers.[3] Polymers having inorganic structure of their chains, and especially siloxane ($\equiv$Si–O–Si$\equiv$),[4] silsesquioxane ($RSiO_{1.5}$), silazane ($\equiv$Si–N$=$), phosphazene ($\equiv$P$=$N–), or borazine ($=$B–N$=$) units, and many other so-called "preceramic polymers" exhibit high thermal stability.[4,5] An application of inorganic fillers and other additives as ingredients of plastic and polymer composites affects substantial improvement of their thermal properties.

### 8.2.2 DETERMINATION OF THE THERMAL BEHAVIOR AND THERMAL STABILITY OF POLYMER COMPOSITES

Very useful tools of investigation of properties of polymers and PNCs are thermal analysis methods (TA). They enable to gain further insight into their structure, especially in the case of montmorillonite nanocomposites. Differential scanning calorimetry (DSC), thermogravimetric analysis (TGA), the integral procedural decomposition temperature (IPDT), dynamic mechanical thermal analysis (DMTA), and thermal mechanical analysis (TMA) are very useful techniques for the characterization of nanocomposite materials.

During heating of polymers occur destruction and crosslinking processes, which have decisive influence on their thermostability—it can be studied by the TGA and DSC. Quantitative characteristics of the thermostability of polymers and polymeric materials can be determined by the TGA method and are given by the different temperatures as thermal stability factors:

$T_o$—at which weight loss begins [initial decomposing temperature (IDT) or $T_{onset}$]
$T_5$—at which weight loss equals 5 wt%
$T_{10}$—at which weight loss equals 10 wt%

$T_{50}$—at which weight loss reaches 50 wt%

$T_{max}$—temperature at the maximum rate of heat loss

$T_{peak}$—the temperature of maximum weight loss rate

IPDT—integral procedural decomposition temperature

and the *char yield* at a certain temperature.

The IPDT, proposed by Doyle,[6] can be calculated from equations given in the literature, based on initial and final temperatures of the TGA measurement and areas from the TGA thermogram, as coefficients.[6–8]

For some polymers [poly(methyl methacrylate) (PMMA), polyethylene, polypropylene, polystyrene] loss of the thermal stability is observed at higher temperature than a decline of the heat resistance—in such cases an upper limit temperature of their use is determined as the heat resistance, but not as the thermostability.[1]

The TGA is very useful for the determination of the decomposition behavior. Impurities may significantly decrease thermal stability of polymers, which could result as extra peaks on thermograms. Melting and boiling points of plasticizers may be observed as well. The TGA results slightly increase with an increasing heating rate (most often it is 10°C/min).[9,10]

The DSC is a thermal analytical technique that allows measuring the difference in the amount of heat necessary to increase the temperature of a sample and reference as a function of temperature. During the experiment nearly the same temperature is maintained for the studied sample and reference. Generally, the linear temperature increase as a function of time is applied. The reference sample should have a well-defined heat capacity over the range of temperatures.[11]

The DSC is widely used for examining polymeric materials to determine their thermal transitions and measurements of characteristic properties of a sample: most often glass transition temperatures ($T_g$) and a crystallization phenomena. Melting points and glass transition temperatures for most polymers are available from standard analytical procedure. For instance, polymer degradation can be observed as the result of the lowering of the expected melting point, $T_m$, which depends on the molecular weight of the polymer and thermal history. The content of a crystalline fraction of polymer can be determined from the crystallization-melting peaks of the DSC graph.[10–13]

The DSC method can also be used to study oxidation, other chemical reactions and the thermal degradation of polymers using a program oxidative onset temperature/time (OOT). The DSC apparatus allows also carrying out differential thermal analysis (DTA), which measures the difference in temperature between the sample and the reference.

Dynamic mechanical thermal analysis (DMTA, dynamic mechanical analysis or dynamic mechanical spectroscopy) is a technique used for characterization of the viscoelastic behavior of polymers and polymeric materials. A sinusoidal stress is applied and the strain in the material is measured, thus the complex modulus is determined. The temperature of the sample or the frequency of the stress is often changed, leading to variations in the complex modulus. This method allows identifying transitions corresponding to other molecular motions and finding the glass transition temperature of the material.

For many plastics can be determined a heat deflection temperature or heat distortion temperature (HDT, HDTUL, or DTUL), at which the sample material is deformed by a desired deflection under a specified load at a speed of temperature increase of 2°C/min. Loading of samples for testing polyolefins is 4.6 kg/cm$^2$, and for the other materials is equal to 18.5 kg/cm$^2$. This property of polymeric materials is applied in engineering and manufacture of products made from thermoplastic polymers.[14]

In the case of plastics used as insulating materials, the maximum temperature of their use is determined as so-called an insulation class. According to the recommendations of the International Electrotechnical Commission (IEC) and according to the PN-63/E-02050 (Insulating materials. Classification), electroinsulating plastic materials are divided into classes (A, B, C, E, F, and H) depending on the maximum continuous operation temperature of electrical and electronic machinery. Usually it is assumed that the maximum temperature of the systematic use of the given class of materials does not exceed the permissible temperature for continuous operation of the next grade materials. For example, composites based on silicone resins belong to the insulating class H, while glass-phenolic laminates are electrically insulating material of class B, with a maximum temperature of use in continuous operation, of 130°C and their maximum operating temperature range is 130–155°C of casual work.

## 8.3 APPLICATIONS OF DIFFERENT POLYMERS AND FILLERS IN POLYMER NANOCOMPOSITES

Polymer nanocomposites (PNCs) are formulated from polymer matrices (thermoplastics, thermosets, or elastomers) which are reinforced with different amounts of nanoparticles (NPs) of inorganic fillers (preferably characterized by high aspect ratios), such as silica, layered silicates, alumina, titania, iron oxide ($Fe_2O_3$), carbon nanotubes, graphene, metal particles (e.g., Ag and Cu), and others. Unexpected synergistic effects caused by two or more ingredients of PNCs were often observed.

### 8.3.1 FILLERS USED IN POLYMER NANOCOMPOSITES

Particles of NPs usually have diameter less than 100 nm.[15] Nanometer-sized particles are prepared from different inorganic or organic components and improve properties of composite materials.[16] For preparation of polymer–inorganic particle nanocomposites (NCs) different nanofillers are used[17]

- Metal oxides ($Al_2O_3$, CaO, ZnO, $Fe_2O_3$, $TiO_2$, etc.)
- Nonmetal oxides ($SiO_2$)
- Metal and nonmetal oxides (present in layered silicates, e.g., in montmorillonite, MMT)
- Metal hydroxides [$Al(OH)_3$, $Ca(OH)_2$, $Mg(OH)_2$, $Zn(OH)_2$, etc.]
- Metal carbonates [$CaCO_3$, $MgCO_3$, $CaCO_3 \cdot MgCO_3$, $BaCO_3$, etc.]
- Metal titanates (e.g., $BaTiO_3$)
- Metals (Al, Fe, Zn, Cu, Ag, Au, etc.)
- Other compounds (e.g., SiC, $Si_3N_4$, BN, AlN, TiN, carbon black, graphite, and fullerenes)
- Carbon nanotubes (CNTs), carbon nanofibers, graphene, POSS derivatives, etc.

The selection of NPs depends on the expected thermal, mechanical, and electrical properties of the NCs. For instance, silicon carbide (SiC) and diamond NPs are applied because of their high hardness, corrosion resistance, and strength.

Carbon nanofibers (CNFs) exhibit physical properties between conventional carbon fibers (5–10 mm) and carbon nanotubes (1–10 nm). CNFs are not concentric cylinders and their length is in a range from about 100 μ to several centimeters, with the larger diameter than CNTs (of the order of 100–200 nm) and an average aspect ratio greater than 100. They have a large surface area and are usually functionalized on a surface.[18] Functionalization of CNTs improves their dispersion and compatibility in polymer matrices.[19] CNTs are most often combined with thermoplastic polymers such as amorphous polymethyl methacrylate (PMMA)[20–23] or semicrystalline PP.[24] Thermal properties of thermosetting epoxy–nanotube composites were very often studied.[25]

### 8.3.2 POLYMER MATRICES APPLIED IN NANOCOMPOSITES

The TGA of numerous PNCs showed that many polymers, for instance PMMA,[26] poly(dimethylsiloxane) (PDMS),[27] polyamide (PA),[28,29] and polypropylene systems,[30] containing different nanofillers exhibited improved thermal properties.

Polyamide 6-clay thermoplastic nanocomposites, containing only 4.2 wt% clay, elaborated by Toyota researchers in the early 1990s,[31–33] showed significantly improved mechanical and thermal properties. The increase of heat distortion temperature (HDT) by 80°C compared to the pristine polymer was also observed.[32,34] These NCs found many industrial applications.[35] In recent years, various polymer–organoclay NCs were prepared from epoxides, polyurethanes, vinyl ester, etc.[36–38] However, the different polymer–MMT NCs: with polyolefins, styrene containing polymers, PMMA, polyamides, polyimides (PI), epoxy resins, polyesters, poly(vinyl chloride) (PVC), polyurethanes (PU), ethylene-propylene-diene terpolymer (EPDM), poly(vinyl alcohol) (PVA), and polylactide (PLA) undergo oxidation and pyrolysis processes during heating, with a formation of volatile products—the incorporation of layered silicates into polymer matrix affects an increase in thermal stability of polymeric matrices.[39–43]

## 8.4 THERMAL PROPERTIES OF POLYMER NANOCOMPOSITES

### 8.4.1 THERMAL PROPERTIES OF POLYMER–LAYERED SILICATE NANOCOMPOSITES

The different experimental methods were used for preparation of polymer-layered silicate (PLS) NCs and their characterization.[44]

The polymer NCs of cyanate ester, epoxy, phenolic, polyamide 11, etc. were very useful for high-temperature applications. These composites were applied as flame retardant coating materials, high strength rocket insulation and ablative materials, etc. Ablative coatings can protect aerospace launching systems against solid rocket exhaust plumes (3600°C) at very high speed. The nanoclay addition also enabled the reduced flammability of coating systems.[45] The NCs prepared from polyamides, epoxides, polystyrene, or vinyl ester exhibited reduced flammability in comparison with neat polymers.[35] Transparent NCs with chemically modified clay, which are lightweight and durable, were developed for a variety of aerospace applications, where some extreme features are required at temperature range −196 to 125°C: higher toughness, dimensional stability (i.e., resistance to microcracking), etc.[46]

#### 8.4.1.1 Nanocomposites of Silicates with Polyolefins and Vinyl Polymers

The thermal stability of polyethylene (PE) NCs was improved with 1.5 and 5 wt% content of organically modified montmorillonite (OMMT) and it was higher than for pure PE. An amphiphilic compound, N-heptaquinolinum, containing both hydrophobic (alkyl and aromatic) and hydrophilic groups was used as the surfactant. The most significant increase $T_{10}$ with 5% filler content was ~36°C in comparison with the neat polymer matrix. It was unexpected that the increase in degradation temperature was not dependent on the increasing amount of char.[47] The effects of treatment with dicumyl peroxide (DCP) and vinyltriethoxysilane (VTES) and nanoclay content were investigated for low-density polyethylene (LDPE)/clay NCs. LDPE was treated with 0.1 phr of DCP and with 1 phr or 3 phr VTES, respectively (System A), and with 0.2 phr of DCP with the same amounts of VTES (System B), and then mixed with different contents (1, 3, and 5 wt%) of modified clay (Cloisite 15A). All the VTES-treated LDPE/clay NCs showed an increase in interlayer spacing, which indicated that the polymer chains were intercalated between the clay layers. TEM micrographs of System B showed some evidence of exfoliated clay layers, indicating a mixed morphology of this system. The LDPE–clay NCs had better thermostability than LDPE. Crosslinking and addition of clay increased the thermal resistance of LDPE in nitrogen. In oxygen, this improvement was much more evident.[48]

At the contents of 5 and 8 wt% the HNTs delayed thermal degradation onset and sped up the thermal degradation of the polypropylene/halloysite nanotubes (PP/HNTs) NCs, prepared by different methods. The results of TGA and DSC analyses showed that, at a low content, the direct stabilizing effect of HNTs on PP caused the significant increase of thermostability of the PP/HNTs NCs. The PP NCs with HNTs were obtained by: (a) water-assisted injection molding (WAIM) method and (b) by compression molding (CM). Both kinds of WAIM and CM PP/HNTs NCs showed increased $T_5$ and $T_{10}$ parameters. The characteristic weight loss temperatures from the TGA curves

**TABLE 8.1**

**TGA Data in Nitrogen of WAIM and CM Processed PP/HNTs NCs**

| Sample | Temperature at 5% Weight Loss (°C) | Temperature at 10% Weight Loss (°C) | Temperature at Maximum Weight Loss (°C) |
|---|---|---|---|
| Pure PP | 389 | 410 | 459 |
| PP/HNT 98/2 | 411 | 429 | 462 |
| PP/HNT 98/2-CM | 391 | 417 | 463 |
| PP/HNT 95/5 | 408 | 419 | 451 |
| PP/HNT 92/8 | 410 | 422 | 453 |

*Source:*  Reprinted from *Polym. Degrad. Stab.*, 98, Wang, B., Huang, H.-X., Effects of halloysite nanotube orientation on crystallization and thermal stability of polypropylene nanocomposites, 1601–1608, Copyright 2013, with permission from Elsevier.

are summarized in Table 8.1. WAIM PP/HNTs 98/2 nanocomposite showed the highest $T_5$ and $T_{10}$ (411°C and 429°C, respectively), which were shifted 21°C and 19°C toward higher temperatures in comparison to neat PP. However, for the PP/HNTs 98/2 NC, prepared by CM method, the $T_5$ and $T_{10}$ were only slightly increased (2°C and 7°C, respectively) as compared to pure PP. The $T_{max}$s of both WAIM and CM 98/2 NCs were slightly higher than that of neat PP, whereas the $T_{max}$s of WAIM 95/5 and 92/8 NC parts were lower than that of neat PP.[49]

The HNTs hindered the thermal degradation in the initial stage, but at the contents of 5 and 8 wt%, the HNTs sped up the degradation in the final step. A catalytic effect of hydroxyl Al–OH groups of HNTs on PP pyrolysis is known from the literature.[50] HNTs may also slow down the liberation of the volatile products in the degradation process, due to the barrier and entrapment effects of the HNTs.[51–53] PP thermally degraded into volatile products above 250°C, in nitrogen atmosphere, probably as a result of the random thermal scissions of C–C chain bonds.[30] The volatile products may be first entrapped into the HNTs, which could effectively delay mass transport and increase thermostability. The effect of the HNTs on the thermal degradation of PP at the contents of 5 and 8 wt% in the final step was explained: (1) HNTs underwent aggregation in the PP matrix at the high contents, which could decrease the thermal stability of the NCs[51,52] and (2) the condensation of the hydroxyl Al–OH groups of the HNTs occurred in the main volatilization stage,[54] which had a catalytic effect on the PP degradation.

NCs of cyclic olefin copolymer (COC) with various contents of organoclay layered silicate NPs were prepared via melt blending. TGA results were strongly dependent on heating rate (2, 5, or 10°C/min). The incorporation of the nanoclay generally increased the thermostability of the COC matrix, particularly at lower contents of the nanoclay, which faded as the nanoclay content was increased. The thermal stability enhancement was a result of the nanoclay accumulation on matrix phase surface, which acted as a physical barrier against pyrolysis. The $T_{onset}$ was in a range of 369–374°C for 2–10 wt% nanoclay content. Under thermooxidative conditions, the penetration of oxygen molecules was hindered. The residual char was lower than the clay content in some cases, probably due to loss of the organic modifier of nanoclay under the imposed processing conditions.[55]

The polymers used in NCs were often grafted using polymerization reactions, and nanoclays were usually surface modified in order to improve compatibility of both ingredients. The poly(vinyl alcohol) (PVA) and a nanoclay filler based on natural bentonite (NBt) were both surface modified by (methacryloxy)methyl(trimethoxy)silane (MAOM-TMOS) and 3-glycidoxypropyl(trimethoxy)-silane (GOP-TMOS). A silylation of nanobentonit (NBt) improved compatibility and enhanced thermostability, which facilitated further processing. Modification of PVA and NBt with these silanes led to a substantial increase of thermal stability of both the modified nanoclay and the modified

PVA. Thermostability was necessary for silane modification and compatibility between the NC ingredients during processing of polymer–clay NCs.[56]

The silylation of PVA with MAOM-TMOS gave more efficient modification of the polymer (the $T_{onset} = 351°C$) than the GOP-TMOS treatment ($T_{onset} = 341°C$) or no treatment (pristine PVA, $T_{onset} = 301°C$). Only Si–O–C bridges were formed between silane and polymer with MAOM-TMOS as the modifier. In the case of GOP-TMOS, less stable C–O–C bridges (716 kJ/mol) than a Si–O–C bridge (802 kJ/mol) were also formed. The $T_{onset}$ of pristine NBt (207°C) was increased to 291°C (for NBt-MAOM-TMOS) and 243°C (for NBt-GOP-TMOS). The MAOM-TMOS can only be attached to the nanoclay surface by Si–O–Si linkages, whereas GOP-TMOS can also form Si–O–C bonds, and Si–O–Si linkage (888 kJ/mol) was more stable than the Si–O–C system.[56]

The thermal stability of nitrile rubber (NBR)–nanoclay composites (3:1, by weight) was improved on addition of nanoclay. The IDT of neat rubber started around 401°C. An addition of nanoclay did not change IDT. However, the $T_{max}$ and the char content of the NCs at 500°C increased with increased nanoclay loading. The enhanced thermostability of NBR–clay NCs could result from the restricted thermal motion of the polymer chains in the silicate layers of the nanoclay.[57] Thermal decomposition of the poly(styrene-co-butyl acrylate)-clay NCs was hindered in the presence of nanoclay, which was exfoliated with (dodecyl)trimethylammonium bromide (DDTMAB) and (vinylbenzyl)trimethylammonium chloride (VBTMAC) surfactants. A small increase in the thermal properties of all the NCs compared with the pure copolymer was observed (IDT only ~200°C). The TGA results were different than results that were obtained for copolymerization in a miniemulsion system.[58] Thermal stability of NCs prepared from copolymers of styrene and butyl acrylate increased with increasing clay loading and it was suggested that the improved thermostability could be caused by clay exfoliation and distribution in these two systems.[59]

### 8.4.1.2 Nanocomposites of Silicates with Epoxy Resins

The thermal properties of the layered silicate nanocomposites were studied very often,[36,60] for example, of epoxy–MMT NCs,[61] and were dependent on the chemical structure of the polymers, the content and type of clays, a dispersion degree of clay layers, and a nature of the purge gas. The formation of a char could hinder the out-diffusion of the volatile decomposition products and the decrease in permeability, leading to the improved thermal stability of the polymer–clay NCs.[38,42] It was often observed in exfoliated NCs that at a high temperature both intercalated and exfoliated MMT layers inhibited the diffusion of gasses evolved during thermal degradation of the samples, while the oxygen diffusion into the NC was inhibited. The so-called "labyrinth effect" of the silicate layers in the polymer matrix could be responsible for the slower evolution of the volatile products,[62,63] which could undergo secondary reactions. The motions of polymer chains strongly interacted with the MMT layers, which reduced heat conduction. A chemical interaction between the polymer matrix and the catalytically active clay surface was probably responsible for more effective char formation during thermal decomposition of polymer–clay NCs.[30,41,42,64] At low clay content (~1 wt%), exfoliation dominated but the amount of exfoliated nanoclay was not enough to enhance the thermal stability through residue formation.[39] In air atmosphere, clay may slow down oxygen diffusion and promote thermooxidative reactions. With increasing the clay content (2–4 wt%) much more exfoliated clay was obtained, and char was formed easier, thus the thermal resistance of the NCs was improved. However, at a higher clay loading level (up to 10 wt%), further increase of thermostability was not observed. A strong insulation effect of clay layers, mass transport barrier to the volatile products of thermal decomposition, and the formation of char were responsible for the improved thermostability and flame retardancy (FR) of different polymer–clay NCs.[65–68]

The heat distortion temperature (HDT) and thermal decomposition for epoxy clay NCs with 5 wt% clay content were increased from 124°C to 133°C and from 348°C to 373°C, respectively, as compared to pristine epoxy resin.[69] The thermostability of unmodified epoxy resins (UME) and different clay modified epoxy resins (CME): bisphenol A diglycidyl ether (BDGE), bisphenol A propoxylate diglycidyl ether (BPDG), bisphenol A-brominated diglycidyl ether (BBDG) and tetraglycidyl of (diaminodiphenyl)methane (TGDDM) were studied by Lakshmi et al. Hexadecyl(trimethyl)ammonium (HDTMA) and hexadecyl(triphenyl)phosphonium (HDTPP) were used as clay modifiers. The initial decomposition temperature (IDT) of the CME was found to be higher than any of the UME systems. The clays modified with HDTTP showed high IDT.[70]

Composites of epoxy resins reinforced with nanoclays and carbon fibers showed improved mechanical and thermal expansion (CTE) characteristics, avoiding microcracking, and they were used for cryogenic storage systems.[71]

Inorganic phases ($SiO_2$, $Al_2O_3$, and MgO) present in clay particles, high temperature resistant groups (phenyl units and/or bromine atoms) present in the chemical structures of epoxides and the triphenylphosphine unit in the HDTPP–MMT clay increased thermal stability of the NCs. A significant improvement of thermostability and fire retardancy (FR) was found for epoxide–clay NCs prepared with organo clay containing 5 wt% of reactive phosphorous as reflected by activation energies and IPDT.[63,72,73]

The thermal degradation behavior of the epoxy clay NCs proceeded in three steps in air and in two steps in nitrogen for the 2 and 10 wt% of clay loading to the epoxide. The improved thermal stability was observed for 2 wt% clay content due to its exfoliated structure, while lowest thermostability was observed for 10 wt% composites with an intercalated structure in NCs.[74] The incorporation of unmodified MMT into the epoxy resin did not affect the $T_g$ value, while the addition of 3 wt% of OMMT increased the $T_g$ by about 15°C due to the better exfoliation of clay in the epoxy matrix.[75]

Epoxy (DGEBA) hybrid nanocomposites were prepared by sol–gel process and modified *in situ* with (3-aminopropyl)trimethoxysilane (APTMS) or (3-glycidoxypropyl)trimethoxysilane (GPTMS), (4,4′-methylenedianiline (DDM) and 1,4-bis(trimethoxysilylethyl)benzene (BTB)). The $T_g$s of all NCs decreased with the increasing BTB content, while good thermostability of NCs was observed even at lower content of 1,4-bis(trimethoxysilylethyl)benzene (BTB) than 10 wt%—the $T_5$ increased from 336°C to 371°C and char yield increased from 27.4% to 30.2%. At 800°C with 30 wt% BTB content the char yield of NCs reached 36.1–37.8%. The structure of inorganic network influenced the $T_5$ of composite. The EP-APTMS-BTB sample with BTB content >20 wt% showed decreased $T_5$ to 350°C, due to the presence of $T_1$ structure in inorganic networks, containing some methoxy or hydroxy groups which degrade easier. The $T_5$ of EP-GPTMS-BTB system was in the temperature range of 377–386°C, depending on crosslinking density of polymer networks.[76]

### 8.4.1.3 Nanocomposites of Silicates with PMMA, Polyesters, Polyurethanes, and Polyethersulfone

Nanocomposites of PMMA with OMMT exhibited homogeneous distribution of spherical particles (~170 nm in diameter) and showed higher thermal stabilities than the neat polymer. The thermal degradation of PMMA occurred in three steps and it was attributed to the presence of head to head linkages and unsaturated chain end, while the major step was assigned to random scission.[77,78] Increased degradation temperature of polymer–nanoclay composites, in comparison with the neat polymers, was assigned to high thermal stability of nanoclay layers, interaction between clay platelets and polymer matrix, and hindrance effect of nanoplatelets on the polymer chains movement and restriction of oxygen permeation by the silicate sheets. By increasing nanoclay content, degradation temperature and a char yield at 600°C increased to 77.7%.[79]

The thermal stability of poly(methyl methacrylate)-*co*-poly[3-tri(methoxy)silylpropyl methacrylate] polymer containing functional (trimethoxy)silyl groups (Figure 8.1) and its nanocomposite intercalated with Cloisite 15A™ were studied.

The NCs were more stable than polymer, and this thermal improvement was proportional to the clay content. Pristine polymer decomposition pathway was dependent on the atmosphere, and proceeded in two steps under air and in three steps under $N_2$. Char formation was also dependent on atmosphere and occurred much easier under air. The experimental results indicated that clay NPs

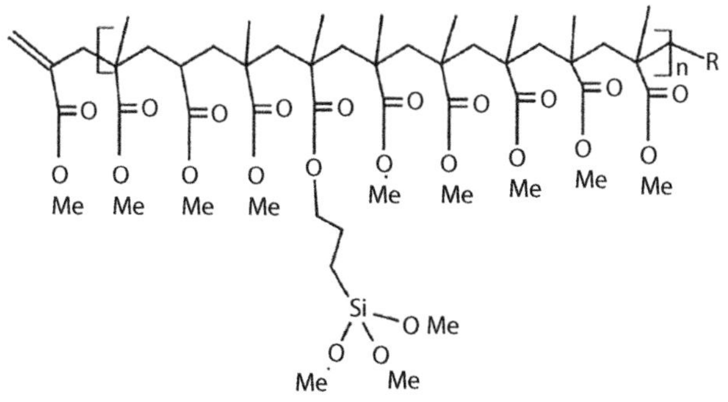

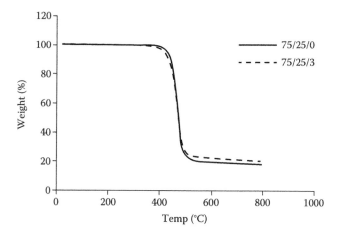

**FIGURE 8.1** Chemical structure of PMMA-*co*-PTMSM. (Carvalho, H.W.P. et al.: Structure and thermal stability of poly(methyl methacrylate)-*co*-poly(3-tri(methoxysilyl)propyl methacrylate)/montmorillonite nano-composites. *Polym. Eng. Sci.* 2013. 53. 1253–1261. Copyright Wiley-VCH Verlag GmbH & Co. KGaA. Reproduced with permission.)

played different functions in polymer stabilization: diffusion barrier, charring, and suppression of degradation steps by chemical reactions between polymer and clay.[80]

The addition of the nanoclay inhibited transesterification reactions and reduced the thermo-stability of poly(ethylene terephthalate) (PET) and poly(ethylene naphthalene 2,6-dicarboxylate) (PEN) blends. PET degraded much faster than PEN in oxygen atmosphere, while an opposite trend was observed in a nitrogen environment. The thermal decomposition behavior of PET–PEN blends was almost similar in oxygen and in nitrogen. Their IDTs were close to 400°C. [1]H-NMR results proved that during the blending process, the transreactions occurred between the two polyesters which led to a formation of crosslinks between their chains—it could improve the thermal stability of the blends. It was concluded that the thermal resistance also increased by increasing PEN content in the blend composition.

The incorporation of MMT into the blends slightly decreased the thermal stability of the NCs (Figure 8.2). Nanoclay inhibited the transesterification reactions between polymeric phases and decreased crosslinking reactions. The reduction of thermostability might be caused as well by the degradation of organic components which were present in the structure of MMT.[81]

**FIGURE 8.2** Effect of nanoclay on thermal decomposition behavior of PET/PEN (75/25) blend. (With kind permission from Springer Science+Business Media: *J. Polym. Res.*, Investigation of thermal behavior and decomposition kinetic of PET/PEN blends and their clay containing nanocomposites, 18, 2011, 1765–1775, Mohammadi, S.R. et al.)

Thermal and mechanical properties of NCs of flexible polyurethanes (PU), filled with 3–9 wt% OMMT, and Fyrol PNX as FR additive, were dependent on the nanoclay content. At a heating rate of 20°C/min the $T_{onset} = 310$–320°C was determined. The effects of the OMMT and MMT/Fyrol PNX on the thermal stability of the NCs foams were expressed by the values: $T_2$, $T_5$, $T_{10}$, $T_{max}$, and the char residue. The modification of elastic PU foams (PUF) with 6–9 wt% of MMT/PNX or with 9 wt% of OMMT caused the decrease in the thermal stability of the NCs in comparison to unmodified flexible PUF. The addition of 3–6 wt% OMMT did not affect the temperature of 2 wt% loss. The PUF with greater content of Fyrol PNX showed lower $T_2$ in the case of both PNX/MMT and OMMT additives, and for unmodified PUF as well.[82]

Polyethersulfone (PES)/halloysite nanotube (HNTs) NCs were prepared by melt compounding. PES with hydroxyl chain ends, PES(OH), were covalently bonded onto the aluminosilicate surface. The $T_5$ and $T_{max}$ values and the char yield at 800°C were determined from TG and DTG curves, under nitrogen or in air atmospheres. The neat polymer was highly stable under both atmospheres with no weight loss up to 400°C. The IDT of PES(OH) under inert environment (475°C) was quite similar to that obtained in air (483°C). The $T_5$ ranged from 502 to 518°C for PES(OH)–HNTs NCs with 6–16 wt% content of the filler. Under nitrogen, neat PES(OH) decomposed mainly through a one-step reaction with the formation of a highly stable residue (36 wt%). Under air the polymer followed a two-step degradation process, which took place at 600–670°C and led to complete mass loss. This latter corresponded to the thermooxidative degradation of the intermediate carbonaceous residue, which was mainly made up of aromatic carbon atoms and was stable under $N_2$. The incorporation of halloysite shifted the start of PES(OH) decomposition to higher temperatures, both in air and in nitrogen. For instance, the $T_5$ of the neat polymer increased by 43 and 19°C in $N_2$ and air atmospheres, respectively, with only 6 wt% HNTs. At higher HNTs content (16 wt%), there was no additional improvement in the first decomposition step of PES(OH), while the second one (i.e., char degradation) was further delayed. The improved thermal and thermooxidative stability of PES in the presence of HNTs was mainly attributed to the labyrinth effect provided by individually dispersed NTs, which was reinforced during the decomposition process by the formation of a protective charred ceramic surface layer. The stabilizing effect of halloysite was even more evident during the degradation of the carbonaceous residue. At higher HNTs content, the thermal stability of PES(OH) was almost unchanged during the initial degradation stage, due to the absence of additional contribution of aggregates to the formation of a tortuous diffusion path, which slowed down the weight loss. The higher thermal resistance of the intermediate carbonaceous residue in air with increasing HNTs fraction was assigned to the formation of an aluminosilicate "skeleton frame" that supported the char structure and enabled the formation of a more thermally stable ceramic residue. The clay aggregates also contributed to a lesser extent to this thermal stabilization mechanism.[83]

### 8.4.1.4 Nanocomposites of Silicates with Poly(Lactic Acid)

In recent years, thermal properties of poly(lactic acid) (PLA) NCs were quite often studied. PLA nanocomposite films, containing 3–5 wt% organoclay (Cloisite 30B), prepared by melt intercalation method, were plasticized with poly(ethylene glycol) (PEG). The addition of 20% PEG to the neat PLA decreased $T_g$ about 30°C. In the presence of 3 wt% clay in the neat PLA, an increase of $T_g$ by 4°C for the plasticized PLA was observed.[84]

PLA-based NCs films composed of an organoclay (Nanofil-2) (1, 3, or 5 wt%) and 5 wt% a poly (ε-caprolactone) (PCL) or without PCL, were exposed to UV radiation. The incorporation of Nanofil-2 into PLA matrix affected photodegradation of studied materials. In the presence of modified MMT, the rate of degradation process was decreased. The TG analysis of all kinds of NCs indicated that MMT modified with quaternary amine delayed the photodegradation process of PLA, whereas the presence of 5 wt% PCL accelerated the decomposition of studied materials during UV radiation.[85]

The PLA/OMMT NCs with 10–30 phr of isopropyl triaryl phosphate ester FR were obtained by a melt compounding technique. The TGA results, carried out under nitrogen, showed that the addition of FR reduced the $T_0$ of PLA/OMMT and only slightly increased char yield of NCs. The $T_0$ was higher for PLA–OMMT (274.6°C) than for neat PLA (259.5°C), but it was decreased with increasing content of FR, from 235°C to 223°C, gradually. The flexibility of the PLA/OMMT NCs increased as a result of plasticization effect, caused by FR. Flexible PLA/OMMT/FR20 and PLA/OMMT/FR30 NCs showed high fire resistance—a V-0 rating was determined during the UL-94 vertical burning test.[86]

NCs of biodegradable PLA with 2 wt% of halloysite nanotubes (PLA/HNTs) were prepared by melt compounding followed by compression molding. Addition of N,N′-ethylenebis(stearamide) (EBS) improved the dispersion of HNTs in the PLA NCs. The decomposition process of PLA/HNTs NCs was dependent on the atmosphere during TGA test as well as the amount of EBS. The thermal properties of PLA/HNTs and PLA/HNT(2 wt%)/EBS NCs were decreased from 329°C to 307°C, with increasing EBS content from 5 wt% to 20 wt%, in comparison to neat PLA, both under nitrogen and oxygen atmosphere. In the nitrogen atmosphere, a single-stage decomposition was observed, while in the oxygen atmosphere a double-stage decomposition process occurred.[87]

Solution casting methods were used for preparation of NCs of PLA with unmodified and organic modified sepiolite. The sepiolite nanofibers were well dispersed in PLA matrix showing a random orientation and a presence of crosslinked moieties. The TGA of the PLA NCs films was carried out under nitrogen atmosphere. The thermostability of PLA was improved. The addition of unmodified sepiolite to the NCs (PLA/Sep) ($T_5 \approx 300$°C) caused improvement of the thermostability of PLA, probably due to a strong interfacial interaction of silanol groups (Si–OH) on sepiolite with the ester groups of PLA. Thus, sepiolite acted as a "crosslinking agent" and retarded the motion of the polymer chains. The thermal stability of PLA NC with sepiolite modified by hexadecyl(trimethyl) ammonium bromide (CTAB) was increased slightly, in comparison with neat PLA. However, it was decreased compared to PLA/Sep NC. This was likely caused by the hydrophobic part of CTAB, which weakened the interaction of Si–OH groups of sepiolite with the ester groups of PLA. Thus, the compatibility between sepiolite and PLA was decreased. The thermal stability of NC, containing sepiolite modified with 12-amino lauric acid (ALA), was even worse in comparison with pure PLA, probably due to the catalytic effect of the organoclay on the decomposition of PLA. The char residue of PLA filled with sepiolite modified with ALA was less than the unmodified sepiolite or CTAB-Sep system, which also confirmed that the ALA modified sepiolite was responsible for decrease of the thermal stability of PLA NCs. Thus, it was found that sepiolite is a very useful inorganic filler for preparation of biodegradable PLA NCs with improved thermal properties.[88]

### 8.4.1.5 Nanocomposites of Silicates with Polyimides

Extremely high thermal resistance exhibit polyimides (PI) and their different nanocomposites. A precursor of PI was synthesized in refluxing methanol medium from a suspension of benzophenone tetracarboxylic acid dianhydride (BTDA), containing organoclay particles (OMMT), and equimolar amounts of diaminodiphenyl ether (ODA). After removal of methanol, foaming and imidization processes of the PI precursor were performed at 100–300°C, within 3 h. The chemical structure of PI foam is shown in Figure 8.3. XRD and TEM results showed that the OMMT NPs were homogeneously dispersed in the foamed PI NC matrix. Their thermal stability was dependent on OMMT contents (0–10 wt%).

The TGA results proved that the addition of organoclay greatly improved the thermal resistance of the PI NC foams. The $T_5$ of PI-OMMT NCs increased from 532°C to 580°C, when OMMT content

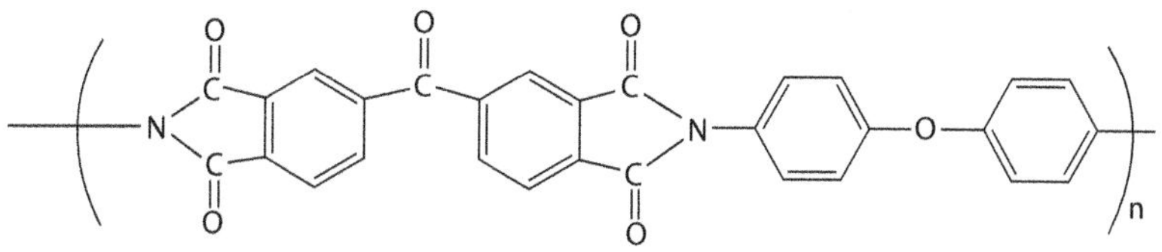

**FIGURE 8.3** Chemical structure of PI foam. (Qi K., Zhang, G: Effect of organoclay on the morphology, mechanical, and thermal properties of polyimide/organoclay nanocomposite foams. *Polym. Compos.* 2014. 35. 2311–2317. Copyright Wiley-VCH Verlag GmbH & Co. KGaA. Reproduced with permission.)

was increased from 2 wt% to 10 wt%. For pure PI $T_5 = 509°C$, and $T_g = 236°C$, while $T_g$ of NCs ranged from 258°C to 282°C with increasing OMMT loading (2–10 wt%). The addition of OMMT caused strong improvement of the thermostability of PI matrix because the decomposition temperatures of the PI NC foams were higher than that of the neat PI foam. Thus, OMMT acted as a thermal barrier for the heat transfer into polymer chains, and it could avoid the quick heat expansion and limit further degradation. The well-dispersed OMMT in the NC foams probably limited the motion of the PI chains and delayed the decomposition by decreasing the mass transfer of the volatile products.[89]

### 8.4.2 THERMAL RESISTANCE OF POLYMER–SILICA NANOCOMPOSITES

Polymer nanocomposites containing silica can be prepared by different methods and they usually have good or very good thermal and mechanical properties.

#### 8.4.2.1 Nanocomposites of Polyethylene with Silica

NCs of high-density polyethylene (HDPE) containing both hydrophilic and hydrophobic fumed silica NPs of different surface areas were often studied. The homogeneous distribution of functionalized fumed silica NPs at low silica loading remarkably improved thermal stability of NCs with respect to neat HDPE. The advantageous effects of fumed silica NPs on the thermal degradation resistance of the NCs demonstrated both $T_2$ and $T_5$ values, while $T_d$ temperature was notably improved only by using two kinds of silica NPs: A380 and Ar974, which caused an increase of $T_d$ of about 70°C. The $T_d$ temperature was strongly dependent on the nanofiller dispersion within the matrix. A strong stabilizing effect of silica NPs on the thermostability of HDPE under oxidizing conditions was accompanied by a single degradation step at temperatures higher than 370°C. Presumably silica NPs acted as thermal insulator and limited the diffusion of oxygen into the polymer matrix and retarded the thermooxidative process. The silica NPs showed tendency to agglomeration to the surface of the melted polymer, and formed a barrier that physically protected the remaining polymer from heat and slowed down the volatilization of the gaseous combustion products. The homogeneous dispersion of functionalized silica reduced the mean distance between the aggregates and inhibited oxygen diffusion.[90]

#### 8.4.2.2 Nanocomposites of Polystyrene with Silica

Polystyrene (PS)/mesoporous silica (SBA-15) NCs with different loading of SBA-15 (2.5–10 wt%) were prepared by *in situ* emulsion polymerization. SBA-15 was well dispersed in PS matrix at a content of 5 wt%. PS/SBA-15 composite showed a slightly higher $T_g$ (5–7°C) than neat PS. The thermal stability of the PS/SBA-15 NCs was improved with the incorporation of SBA-15 into the polymer matrix. The $T_{onset}$ (~350°C), $T_{max}$, and the end decomposition temperature ($T_{end}$) were all increased. The $T_5$ values were close to 370°C for pure PS and PS/SBA-15 NCs. The increase of char yields at 600°C suggested that the SBA-15 silica was incorporated with polymer, where the strong interaction between the SBA-15 particles and the polymer matrix stabilized the polymer and increased its thermal resistance. The mesoporous SBA-15 with a large pore size and pore volume was more effective in trapping radicals which led to increase of the decomposition temperature.[91]

#### 8.4.2.3 Nanocomposites of PMMA with Silica

Incorporation of 1–9 wt% of pristine or modified silica NPs into PMMA/silica NCs resulted in better thermal and mechanical properties than those of pristine NPs. Silanol groups of the silica were functionalized with methyl methacrylate groups from (3-methacryloxypropyl)dimethylchlorosilane. The best improvement of thermophysical properties was achieved for NCs with 7 wt% content of pristine silica NPs ($T_5 = 252.4°C$, related to 236.1°C for PMMA). A surface modification resulted in enhanced thermal stability of NCs, which was observed with increasing modified silica NPs contents, except for one sample. The thermal stabilities of all the NCs were higher than the neat PMMA, whereas modified NPs improved thermostability more than pristine silica. This could be ascribed to a better dispersion of modified NPs and lower monomer conversion values, which led to higher contents of silica in the NCs. Two stages of degradation were observed in all PMMA/pristine silica and PMMA/modified silica NCs. The first step of degradation was related to the decomposition of thio-groups, which was delayed by silica NPs. The main stage of degradation was ascribed to random chain scission, which occurred from 280°C to 360°C and higher amounts of silica increased degradation temperature.[92]

### 8.4.2.4 Nanocomposites of Phenolic Resin with Mesoporous Silica

The mesoporous silica/phenolic resin (SBA-15/PR) NCs were prepared via *in situ* polymerization. The surface of SBA-15 was modified using 3-glycidyloxypropyl(trimethoxy)silane (GOTMS) as a coupling agent. The sample SBA-15-GOTMS contained SBA-15 surface treated with GOTMS and E-SBA-15/PR NC sample was extracted with ethanol. The TGA, determined under a nitrogen atmosphere, revealed that only 1.8 wt% weight loss was found for unmodified SBA-15 below 800°C, but 13.5 and 34.0 wt% between 50°C and 800°C for SBA-15-GOTMS and E-SBA-15/PR, respectively. The 11.7 wt% of GOTMS was grafted onto the surface of the SBA-15 and the chemical reaction occurred between SBA-15-GOTMS and the phenolic resin. The $T_g$s and thermal stability of the PR NCs were improved in comparison with the pure PR at silica loading 1–3 wt%, presumably due to the homogeneous dispersion of the modified SBA-15 in the PR matrix. The $T_5$ and $T_{10}$ of the NCs increased with respect to the amount of SBA-15, for instance, the $T_5$ and $T_{10}$ of SBA-15/PR NCs with 1 and 3 wt% SBA-15 increased from 263 and 312°C (for neat PR) to 266°C and 317°C (1 wt%), and 364°C and 380°C (3 wt%), respectively. The char yield of the NCs also increased with the increase of SBA-15 content, which hindered the formation of gaseous products. The improvement of the thermostability of the NCs was attributed to the presence of strong interaction between SBA-15 and polymer matrix.[93]

### 8.4.2.5 Nanocomposites of Polycarbonate with Silica

The high thermal degradation temperature was observed for polycarbonate NCs filled with 1 wt% pristine and modified silica with 3-glycidyloxypropyl(trimethoxy)silane (KH560-SiO$_2$), which were prepared by melt compounding. Samples of PC-silica NCs were heated from 50°C to 700°C at 10°C/min under nitrogen atmosphere and held for 4 min at 700°C. The TGA confirmed that alcoholysis reaction between PC and silica NPs occurred during the thermal treatment. The TGA/FTIR results proved that no new degradation volatile products were formed during the thermal degradation of NCs, but the total amounts of all gaseous products decreased by adding silica NPs. The degradation activation energies of both NCs increased significantly relative to neat PC, especially for the composite with modified silica. The thermal degradation temperatures of both PC/SiO$_2$ and PC/KH560-SiO$_2$ NCs shifted to higher values relative to neat PC. For instance, the values of $T_{onset}$ (461–472°C) and $T_{max}$ (~515°C) increased 25–35°C and 10°C, respectively, with respect to PC. It seemed that the enhancement of thermostability fits the barrier model. It was considered that the thermal degradation mechanism of composites was different from that of pure PC.[94]

### 8.4.2.6 Nanocomposites of Polyamide-Imides and Polyimides with Silica

Hybrid organic–inorganic membrane materials, based on a high molecular weight sulphonated polyamide-imide resin (PA-I) and silica were prepared by the sol–gel method using TEOS and (aminopropyl)triethoxysilane (APTS) as the silica precursors. Incorporation of APTS greatly increased the compatibility of organic and inorganic phases. Silica NPs (<50 nm) were well dispersed in the polymer matrix and chemically bonded with the polymer phase. The thermal stability and hydrophilic properties of hybrid materials were also significantly improved.[95] PA-I and the hybrid NCs showed a good thermooxidative stability and a high $T_g$ up to 290°C. The incorporation of silica particles improved both $T_5$ and $T_{10}$ values (as compared to the pure PA-I sample) and the char yield increased with increasing silica content, indicating a general improvement of the thermostability. The positive effect of silica on the thermostability of hybrid materials was also confirmed by the increase of $T_g$—the pure PA-I exhibited a $T_g$ value of 194°C, while the addition of silica, increased the $T_g$ to 221°C, when the silica content was 7 wt%, and to 234°C, when the silica content was further increased to 15 wt%. This was explained by the rigidity of the hybrid films. The incorporation of silica into the polymer matrix decreased the flexibility of the polymer chain and increased the $T_g$s of the hybrid NC.[95]

The other hybrid films of well-dispersed silica NPs in polyimide (PI/SiO$_2$) were also prepared through the sol–gel processing. The pure PI was synthesized from 4,4′-oxydianaline (ODA) and pyromellitic dianhydride (PMDA), and the residual amino groups reacted with the NCO groups of coupling agent (isocyanatopropyl)triethoxysilane (ICTOS). Moreover, addition of different amounts of tetraethoxysilane (TEOS) led to chemically bonded hybrid NC films with different silica contents (5, 8, 10, 13, and 15 wt%). The chemical interaction between the two phases resulted in the formation of well-dispersed silica NPs (20–30 nm) in the PI matrix, and excellent thermal stability of organic–inorganic films. Under nitrogen atmosphere, the decomposition temperature of the pure PI films and unbonded hybrids began around 550°C, while for hybrid NCs it began around 570°C. The char yields of the NCs films at 800°C increased with the increased silica content.[96]

Reaction of polyamic acids (PAA) with pendant hydroxyl groups with (isocyanatopropyl) triethoxysilane led to novel PI-silica NCs with interphase chemical bonding and reactive alkoxysilane groups in the polymer structure (Figure 8.4). The reinforcement of PAA matrices with or without alkoxysilyl groups attached to the PI chain was carried out by addition of TEOS, which underwent hydrolysis and condensation reactions in a sol–gel process. The three thin hybrid films, without OH groups and with OH groups, with various SiO$_2$ contents (0–40 wt%) were further imidized by heating up to 300°C. The cocondensation of alkoxysilane groups on the polymer chain with TEOS gave the silica network which was bound chemically with the PI matrix. Higher thermal stability and mechanical strength, improved transparency, and low values of thermal coefficient of expansion were observed in the case of chemically bonded PI NCs.[97]

The $T_d$ of the pure PI was above 500°C. The PI with hydroxyl groups, used for preparation of the bonded hybrids, showed slightly higher $T_d$ than the PI used for the unbonded hybrids. It seemed that the presence of OH groups on the polymer chain acted as a free radical quencher during the oxidative degradation process. The chemically bonded PI-silica hybrid NCs showed slightly better thermal resistance than the PI matrix. The char yields above 770°C in all the hybrids were proportional to the silica contents used in the matrix. The PI–silica hybrids showed that interphase chemical links between the silica network and the polymer chain avoided the agglomeration of silica particles and reduced the particle size to nanolevel, thus caused more homogenous distribution in the matrix. The values of CTE were also greatly reduced.[97]

The thermostability of polyimide (Figure 8.5) (PI)/mesoporous silica nanospheres (MSNs) NC films with different contents of MSNs were improved by the presence of MSNs. The addition of the

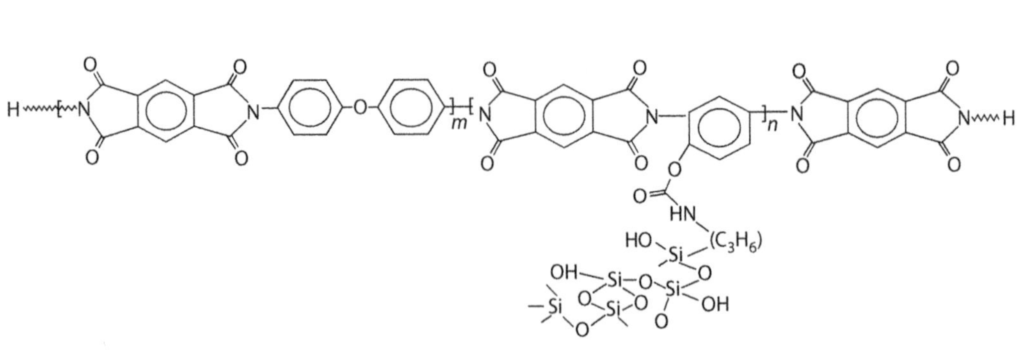

**FIGURE 8.4**  The chemical structure of polyimide–silica hybrids. (Reprinted with permission from Al Arbash, A. et al., *J. Nanomater.*, Article ID 58648, 1–9. Copyright 2006, Hindawi Publishing Corporation.)

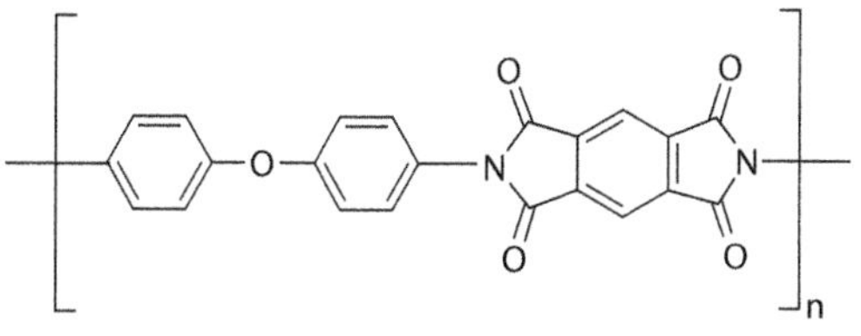

**FIGURE 8.5** Chemical structure of imidized PI.

MSNs caused increase of a $T_g$ of PI/MSNs NCs—the NCs with 5 wt% content of MSNs showed the $T_g$ of 379.6°C, which was apparently higher than for the neat PI (346.2°C).

The residues of 20 wt% appeared at 673°C for neat PI, at 681°C for 3 wt% content of MSNs, at 691°C for 5 wt% of MSNs, and at 697°C for 7 wt% of MSNs. The thermal resistance was probably improved by formation of a crosslink between the PI and the silica network.[98]

Porous polyimide (PI) microspheres, having excellent heat resistance, were prepared in two steps.[99] First, novel PI/nanosilica composite microspheres were prepared via the self-assembly structures of poly(amic acid) (PAA, precursor of PI)/nanosized $SiO_2$ blends after *in situ* polymerization, followed by the two-step imidization. Next, the encapsulated NPs were treated with hydrofluoric acid, leading to the pores.

The thermal properties of the pure PI spheres (Figure 8.6 and Table 8.2), the PI/$SiO_2$ spheres, and the porous PI spheres were analyzed in the range of 40–900°C under nitrogen.

The porous PI spheres showed still a very high heat resistance. The observed $T_5$, $T_{10}$, and $T_{max}$ values indicated that the incorporation of nanopores did not substantially affect the thermal stabilities of PI porous microspheres. No apparent weight loss up to 455°C for the porous PI was observed, and the distinct weight loss occurred at 562°C.[99]

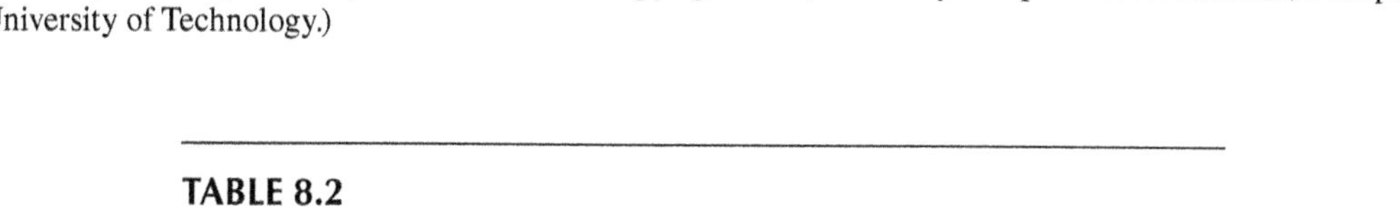

**FIGURE 8.6** Chemical structure of the porous polyimide microspheres. (Reprinted with permission from Liu, M.Q. et al. *Express Polym. Lett.* 9:14–22. Copyright 2015, Directory of Open Access Journals, Budapest University of Technology.)

### TABLE 8.2
### Thermal Stabilities of the Three Type Spheres

| Sample Code | $T_5$ (°C) | $T_{10}$ (°C) | $T_{max}$ (°C) | Char Yield at 900°C (°C) |
|---|---|---|---|---|
| Pure PI | 499 | 566 | 601 | 48.11 |
| PI/$SiO_2$ composites | 499 | 552 | 569 | 73.97 |
| Porous PI | 455 | 519 | 562 | 52.49 |

*Source:* Reprinted with permission from Liu, M.Q. et al. *Express Polym. Lett.* 9: 14–22. Copyright 2015, Directory of Open Access Journals, Budapest University of Technology.

### 8.4.3 Thermal Properties of Polymer Nanocomposites with Other Inorganic Fillers

Novel poly(amide-imide)/TiO$_2$ NC (PAI/TiO$_2$ NC) was synthesized by direct polycondensation reaction of N-trimellitylimido-L-valine with diaminediphenylsulphone in the presence of triphenyl phosphite and the molten tetra-n-butylammonium bromide, by means of an ultrasonic irradiation. The surface of TiO$_2$ NPs was modified by N-trimellitylimido-L-valine as a bioactive coupling agent, providing environmentally friendly TiO$_2$NPs. A considerable enhancement of thermal properties of these new NCs was achieved with 15 wt% addition of TiO$_2$ NPs. The T$_5$ and T$_{10}$ for samples containing 5 and 10 wt% of TiO$_2$ were decreased, which could be explained as a result of the oxidative degradation of polymer, while in NC with 15 wt% of TiO$_2$, T$_5$ and T$_{10}$ were shifted to a higher temperature range, 403°C and 470°C, respectively, in comparison with that of neat PAI (368°C and 439°C, respectively). Thus, TiO$_2$ acted as filler and thermal insulator at high contents. Although, the TiO$_2$ NPs caused a decrease of thermostability, the char yield at 800°C was slightly increased with the increase of TiO$_2$ content.[100]

Polybenzoxazine (PBA-a) composites based on bisphenol-A and aniline, with a very high loading of alumina particles (83% by weight, 60% by volume) showed significantly increased T$_g$, T$_5$, and char yield with increasing the alumina contents. A good distribution of the alumina particles in the PBA-a matrix was confirmed by SEM method. Alumina exhibited very high thermal stability—no weight loss up to 800°C was observed during the TG analysis. At 800°C, only the PBA-a fraction was decomposed thermally and formed the char. The PBA-a matrix showed the T$_5$ of 317°C and the 25% char residue at 800°C. The T$_5$ at the highest alumina loading of 83% in the PBA-a was found to be ~389°C, which is 72°C greater than that of the neat PBA-a. This crucial improvement in the thermal properties of the PBA-a composites with high alumina content probably resulted from the barrier effect of Al$_2$O$_3$ as well as the strong interfacial interaction of the PBA-a resin to the alumina. The (PBA-a)-alumina composites may be recommended for applications that require high modulus and hardness as well as high thermostability.[101]

Novel polyimide/α-Fe$_2$O$_3$ (PI/α-Fe$_2$O$_3$) hybrid nanocomposite films were prepared by thermal curing of poly(amic acid) (obtained from 3,5-diamino-N-(9H-fluoren-2-yl)-benzamide and (3,3′,4,4′-benzophenone)tetracarboxylic dianhydride) with different amounts of silane-modified NPs of a filler (α-Fe$_2$O$_3$). The organophilic surface modification of α-Fe$_2$O$_3$ NPs with 3-aminopropyl(triethoxy)silane enabled achieving the homogeneous dispersion of α-Fe$_2$O$_3$ on nanoscale in a polymeric matrix. The thermal properties of pure PIs and PI hybrid NC with 15 wt% of the filler were studied by TGA method, under nitrogen atmosphere. The TGA data (Figure 8.7) showed a great improvement of thermal resistance of novel NC films in comparison with the pure polymer.[102]

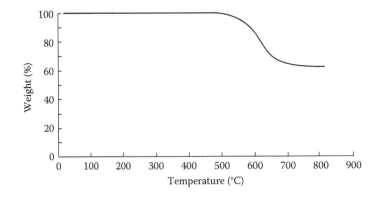

**FIGURE 8.7** TGA thermogram of PI nanocomposite containing 15 wt% α-Fe$_2$O$_3$. (With kind permission from Springer Science+Business Media: *J. Polym. Res.*, Polyimide nanocomposite films containing α-Fe$_2$O$_3$ nanoparticles, 22, 2015, 630, Rafiee, Z., Golriz, L.)

NPs of $\alpha$-$Fe_2O_3$ have high thermal stability, not only due to their larger surface area, so the incorporation of $\alpha$-$Fe_2O_3$ NPs into a PI matrix improved the thermal stability and flame retardant properties of the PI/$\alpha$-$Fe_2O_3$ NCs. The thermostability of NC was higher than pure PI, due to the good compatibility of modified $\alpha$-$Fe_2O_3$ particles with polymer matrix. These NCs were extremely thermally stable—up to 500°C. The $T_{10}$ of the pure PI and NC with 15 wt% content of $\alpha$-$Fe_2O_3$ ranged from 526°C to 585°C. The amount of residue (char yield) of this NC was more than 63% at 800°C. The DSC thermogram of NC showed very high $T_g$ (294°C). The calculated values of limiting oxygen index (LOI) derived from char yield of PI's NCs containing $\alpha$-$Fe_2O_3$ exceeded 42. Estimated values of LOI of the polymers were calculated from Van Krevelen and Hoftyzer equation:[103]

$$LOI = 17.5 + 0.4CR, \quad where: CR = char\ yield$$

Thus, these NCs belong to self-extinguishing materials.

A series of polyoxymethylene polyacetal copolymers (POM) NCs with 5–10 wt% of nanohydroxyapatite (n-HAp) were prepared by melt processing. POM has a tendency to a thermal depolymerization into formaldehyde. TGA results indicated that with an increase in n-HAp loading the thermal stability of POM matrix decreased significantly, depending on the POM molecular weight. The difference between the thermostability of pure POM and POM/10.0 wt% n-HAp was 30–60°C, depending on the molecular composition of copolymers. The thermal stability of POM decreased with a decrease of molecular weight: this effect was assigned to the higher concentration of end groups in the POM matrix with low molecular weight. It was proposed that POM degradation was initiated either by bond dissociation at the chain end (–OH groups), or by random main-chain scission for the end-capped polymer followed by unzipping. If the hydroxylic groups of POM were not blocked, rapid depolymerization took place at elevated temperatures. The lower thermostability of POM/n-HAp NCs resulted from the high sensitivity of the POM main chain to attack by base under heating. A general formula of $[Ca_{10}(PO_4)_6(OH)_2]$ corresponds to a structure $3Ca_3(PO_4)_2 \cdot Ca(OH)_2$, which contains calcium hydroxide. Nano-HAp particles with a very large surface area were more reactive at higher temperature than microsized Hap.[104]

Novel poly(amideimide) (PAI)/$TiO_2$ hybrid bionanocomposites (BNCs) containing L-isoleucine and diphenylsulfone moieties in the main chain were prepared via an ultrasonic irradiation process. The surface of titania NPs was modified with (3-aminopropyl)triethoxysilane. The surface modified $TiO_2$ NPs, with a diameter of benzimidazole less than 40 nm, were uniformly dispersed in the PAI matrix. The BNCs showed much higher thermal stability than pure PAI. The TGA results (in nitrogen atmosphere) of the PAI and PAI/$TiO_2$ BNCs included the $T_5$, $T_{10}$, and char yield at 800°C. The IDT for PAI/$TiO_2$ BNCs were quite high (>360°C), when content of $TiO_2$ reached 15–20 wt%. The char yield of hybrid BNCs ranged from 49% to 62% at 800°C. The thermostability increased presumably due to excellent insulating features of the $TiO_2$ NPs, which also acted as a mass transport barrier to the volatile products formed during decomposition.[105]

New thermally stable aromatic poly(benzimidazole-amide)s PBIAs were prepared by polycondensation of bis-benzimidazole diamines [1,4-bis(5-amino-1H-benzimidazol-2-yl)-benzene and 1,3-bis(5-amino-1H-benzimidazol-2-yl)benzene] with two different diacid chlorides. Next, a series of novel NCs, complexed through C–H or N–H protons of benz-imidazole moieties with $Ag^+$ or $Cu^{2+}$ was prepared. The FE-SEM, TEM, and SEM–EDX results indicated that Cu and Ag nanoparticles were dispersed homogenously in the polymer matrix. The thermostability of the PBIAs NCs was good: $T_5$ was in the range of 310–350°C, $T_{10}$—in the range of 315–380°C, char yields: 56–64 wt%. High LOI values were determined in inert atmosphere: 40–43%.[106]

High-density polyethylene (HDPE) NCs with different volume fractions (1–20 vol%) of aluminum nitride (n-AlN) NPs were prepared by melt mixing. HR-TEM micrographs and AFM pictures confirmed homogeneous dispersion of AlN NPs, as well as the existence of long interconnected chain-like aggregates.

The thermal decomposition of the pristine HDPE and HDPE/n-AlN NCs were analyzed at different heating rates (5, 10, 15, and 20°C/min) from ambient temperature to 800°C in a pure nitrogen. The TGA data (Figure 8.8) indicated only a small increase in the thermal stability of the composites, $T_{10}$ and $T_{peak}$ shifted to higher values with growing heating rate. The $T_{10}$ values were in a range of 438–458°C, both for neat HDPE and all HDPE/n-AlN NCs. However, with increasing content of n-AlN in HDPE, only a marginal increase of $T_{10}$ and $T_{30}$ was observed. It appeared that both branching and random chain scission occurred simultaneously, resulting in rapid single-stage decomposition. The $T_{max}$ increase with increase in nano-AlN content was due to the reduction of the chain mobility of the polymer matrix. The char yield was enhanced appreciably for HDPE/n-AlN composites—from 0.385 for HDPE to 35.1% for HDPE/20 vol% n-AlN.[107]

Thermal and mechanical properties of polyphenylene sulfide (PPS) composites containing 1–20 wt% of volcanic ash (VA) were significantly improved by VA. The PPS and its composites with different VA contents showed a relatively good thermostability under a nitrogen atmosphere because no significant mass loss (<0.5%) occurred until 450°C. A single degradation stage of neat PPS was initiated at ~450°C (at a heating rate of 20°C/min) and showed the maximum rate of weight loss at ~540°C. The $T_{10}$ of neat PPS was 508°C. The char yield of ~44% was observed at

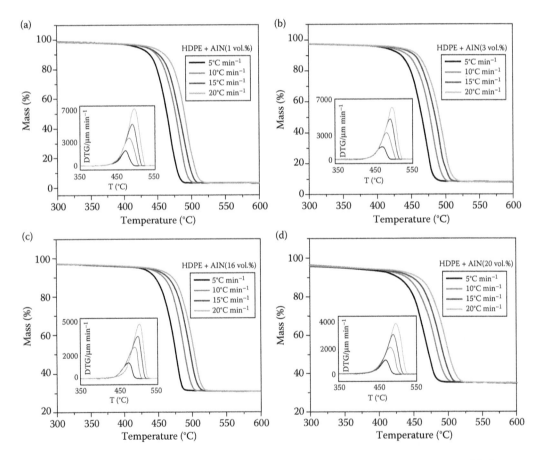

**FIGURE 8.8** **(See color insert.)** TGA mass loss (%) and inset images are derivative mass loss (DTG) versus temperature curves, at four different heating rates: 5°C/min, 10°C/min, 15°C/min, and 20°C/min, for different volume fraction loadings of nanocomposites: (a) HDPE/1 vol.% n-AlN, (b) HDPE/3 vol.% n-AlN, (c) HDPE/16 vol.% n-AlN, and (d) HDPE/20 vol.% n-AlN. (With kind permission from Springer Science+Business Media: *J. Therm. Anal. Calorim.,* Structural and thermal properties of HDPE/n-AlN polymer nanocomposites, 118, 2014, 1513–1530, Rajeshwari, P., Dey, T.K.)

700°C and it also increased with increase of the filler content. No significant changes at $T_{10}$ values occurred up to 20 wt% VA content in PPS–VA composites. The 10 wt% VA loading was enough for the highest thermal stability of VA/PPS composites. Further filler content led to no significant change in thermostability. The degradation mechanism involved random chain scission followed by disproportionation, cyclization, and depolymerization from radical chain ends. The higher thermal resistance could be attributed to the higher thermal conductivity of VA particles than PPS matrix.[108]

### 8.4.4 Thermal Stability of Polymer Nanocomposites with POSS Fillers

Organic–inorganic hybrids comprising epoxy resin and polyhedral oligomeric silsesquioxanes (POSSs) were prepared via *in situ* polymerization of the diglycidyl ether of bisphenol A (DGEBA) and (4,4′-diaminodiphenyl)methane (DDM). The phenyl POSSs had active functional tetrasilanol groups that took part in the ring-opening reaction with the oxirane moiety. The organic and inorganic moieties were connected by covalent bonds which enhanced the compatibility of the organic and inorganic phases. TGA showed that the incorporation of the POSSs into epoxy resin improved the thermal stability of the hybrids. For the pure epoxy resin, $T_g$ of 152°C was found from DSC curves. The presence of a single $T_g$ indicated that the hybrids were homogeneous. The hybrids containing 10, 20, 30, and 40 wt% POSSs had very similar $T_g$s, which were slightly lower than that of the neat epoxide, and they did not change significantly as the POSS content increased.[109] The thermostability of the hybrids increased with POSSs content; in a nitrogen atmosphere it was higher than that of the pure polymer. The values of the char yields of the hybrids also increased as the filler content increased. The char yields were higher than for the copolymers at high temperatures and were higher than the amounts of inorganic filler in different samples. The integral procedural decomposition temperature (IPDT) was correlated with the volatile parts of polymeric materials, and was used to evaluate the inherent thermal stabilities of polymeric materials. The IPDT of pure epoxide was 702°C. The IPDTs values of the hybrids with different POSS contents were higher than that of the pure polymer: 802, 1009, 1306, and 1559°C, respectively. The addition of POSS which formed silica, which cannot degrade further, influenced the high-temperature thermostability of NCs.[109]

Epoxy resin NCs based on tetramethylbiphenyl diglycidyl epoxy resins (TMBP) were prepared through *in situ* copolymerization with (4,4′-diaminodiphenyl)sulfone (DDS) and octapropylglycidylether silsesquioxane (OGPOSS). The thermal properties ($T_5 = 363$–367°C and $T_{10} = 378$–386°C) of the TMBP/OGPOSS NCs were almost the same as for pure epoxide, but their $T_g$s (185–219°C) were slightly higher for samples with increased OGPOSS content. The presence of OGPOSS did not change the mechanism of the thermal decomposition of TMBP epoxy matrix, which occurred in one step. The presence of rigid biphenyl groups and bulky methyl substitutions of TMBP resulted in the high $T_5$ of the TMBP systems. Nanoscale dispersion of OGPOSS cages in the epoxy matrix was considered as an important factor responsible for the increased thermal stability.[110] Polyhedral oligomeric 9,10-dihydro-9-oxa-10-phosphaphenanthrene-10-oxide (DOPO) or (octadiphenylsulfonyl)silsesquioxane (ODPSS) (1.25–5.00 wt%) were used as flame retardant additives to an epoxy resin (EP) of diglycidyl ether of bisphenol A (DGEBA), which were cured with (4,4′-diaminodiphenyl)sulphone (DDS).[111]

EP-1 with 5 wt% ODPSS exhibited similar $T_5$ and $T_{max}$, but its char residue was higher than in the case of the EP-neat (Figure 8.9 and Table 8.3). It was explained as the effect of the addition of the ODPSS with excellent thermal stability.

The incorporation of ODPSS caused only slight decrease of the decomposition temperature of EP composites. However, with the increasing content of DOPO, both $T_5$ and $T_{max}$ of EP composites were decreased, presumably due to the lower thermostability of DOPO in comparison with neat EP and ODPSS. A distinct synergistic effect of the mixture of ODPSS and DOPO on fire retardancy of the EP composites was observed.[111]

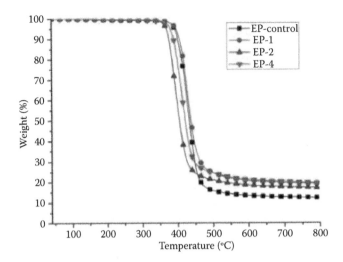

**FIGURE 8.9** TGA curves of EP composites in nitrogen. (Reprinted from *Polym. Degrad. Stab.*, 109, Li, Z., Yang, R. Study of the synergistic effect of polyhedral oligomeric octadiphenylsulfonyl-silsesquioxane and 9,10-dihydro-9-oxa-10-phosphaphenanthrene-10-oxide on flame-retarded epoxy resins, 233–239, Copyright 2014, with permission from Elsevier.)

**TABLE 8.3**
**TGA and DSC Data of EP Composites**

| Samples | $T_g$ (°C) | $T_5$ (°C) | $T_{10}$ (°C) | Residues at 800°C (%) |
|---------|-----------|-----------|--------------|----------------------|
| EP-neat | 178 | 392 | 430 | 12 |
| EP-1 | 182 | 396 | 429 | 19 |
| EP-2 | 165 | 369 | 393 | 17 |
| EP-3 | 170 | – | – | – |
| EP-4 | 175 | 380 | 414 | 20 |
| EP-5 | 178 | – | – | – |

*Source:* Reprinted from *Polym. Degrad. Stab.*, 109, Li, Z., Yang, R., Study of the synergistic effect of polyhedral oligomeric octadiphenylsulfonyl-silsesquioxane and 9,10-dihydro-9-oxa-10-phosphaphenanthrene-10-oxide on flame-retarded epoxy resins, 233–239. Copyright 2014, with permission from Elsevier.

The melt viscosity of the poly(lactic acid) (PLA) composites plasticized with poly(ethylene glycol) of $M_w = 8000$ g/mol was decreased with the incorporation of the reactive and nonreactive polyhedral oligomeric silsesquioxane derivatives (POSSs)—the lowest viscosity showed PLA NCs modified with epoxy substituted POSS (G-POSS, see Figure 8.10), which was liquid at processing temperature. Addition of 10 wt% PEG and 1–10 wt% POSS did not change the melting point of PLA, while the $T_g$ of the PLA decreased in the presence of POSSs.

The thermal stability of the composites, determined under nitrogen atmosphere at a heating rate of 20°C/min, was improved in the presence of all POSSs particles, for example, for 3 wt% POSSs content the $T_5$ and $T_{10}$ increased 7–23°C (Table 8.4).[112] POSS particles played a role in the physical barrier in the PLA/PEG composites, which limited the heat flux to the matrix.[113]

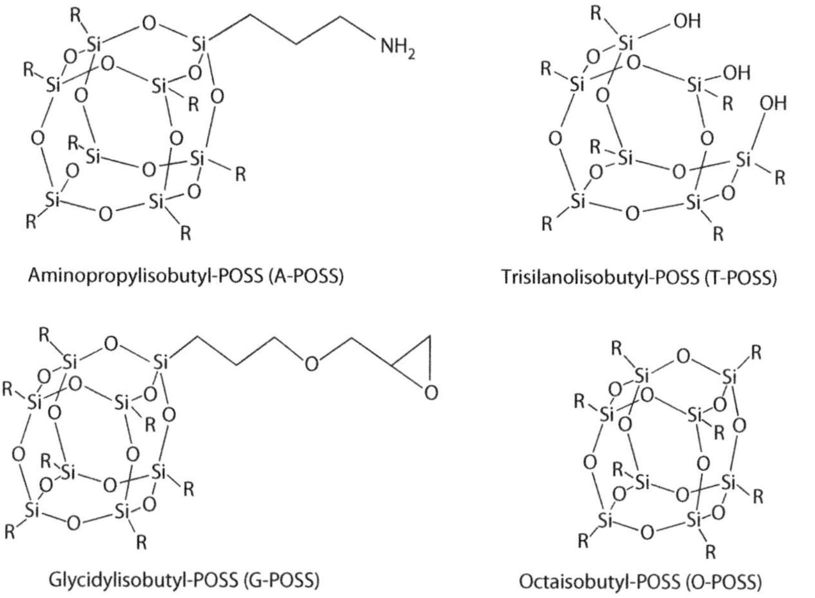

Aminopropylisobutyl-POSS (A-POSS)

Trisilanolisobutyl-POSS (T-POSS)

Glycidylisobutyl-POSS (G-POSS)

Octaisobutyl-POSS (O-POSS)

**FIGURE 8.10** Chemical structures of POSSs. (Kodal, M., Sirin, H., Ozkoc, G.: Effects of reactive and nonreactive POSS types on the mechanical, thermal, and morphological properties of plasticized poly(lactic acid). *Polym. Eng. Sci.* 2014. 54. 264–275. Copyright Wiley-VCH Verlag GmbH & Co. KGaA. Reproduced with permission.)

**TABLE 8.4**
**TGA Results of PLA, PLA/PEG, and PLA/PEG/POSS Composites**

| Materials | $T_5$ (°C) | $T_5$ (°C) | $T_{max}$ (°C) |
|---|---|---|---|
| PLA | 341.35 | 349.76 | 377.18 |
| PLA/PEG | 340.64 | 349.89 | 378.69 |
| PLA/PEG/A-POSS3 | 363.25 | 375.01 | 407.79 |
| PLA/PEG/T-POSS3 | 354.54 | 371.20 | 406.00 |
| PLA/PEG/G-POSS3 | 347.37 | 364.50 | 403.85 |
| PLA/PEG/O-POSS3 | 353.71 | 371.21 | 406.05 |

*Source:* Reprinted from Kodal, M., Sirin, H., Ozkoc, G.: Effects of reactive and nonreactive POSS types on the mechanical, thermal, and morphological properties of plasticized poly(lactic acid). *Polym. Eng. Sci.* 2014. 54. 264–275. Copyright Wiley-VCH Verlag GmbH & Co. KGaA. Reproduced with permission.

### 8.4.5 Thermal Properties of Polymer Nanocomposites with Carbon Fillers

In recent decades, a continuously growing interest has been observed in applications of carbonaceous fillers for fabrication of PNCs.

The $T_{onset}$ and the $T_{peak}$ were higher in the nanotube/polymer NCs. For example, addition of 5 wt% multiwall carbon nanotubes (MWCNT) caused a 24°C increase in $T_{onset}$, in comparison to that of the neat PAN.[42,114–116] It was assumed that the higher thermal conductivity in the NT/polymer composites could facilitate heat transfer within the composite and caused their improved thermostability. Dispersed NTs probably hindered the diffusion of degradation products, delayed the onset of degradation, and shifted $T_{peak}$ to higher temperatures.[117,118] A significant improvement of the thermal stability was found for poly(ethylene-*co*-vinyl acetate) (EVA), filled with the MWCNT as compared to neat EVA. First, the deacetylation of the EVA matrix occurred, then the volatilization of degradation product (the acetic acid) and the formation of a stable char occurred at higher temperatures.[118,119]

Functionalization of CNTs with carboxylic groups makes their dispersion properties in PNCs easier, and in a case of the single wall carbon nanotubes (SWCNTs) reduced enhancement of the thermal conductivity. Polymer composites with functionalized MWCNTs showed improved thermal properties. An addition of ~3 wt% MWCNTs caused the increased thermal conductivity of NCs because surface functionalization improved the bonding between polymer matrix and CNTs. The obtained results confirmed an important role of CNTs (and graphene flakes as well) as fillers in applications directed to improvement of thermal, mechanical, and electrical properties of NCs.[120]

The effect of montmorillonite (MMT) as a secondary filler in CNT/HDPE NCs on the dispersion of carbon nanotube reinforced high-density polyethylene (HDPE) nanocomposites (CNT/HDPE) was studied. The TGA results showed improved thermal stability of the NCs having 3 wt% of MMT and CNT. Addition of MMT avoided formation of CNTs aggregates and increased thermomechanical properties of NCs.[121]

The polyimide (PI) NCs containing up to 2.0 wt% of MWCNT exhibited especially good thermal properties. The dispersion of MWCNT in the PI matrix enhanced the thermal resistance of PI/MWCNT because the MWCNT bond chemically or physically interacts with the PI matrix. The acid treated MWCNT showed a better adhesion to the polymer matrix than the untreated MWCNT and may form hydrogen bonds with the C=O groups of the polyimide molecules.[122] The $T_g$ values of the PI/MWCNT NCs were gradually increased with the increasing loading of acid-treated MWCNT. The $T_g$ of the 2.0 wt% acid-treated PI/MWCNT nanocomposite was very high (282.7°C). The thermal properties of the PI NCs, under nitrogen, improved with increasing MWCNT content. The crosslinking of PI/MWCNT NCs was increased by good distribution of the MWCNT in the PI resins, and resulted in the increase of the thermal and electrical properties of the NCs. With increased MWCNT content from 0.5 to 2.0 wt%, the IDT of the PI/MWCNT NCs increased from 533°C to 555°C, and it was dependent on the degree of dispersion of the MWCNT in the PI matrix. The IDT and the $T_{max}$ increased with higher MWCNT contents. An IPDT reached maximum value (2586°C) for 1 wt% of MWCNT.[122]

Polyimide containing diphenyl sulfone moiety was synthesized by the polycondensation reaction of 4,4′-diamino diphenyl sulfone with pyromellitic anhydride in the presence of *iso*-quinoline in *m*-cresol solution. A nanocomposite of this polyimide in the main chain with 9.3–10.9 wt% of homogeneously dispersed silver nanoparticles was prepared. The thermal properties of neat PI and PI–silver NCs film were studied by TGA (in a nitrogen atmosphere) and DSC experiments. The $T_5$ and $T_{10}$ of PI–Ag NC reached 375°C and 400°C, respectively, and were higher than for the pure PI (330°C and 375°C, respectively). The higher thermostability of NCs arose from the presence of inorganic Ag NPs in the PI matrix. The char yield at 600°C also increased from 53% to 61%.[123]

Small CB agglomerates adsorbed onto graphene surfaces were present in a composite of styrene-butadiene rubber (SBR) with novel carbon black–reduced graphene (CB–RG) hybrid filler. The high surface area of graphene and the strong interfacial p–p interaction between SBR and graphene resulted in the higher $T_g$ values of the SBR/CB–RG blend than in a case of the SBR/CB blend. It was observed that the decomposition of graphene oxide (GO) took place in two steps. In the temperature range less than 150°C, the weight loss of GO was mainly assigned to the evaporation of water absorbed on the graphene sheets. Labile oxygen-containing groups, such as −COOH, −OH, etc. easily undergo the pyrolysis, resulting in the weight loss of GO at 200–300°C. The thermostability of SBR composites filled with CB and CB–RG (10:2) showed no significant difference. The TGA curves of SBR/CB–RG composites slightly shifted toward higher temperatures in the range from 400°C to 460°C, in comparison to SBR/CB composites. The emission of small gaseous molecules, leading to better thermal stability, was probably inhibited as a result of the high aspect ratio of monodispersed graphene layers which acted as a barrier.[124]

Reinforced composites materials based on a blend (70:30) of poly(ethylene terephthalate) (PET) and polypropylene (PP) with exfoliated graphite nanoplatelets (GNP) were prepared by melt extrusion followed by injection molding. Styrene-ethylene-butylene-styrene-$g$-maleic anhydride (10 phr) served as a compatibilizing additive in PET/PP/GNP composites. A dispersion of 3 phr GNPs content in PET/PP was homogenous and exhibited the highest thermal stability, which generally improved with increasing GNP content. Any chemical interaction between GNP and the polymer matrix was not confirmed on the basis of the analysis of FTIR spectra. GNP showed high thermal resistance. A single step decomposition process of the blends and composites in a nitrogen atmosphere was observed, which indicated effective compatibilization of the blends. The $T_{10}$, $T_{50}$, and $T_{max}$ were high for all the composites compared with the neat PET/PP blend. The increased thermal stability of the composites was assigned to the high aspect ratio of GNPs, which served as a barrier and prevented the emission of gaseous molecules during thermal degradation. The great improvement of thermostability at 3.0 phr GNP was attributed to the homogenous dispersion of the GNPs in the matrix at this filler level, which hindered the oxygen diffusion by formation of char layers on the composites' surfaces, improving their thermal stability. The highest char yield was obtained with the 5.0 phr GNP content, due to very high thermostability of the graphene platelets.[125]

The poly(ethylene-$co$-vinyl acetate) (EVA) composites filled with expanded graphite (EG) platelets, with and without anionic sodium dodecyl sulfate (SDS) modification, showed very good distribution of the modified EG platelets in the EVA matrix and an improved interfacial adhesion between the polymer and the SDS–EG particles. The addition of EG improved the thermostability of EVA, and this stabilizing effect was further enhanced for the EG treated with SDS. The $T_g$s were higher with increasing both the EG and modified EG content. The incorporation of EG improved the thermal stability of EVA because the EG was thermally stable up to 850°C. The $T_{max}$ for the second degradation step of the EVA/EG samples increased with an increase of EG content. It was proposed that the EG layers affected the diffusion of oxygen and volatile products through the composites. The thermostability of EVA was much better, even in the presence of 2 wt% content of SDS-EG, when the $T_{max}$ was improved by 17°C, while in the case of unmodified EG the increase was only 3°C. The $T_{max}$ values were almost the same for the samples containing 10 wt% EG and EG–SDS. The thermal properties of EVA/EG and EVA–SDS–EG composites mainly depend on two parameters: (1) a strength of interaction between the polymer free radical chains and volatile degradation products and the filler and (2) the amount of filler that may interact with the free radical chains and volatile degradation products, and delay the degradation of the polymer and/or the diffusion of the volatile degradation products out of the polymer.[126]

The core–shell hybrid materials of polyarylene nitrile ethers (PEN) with phthalocyanine copper (CuPc) and MWCNTs were prepared via solution casting method. These new CuPc/MWCNTs hybrid materials showed good dispersion and compatibility in the PEN polymer matrix, and led to the excellent thermal, mechanical, and dielectric properties.[127]

An observed improvement of $T_g$ for CuPc/MWCNTs/PEN composite films, compared with pure PEN (Table 8.5), was ascribed to the interaction of nitrile groups between fillers and PEN chains, which can enhance the stability of polymer structure. The $T_5$ value of composite films, determined in nitrogen atmosphere (at a heating rate of 20°C/min) was 518°C (see Table 8.5), when the CuPc/MWCNTs content was 3 wt%, which was higher than that of pure PEN. This result indicated that the thermostabilities of PEN matrix began to improve as the CuPc/MWCNTs content reached 3 wt%. The $T_{10}$ values of CuPc/MWCNTs/PEN composites increased from 534°C up to 588°C, and the char yields at the temperature of 800°C rose from 63.3% to 76.7%. These enhancements were mainly assigned to homogeneous dispersion of NPs and good compatibility (good matrix–fillers interaction), thermal conductivity and the barrier effect of the NTs.[127]

Plasticized PLA-based nanocomposites (PLA–EPO–xGnP NCs), prepared by melt blending of the matrix with 5 wt% of epoxidized palm oils (EPO) and different amounts of graphene nanoplatelets (xGnP) as a filler, showed a substantial increase in thermal stability with increasing content of xGnP. A significant decrease of $T_g$ was observed up to 0.3 wt% incorporation of xGnP into p-PLA,

**TABLE 8.5**
**Thermal Properties of Pure PEN Film and CuPc/MWCNTs/PEN Composite Films**

| Mass Fraction (wt%) | $T_g$ (°C) | $T_5$ (°C) | $T_{10}$ (°C) | Char Yield (wt%) |
|---|---|---|---|---|
| 0 | 195 | 514 | 534 | 63.31 |
| 1 | 200 | 508 | 564 | 74.53 |
| 2 | 202 | 512 | 577 | 75.16 |
| 3 | 203 | 518 | 582 | 74.79 |
| 4 | 205 | 518 | 587 | 75.82 |
| 5 | 207 | 522 | 588 | 76.68 |

*Source:* With kind permission from Springer Science+Business Media: *J. Polym. Res.*, Effect of CuPc@MWCNTs on rheological, thermal, mechanical and dielectric properties of polyarylene ether nitriles (PEN) terminated with phthalonitriles, 21, 2014, 525, Long, Y. et al.

which improved thermostability due to an excellent insulating ability and mass transport barrier to the volatile decomposition products, even at extremely low contents.[128–130] The addition of xGnP increased the initial decomposition temperature. With 1 wt% of xGnP the $T_5$ of NC increased to 421°C, compared to 397.6°C for pure PLA. A substantial shift from 326 to 361 was observed for the $T_{95}$, while $T_{max}$ decreased from 429°C to 395°C of PLA–EPO/1.0 wt% xGnP. The xGnP formed a heat barrier, which enhanced the overall thermal stability of the polymer NCs, as well as took a part in the formation of char during thermal decomposition. The xGnP layers help to accumulate heat, which could accelerate the decomposition process. The improvement in thermostability could also be assigned to the "tortuous path" effect of xGnPs, which delayed the liberation of volatile degradation products. The high content of xGnP or well-dispersed xGnP in polymer matrix was expected to force the degradation products to go through more tortuous path and hence improved the thermal stability. Similar results were reported for other graphene-based nanocomposites.[131]

The synergistic effect of adipic acid ester of pentaerythritol (AAPE) with zinc and calcium stearates ($CaSt_2/ZnSt_2$) on thermostability of poly(vinyl chloride) (PVC) was studied by TGA, thermal aging test, and the conductivity measurements, which was carried out in an argon atmosphere. The addition of $CaSt_2/ZnSt_2$ and AAPE slightly improved $T_{onset}$ stability (from 220°C for PVC with 4 wt% $CaSt_2/ZnSt_2$, to 224.8–231.5°C with 4 wt% of both additives: 1:19 ÷ 1:5.7 wt/wt) and long-term thermal resistance of PVC. The results of UV–visible spectroscopy revealed that $CaSt_2/ZnSt_2/$ AAPE may avoid the formation of conjugated double bonds. Probably AAPE chelated $ZnCl_2$ and avoided a dehydrochlorination of PVC, increasing its thermal stability. The synergistic effect of $CaSt_2/ZnSt_2$ and AAPE was stronger than in a case of polyethylene and $CaSt_2/ZnSt_2$, which resulted in better compatibility with PVC and lower melting point of AAPE.[132]

A novel material (CB–PI) was obtained from polyimide (PI) and carbon black (CB). The CB and PI were chemically crosslinked in the synthesized composite CB–PI (Figure 8.11), leading to the extremely high thermal and mechanical properties of the high-performance CB–PI composite. The huge increase of $T_g$ of CB–PI, by 204°C from 379°C to 583°C, in comparison to that of neat PI was observed. In a nitrogen atmosphere, the thermal decomposition temperature ($T_d$) of CB–PI increased significantly, by 76°C—from 508°C to 584°C, in comparison with PI. At a temperature of 508°C PI showed only a 1% weight loss. The $T_d$ of PI–CB was almost independent of the CB content (10–40 wt%) and occurred at 573–584°C, while $T_{d10}$ ranged from 620°C to 628°C. The chemical crosslinking of CB and PI occurred at higher temperature than was expected (around 350°C), that is, during imidization process.[133]

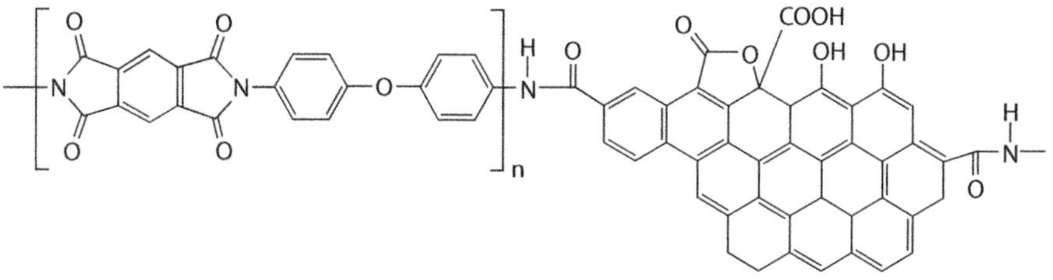

**FIGURE 8.11**    Chemical structure of PI–CB composites. (Kwon, J. et al.: Fabrication of polyimide composite films based on carbon black for high-temperature resistance. *Polym. Compos.* 2014. 35. 2214–2220. Copyright Wiley-VCH Verlag GmbH & Co. KGaA. Reproduced with permission.)

The homogeneous NCs films of poly(methyl methacrylate) (PMMA) with multiwall carbon nanotubes (MWCNTs), amino modified, and silver complexed, were synthesized through *in situ* free radical polymerization. Silver metal nanohybrids (Ag/MWCNTs) were prepared by two strategies: (1) reduction of metal salt in the presence of sodium dodecyl sulfate and (2) *in situ* growth from $AgNO_3$ aqueous solution. The amino functionalization which was carried out by ball milling with ammonium bicarbonate ($NH_4HCO_3$) improved the dispersion of MWCNT in monomer and provided a new class of radiation resistant NCs. Microscopic studies confirmed the homogeneous dispersion of amino functionalized and metal doped MWCNTs in polymer matrix, increasing the thermal stability, thermomechanical strength, $T_g$s, and thermal conductivity of NCs even at 0.25 wt% addition of modified nanofiller. At ~50°C, the thermal stability and thermal conductivity of NCs film was increased up to 63% at 0.25 wt% content of Ag/MWCNTs, as compared to the neat polymer. The increase of thermomechanical properties of NCs was observed up to 85% at 100°C in the presence of adsorbed surfactant. The NC films also showed a very good resistance against UV/ $O_3$ radiations in comparison with pure PMMA.[134]

Surface functionalized carbon black (FCB) reinforced polybenzoxazine (PBZ) NCs, with a changing content of FCB (0.5, 1.0, and 1.5 wt%), were prepared by ring opening polymerization and thermal curing. The FCB NPs were homogeneously dispersed in the PBZ matrix. The covalent and hydrogen bonding interactions between the FCB and the PBZ matrix hindered the movement of the polymer chain segments and slowed down molecular motions of FCB–PBZ NCs. The $T_g$ values of the NCs were increased from 180°C for neat PBZ to 198°C for 1.5% FCB–PBZ. A significant increase of the thermal stability of FCB–PBZ NCs was observed, as compared to neat PBZ. No apparent weight loss below 300°C was noticed. The $T_5$ increased to 325°C, as compared to 312°C for pure PBZ, and $T_{10}$ increased to 396°C. The incorporation of highly thermally stable FCB into the PBZ improved thermal resistance of FCB–PBZ NCs.[135] It was assumed that the greater interfacial interaction and compatibility between the two phases strengthened dispersion of FCB into the polymer matrix at a molecular level. The presence of FCB delayed the decomposition of the PBZ, which usually resulted from the cleavage of the oxazine rings. The residue of ash from FCB and the carbonized PBZ after the volatilization of the ester, $CO_2$, or $H_2O$ (formed by the decomposition of PBZ) also affected the thermostability of NCs. The final residue of char formed from the NCs at 800°C increased with the increase of the FCB content.[135]

Thermal and thermooxidative degradation high-density polyethylene (HDPE)/fullerene (C60) composites were studied. The influence of C60 on the thermal stability of HDPE was different under nitrogen and air atmosphere. TG–FTIR and pyrolysis–gas chromatography–mass spectrometry (Py–GC–MS) showed that in $N_2$ the addition of C60 increased the $T_{onset}$ by ~10°C (up to 464°C), as compared to neat HDPE. The addition of C60 also substantially improved the thermostability of HDPE in air. With the 2.5 wt% content of C60 the $T_{onset}$ increased by ~91°C (up to 402°C), in comparison to 311°C for pure HDPE.[136] The results of viscoelastic behavior and gel content suggested that C60 trapped the alkyl radicals and alkyl peroxide radicals, inhibited hydrogen abstraction to reduce the scission of polymer chains, and slowed down their thermal degradation. It was also found that in the absence of C60 or with low C60 concentration, hydrogen abstraction occurred and led to the formation of alkyl and alkyl peroxide radicals, which played an important role in the thermal oxidative degradation, accelerating the chain scission.[136]

The thermal properties of polypropylene NCs containing SWNTs, SWNTs functionalized with dodecyl units, and vapor-grown carbon fibers were studied. The thermal decomposition temperature of NCs was significantly dependent on extruder processing conditions, on the fraction of polymer chains stabilized in the interphase, and on the type of NTs incorporated.[137]

Nanocomposite polymer thermal interface materials (TIMs) containing graphene and multilayer graphene (MLG) showed the extremely high thermal conductivity enhancement (TCE) factors at low filler contents. The TCE of 2300% at 10% filler content was the highest ever reported. A possibility of achieving K ~14 W/mK in the commercial thermal grease via the addition of only 2% of the optimized graphene—MLG nanocomposite mixture was also described. The graphene-based TIMs had thermal resistance RTIM reduced by orders-of-magnitude and can be produced in large quantities on an industrial scale cheaply.[138]

Thermally conductive NCs were prepared from epoxy resins and different contents of multilayer graphitic nanoplatelets. A linear increase in the thermal conductivity (TC) of the NC was observed for higher graphene loading.[139] The incorporation of 5 wt% graphene oxide (GO) into the epoxy matrix caused a four times higher TC than the neat resin, and was much more increased by incorporation of 20 wt% GO, or up to 20 times for a graphene content of 40 wt%. At similar contents, functionalized graphite was a more effective filler in improving the TC of the PNCs. The thermal conductivity of PNCs was further improved with silane-crosslinked GO–NCs[139] and in NCs with strong interfacial interactions.[140,141] Preparation of silicone–graphene NCs with the increased TC and dimensional stability[140] and dielectric epoxy thermosets with the increased TC[142] were also described. Stable GO-encapsulated carbon nanotube hybrid NCs showed a high TC and dielectric constant.[143]

Recently, the improved thermal properties of polyimides and controlling thermoelectrical properties of polyaniline (PANI) films were reported.[144–146]

### 8.4.6 THERMAL STABILITY OF POLYMER NANOCOMPOSITES WITH ORGANIC FILLERS

Usually, the addition of cellulosic fillers led to a decrease of thermostability of composites with commodity thermoplastics (polyethylene, polypropylene, etc.). However, modification of polyvinyl alcohol (PVA) with cellulose nanofibers caused an increase of thermal stability of PVA. The results of $T_{10}$ and $T_{max}$ for PVA and PVA/cellulose NCs are described in Table 8.6.

**TABLE 8.6**
**TGA Results for PVA and PVA/Cellulose Nanocomposites**

| Sample | $T_{max}$ (°C) | $T_{10}$ (°C) |
|---|---|---|
| PVA | 252.2 | 235.9 |
| PVA/NC(2) | 265.7 | 243.7 |
| PVA/NC(5) | 282.5 | 248.7 |
| PVA/NC(7) | 293.2 | 253.0 |
| PVA/NC(10) | 295.6 | 256.1 |

*Source:* Adapted from Kakroodi, A.R., Cheng, S., Sain, M., Asiri, A., *J. Nanomater.*, 7 pages, Article ID 903498. doi: 10.1155/2014/903498. Copyright 2014, Hindawi Publishing Corporation. With permission.

In the case of an inclusion of cellulose nanofibers to PVA, an enhancement in degradation temperature of NCs was observed, which was ascribed to (1) higher thermal stability of nanofibers (due to separation of less stable components such as lignin and hemicellulose) and (2) reduction of the mobility of polymer chains (in PVA) caused by homogenous distribution of cellulose nanofibers in the matrix. For instance, $T_{10}$ of PVA was 235.9°C, while it was improved to 243.7°C in the nanocomposite with 2% cellulose nanofiber. The $T_{10}$ of PVA/NC(10%) was 256.1°C. Thermal degradation of PVA and NCs occurred in two stages: (1) 230–330°C and (2) 390–450°C. It was shown that decomposition temperature of PVA modified with cellulose nanofiber was increased, while decomposition rates decreased.[147]

Melt blending was used for preparation of intumescent flame retardant polyurethane–starch (IFR–PU–starch) NCs. Microencapsulated ammonium polyphosphate (MCAPP) served as compatibilizer, fire retardant, and water resistant additive to NCs. SEM confirmed homogeneity of starch distribution in the PU matrix. In the presence of 5–10 wt% of starch, the slightly decreased thermal stability of NCs was determined by TGA, under a nitrogen. The $T_5$ values: 281°C (with 5 wt% of starch and 25 wt% IFR) and 275°C (with 10 wt% of starch and 20 wt% IFR) were lower, as compared to 297°C for unmodified PU. However, char yields were increased from 3 wt% (for neat PU) to 24.7–29.3 for these NCs. For PU sample, containing 20 wt% IFR and 10 wt% starch limiting oxygen index (LOI) increased to 40.0 (in comparison to 22.0 for neat PU) and UL94 V0 rating was observed.[148]

## 8.5  SUMMARY AND CONCLUSIONS

Many examples of thermally stable PNCs, described in this chapter, confirm their good heat resistance. For instance, excellent thermal and flame resistant properties exhibit different polymer nanocompsites filled with silica or POSS, or modified with silicones,[149,150] but outstanding, extremely high thermal resistance (above 500°C) was achieved especially in the case of different polyimide nanocomposites. The improved thermal stability of PNCs was achieved by the addition of thermally stable modifiers (including oligomeric compounds), which provided good compatibility, due to low migration characteristics.[41,42] Obviously, literature cited in this chapter doesn't cover all world achievements in this field.

A few years ago it was predicted[23] that in a nearer future a new collection of materials might be offered by PNCs technology in many market sectors: automotive, packaging, household goods, electrical and electronic goods, aerospace, etc. In the future, the new commercial products may also include new applications, for example, engine covers, lighting accessories, batteries, computer chips, memory devices, catalytic converts, nanoparticles in inkjets and cosmetics, biosensors for diagnostics, etc. In the far future, new applications of PNCs in automotive, bionanotechnology, and aerospace fields can be expected.

## REFERENCES

1. Fejgin, J. 1997. Chapter 1: Polimery termoplastyczne. In *Chemia Polimerów, vol. 2*, ed. Florjańczyk Z., Penczek S., 23–24. Warsaw (Poland), Oficyna Wydawnicza Politechniki Warszawskiej.
2. Whelan, T. 1994. *Polymer Technology Dictionary*, 237. London, Chapman & Hall.
3. Sęk, D. 1998. Chapter 6: Polimery termoodporne. In *Chemia Polimerów, vol. 3*, ed. Florjańczyk, Z., Penczek, S., 89–119. Warsaw (Poland), Oficyna Wydawnicza Politechniki Warszawskiej.
4. Matyjaszewski, K. 1998. Chapter 7: Polimery nieorganiczne i organometaliczne. In *Chemia Polimerów, vol. 3*, ed. Florjańczyk, Z., Penczek, S., 124–134. Warsaw (Poland), Oficyna Wydawnicza Politechniki Warszawskiej.
5. Dvornic, P.R. 2008. High temperature stability of polysiloxanes. In *Gelest Catalog 4000–A. Silicon Compounds: Silanes & Silicones: A Survey of Properties and Chemistry*, ed. Arkles, B., Larson, G., 441–454. Morrisville, PA, Gelest, Inc.
6. Doyle, C.D. 1961. Estimating thermal stability of experimental polymers by empirical thermogravimetric analysis. *Anal. Chem.* 33:77–79.

7. Zhang, X.H., Huang, L.H., Chen, S., Qi, G.R. 2007. Improvement of thermal properties and flame retardancy of epoxy-amine thermosets by introducing bisphenol containing azomethine moiety. *Express Polym. Lett.* 1:326–332.
8. Guo, B., Jia, D., Cai, C. 2004. Effects of organo-montmorillonite dispersion on thermal stability of epoxy resin nanocomposites. *Eur. Polym. J.* 40:1743–1748.
9. Wunderlich, B. 1990. *Thermal Analysis*, 137–140. New York, Academic Press.
10. Dean, J.A. 1995. *The Analytical Chemistry Handbook*, 15.1–15.5. New York, McGraw Hill, Inc.
11. Pungor, E. 1995. *A Practical Guide to Instrumental Analysis*, 181–191. Boca Raton, FL, CRC Press/Taylor & Francis.
12. O'Neill, M.J. 1964. The analysis of a temperature-controlled scanning calorimeter. *Anal. Chem.* 36:1238–1245. doi:10.1021/ac60213a020.
13. Wunderlich, B. 1980. *Macromolecular Physics, Vol. 3*, Chapter 8, Table VIII.6. New York, Academic Press.
14. Biswas M., Sinha R.S. 2001. Recent progress in synthesis and evaluation of polymer–montmorillonite nanocomposites. *Adv. Polym. Sci.* 155:167–221.
15. Report: Nanoscience and Nanotechnologies. 2004. London, England, The Royal Society & the Royal Academy of Engineering.
16. Tang, J., Wang, Y., Liu, H., Xia, Y., Schneider, B. 2003. Effect of processing on morphological structure of polyacrylonitrile matrix Nano–ZnO composites. *J. Appl. Polym. Sci.* 90:1053–1057.
17. Zheng, Y.P., Zheng Y., Ning, R.C. 2003. Effects of nanoparticles $SiO_2$ on the performance of nanocomposites. *Mater. Lett.* 57:2940–2944.
18. Jayaraman, K. 2004. Recent advances in polymer nanofibers, *J. Nanosci. Nanotechnol.* 4:52–65.
19. Han, J., Jaffe, R.J. 1998. Energetics and geometries of carbon nanocone tips. *Chem. Phys.* 108:2817–2823.
20. Gorga, R.E., Cohen, R.E. 2004. Toughness enhancements in poly(methyl methacrylate) by addition of oriented multiwall carbon nanotube. *J. Polym. Sci. B: Polym. Phys.* 42:2690–2702.
21. Jin, Z., Pramoda, K.P., Xu, G., Goh, S.H. 2001. Dynamic mechanical behavior of melt processed multiwalled carbon nanotube/poly(methyl methacrylate) composites. *Chem. Phys. Lett.* 337:43–47.
22. Zeng, J., Saltysiak, B., Johnson, W.S., Schiraldi, D.A., Kumar, S. 2004. Processing and properties of poly (methyl methacrylate)/carbon nano fiber composites. *Composites, B: Engineering* 35:173–178.
23. F. Hussain, M. Hojjati, M. Okamoto, Gorga, R.E. 2006. Review article: Polymer–matrix nanocomposites, processing, manufacturing, and application: An overview. *J. Compos. Mater.* 40:1511–1575.
24. Węgrzyn, M., Galindo, B., Benedito, A., Gimenez, E. 2015. Morphology, thermal, and electrical properties of polypropylene hybrid composites co-filled with multi-walled carbon nanotubes and graphene nanoplatelets. *J. Appl. Polym. Sci.* 132: 42793.
25. Lau, K.T., Lu, M., Lam, C.K., Cheung, H.Y., Sheng, F.L., Li, H.L. 2005. Thermal and mechanical properties of single-walled carbon nanotube bundle-reinforced epoxy nanocomposites: The role of solvent for nanotube dispersion. *Compos. Sci. Technol.* 65:719–725.
26. Blumstein, A. Polymerization of adsorbed monolayers. II. Thermal degradation of the inserted polymer. *J. Polym. Sci. A* 3:2665–2672.
27. Burnside, S.D., Giannelis, E.P. 1995. Synthesis and properties of new poly(dimethylsiloxane) nanocomposites. *Chem. Mater.* 7:1597–1603.
28. Qin, H., Su, Q., Zhang, S., Zhao, B., Yang, M. 2003. Thermal stability and flammability of polyamide 66/montmorillonite nanocomposites. *Polymer* 44:7533–7538.
29. Ide, F., Hasegawa, A. 1974. Studies on polymer blend of nylon 6 and polypropylene or nylon 6 and polystyrene using the reaction of polymer. *J. Appl. Polym. Sci.* 18:963–974.
30. Zanetti, M., Camino, G., Peichert, P., Mülhaupt, R. 2001. Thermal behaviour of poly(propylene) layered silicate nanocomposites. *Macromol. Rapid Commun.* 22:176–180.
31. Oriakhi, C.O. 1998. Nano sandwiches. *Chem. Britain* 34:59–62.
32. Okada, A., Usuki, A. 1995. The chemistry of polymer–clay hybrids. *Mater. Sci. Eng.* C3:109–115.
33. Kozima, Y., Usuki, A., Kawasumi, M. et al. 1993. Mechanical properties of nylon-6 clay hybrid. *J. Mater. Res.* 8:1185–1189.
34. Usuki, A., Kawasumi, M., Kojima, Y., Okada, A., Kurauchi, T., Kamigaito, O.J. 1993. Swelling behavior of montmorillonite cation exchanged for ω-amino acids by ε-caprolactam. *Mater. Res.* 8:1174–1178.
35. Christopher, O.O., Lerner, M. 2001. Nanocomposites and Intercalation Compound, *Encyclopedia of Physical Science and Technology, Vol. 10,* 3rd ed., San Diego, Academic Press.
36. Kornmann, X., Linderberg, H., Bergund, L.A. 2001. Synthesis of epoxy–clay nanocomposites: Influence of the nature of the clay on structure. *Polymer* 42:1303–1310.

37. Zhang, K., Wang, L., Wang, F., Wang, G., Li, Z. 2004. Preparation and characterization of modified-clay-reinforced and toughened epoxy–resin nanocomposites. *J. Appl. Polym. Sci.* 91:2649–2652.

38. Ingram, S., Rhoney, I., Liggat, J.J., Hudson, N.E., Pethrick, R.A. 2007. Some factors influencing exfoliation and physical property enhancement in nanoclay epoxy resins based on diglycidyl ethers of bisphenol A and F. *J. Appl. Polym. Sci.* 106:5–19.

39. Leszczyńska, A., Njuguna, J., Pielichowski, K., Banerjee, J.R. 2007. Polymer/montmorillonite nanocomposites with improved thermal properties: Part I. Factors influencing thermal stability and mechanisms of thermal stability improvement. *Thermochim. Acta* 453:75–96.

40. Alexandre, M., Dubois, P. 2000. Polymer-layered silicate nanocomposites: Preparation, properties and uses of a new class of materials. *Mater. Sci. Eng.* 28:1–63.

41. Leszczyńska, A., Njuguna, J., Pielichowski, K., Banerjee, J.R. 2007. Polymer/montmorillonite nanocomposites with improved thermal properties: Part II. Thermal stability of montmorillonite nanocomposites based on different polymeric matrixes. *Thermochim. Acta* 454:1–22.

42. Ge, J., Hou, H., Li, Q. et al. 2004. Assembly of well-aligned multiwalled carbon nanotubes in confined polyacrylonitrile environments: Electrospun composite nanofiber sheets. *J. Am. Chem. Soc.* 126:15754–15761.

43. Corcione, C.E., Frigione, M. 2012. Characterization of nanocomposites by thermal Analysis. *Materials* 5:2960–2980.

44. Nguyen, Q.T., Baird, D.G. 2006. Preparation of polymer–clay nanocomposites and their properties. *Adv. Polym. Technol.* 25:270–285.

45. Koo, J., Pilato, L. 2005. Polymer nanostructured materials for high temperature applications. *SAMPE J.* 41(2):7.

46. Transparent Nanocomposites for Aerospace Applications. 2004. *Advanced Composites Bulletin*, February 2004:5.

47. Olewnik, E., Garman, K., Piechota, G., Czerwiński, W. 2012. Thermal properties of nanocomposites based on polyethylene and n-heptaquinolinum modified montmorillonite. *J. Therm. Anal. Calorim.* 110:479–484.

48. Sibeko, M.A., Luyt, A.S. 2014. Preparation and characterisation of vinylsilane crosslinked low-density polyethylene composites filled with nano clays. *Polym. Bull.* 71:637–657.

49. Wang, B., Huang, H.-X. 2013. Effects of halloysite nanotube orientation on crystallization and thermal stability of polypropylene nanocomposites. *Polym. Degrad. Stab.* 98:1601–1608.

50. Marcilla, A., Gomez, A., Menargues, S., Ruiz, R. 2005. Pyrolysis of polymers in the presence of a commercial clay. *Polym. Degrad. Stab.* 88:456–460.

51. Du, M.L., Guo, B.C., Jia, D.M. 2006. Thermal stability and flame retardant effects of halloysite nanotubes on poly(propylene). *Eur. Polym. J.* 42:1362–1369.

52. Lecouvet, B., Sclavons, M., Bourbigot, S., Devaux, J., Bailly, C. 2011. Water-assisted extrusion as a novel processing route to prepare polypropylene/halloysite nanotube nanocomposites: Structure and properties. *Polymer* 52:4284–4295.

53. Lecouvet, B., Gutierrez, J.G., Sclavons, M., Bailly, C. 2011. Structure–property relationships in polyamide 12/halloysite nanotube nanocomposites. *Polym. Degrad. Stab.* 96:226–235.

54. Handge, U.A., Hedicke-Hoechstoetter, K., Altstaedt, V. 2010. Composites of polyamide 6 and silicate nanotubes of the mineral halloysite: Influence of molecular weight on thermal, mechanical and rheological properties. *Polymer* 51:2690–2699.

55. Jafari, S.H., Hesabi, M.N., Khonakdar, H.A., Asadinezhad, A. 2012. Cyclic olefin copolymer/layered silicate nano-composite: Solid and melt viscoelastic properties and degradation behavior. *J. Polym. Res.* 19:9911.

56. Geyer, B., Roehner, S., Lorenz, G., Kandelbauer, A. 2015. Improved thermostability and interfacial matching of nanoclay filler and ethylene vinyl alcohol matrix by silane-modification. *J. Appl. Polym. Sci.* 132:41227.

57. Balachandran, M., Bhagawan, S.S. 2012. Mechanical, thermal and transport properties of nitrile rubber (NBR)-nanoclay composites. *J. Polym. Res.* 19:9809.

58. Hatami, L., Haddadi-Asl, V., Roghani-Mamaqani, H., Ahmadian-Alam, L., Salami-Kalajahi, M. 2011. Synthesis and characterization of poly (Styrene-*co*-Butyl Acrylate)/clay nanocomposite latexes in miniemulsion by AGET ATRP. *Polym. Compos.* 32:967–975.

59. Ahmadian-Alam, L., Haddadi-Asl V., Roghani-Mamaqani, H., Hatami, L., Salami-Kalajahi, M. 2012. Use of clay-anchored reactive modifier for the synthesis of poly (styrene-*co*-butyl acrylate)/clay nanocomposite via *in situ* AGET ATRP. *J. Polym. Res.* 19:9773.

60. Hussain, F., Hojjati, M., Okamoto, M., Gorga, R.E. 2006. Review article: Polymer–matrix nanocomposites, processing, manufacturing, and application: An overview. *J. Compos. Mater.* 40:1511–1575.
61. Chen, J.S., Christopher K.O., Weisner, U., Giannelis. 2002. Study of the interlayer expansion mechanism and thermal-mechanical properties of surface-initiated epoxy nanocomposites. *Polym. Sci.* 43:4895–4904.
62. Camino, G., Sgobbi, R., Colombier, S., Scelza, C. 2000. Investigation of flame retardancy in EVA. *Fire Mater.* 24:85–90.
63. Azeez, A.A., Rhee, K.Y., Park, S.J., Hui, D. 2013. Epoxy clay nanocomposites—Processing, properties and applications: A review. *Compos. B: Eng.* 45:308–320.
64. Qin, H., Zhang, S., Zhao, C. et al. 2004. Thermal stability and flammability of polypropylene/montmorillonite composites. *Polym. Degrad. Stab.* 85:807–813.
65. Ray S.S., Okamoto M. 2003. Polymer/layered silicate nanocomposites: A review from preparation to processing. *Prog. Polym. Sci.* 28:1539–1641.
66. Ray, S.S., Bousima, M. 2005. Biodegradable polymers and their layered silicate nanocomposites. In: Greening the 21st century materials world, *Prog. Mater. Sci.* 50:962–1079.
67. Becker O., Varley R.J., Simon G.P. 2004. Thermal stability and water uptake of high performance epoxy layered silicate nanocomposites. *Eur. Polym. J.* 40:187–195.
68. Zhu J., Uhl F.M., Morgan A.B. et al. 2001. Studies on the mechanism by which the formation of nanocomposites enhances thermal stability. *Chem. Mater.* 13:4649–4654.
69. Zhang K., Wang L., Wang F., Wang G., Li Z. 2004. Preparation and characterization of modified-clay-reinforced and toughened epoxy–resin nanocomposites. *J. Appl. Polym. Sci.* 91:2649–2652.
70. Lakshmi, M.S., Narmadha, B., Reddy, B.S.R. 2008. Polymer degradation and stability. *Polym. Degrad. Stabil.* 93:201–213.
71. Timmerman, J., Hayes, B., Seferis, J. 2002. Nanoclay reinforcement effects on the cryogenic micro cracking of carbon fiber/epoxy composites. *Compos. Sci. Technol.* 62:1249–1258.
72. Wang, W.S., Chen, H.S., Wu, Y.W., Tsai, T.Y., Chen-Yang, Y.W. 2008. Properties of novel epoxy/clay nanocomposites prepared with a reactive phosphorus-containing organoclay. *Polymer* 49:4826–4836.
73. Blumstein, A. 1965. Polymerization of adsorbed monolayers. II. Thermal degradation of the inserted polymer. *J. Polym. Sci. A* 3:2665–2673.
74. Gu, A., Liang, G. 2003. An investigation on post-fire behavior of hybrid nanocomposites under bending loads. *Polym. Degrad. Stab.* 80:383–391.
75. Kaya, E., Tanoglu, M., Okur, S. 2008. Layered clay/epoxy nanocomposites: Thermomechanical, flame retardancy, and optical properties. *J. Appl. Polym. Sci.* 109:834–840.
76. Chiu, Y.C., Huang, C.C., Tsai, H.C. 2014. Synthesis, characterization, and thermo mechanical properties of siloxane-modified epoxy-based nano composite. *J. Appl. Polym. Sci.* 131:40984.
77. Akat, H., Tasdelen, M.A., Prez, F.D., Yagci, Y. 2008. Synthesis and characterization of polymer/clay nanocomposites by intercalated chain transfer agent. *Eur. Polym. J.* 44:1949–1954.
78. Costache, M.C., Wang, D., Heidecker, M.J., Manias, E. 2006. The thermal degradation of poly(methyl methacrylate) nanocomposites with montmorillonite, layered double hydroxides and carbon nanotubes. *Polym. Adv. Technol.* 17:272–280.
79. Khezri, K., Haddadi-Asl, V., Roghani-Mamaqani, H., Salami-Kalajahi, M. 2012. Encapsulation of organo-modified montmorillonite with PMMA via *in situ* SR&NI ATRP in miniemulsion. *J. Polym. Res.* 19:9868.
80. Carvalho, H.W.P., Suzana, A.F., Santilli, C.V., Pulcinelli, S.H. 2013. Structure, and thermal stability of poly(methyl methacrylate)-*co*-poly(3-tri(methoxysilyl)propyl methacrylate)/montmorillonite nanocomposites. *Polym. Eng. Sci.* 53:1253–1261.
81. Mohammadi, S.R., Khonakdar, H.A., Ehsani, M., Jafari, S.H., Wagenknecht, U., Kretzschmar, B. 2011. Investigation of thermal behavior and decomposition kinetic of PET/PEN blends and their clay containing nanocomposites. *J. Polym. Res.* 18:1765–1775.
82. Piszczyk, Ł., Danowska, M., Mietlarek-Kropidłowska, A., Szyszka, M., Strankowski, M. 2014. Synthesis and thermal studies of flexible polyurethane nanocomposite foams obtained using nanoclay modified with flame retardant compound. *J. Therm. Anal. Calorim.* 118:901–909.
83. Lecouvet, B., Sclavons, M., Bourbigot, S., Bailly, C. 2013. Thermal and flammability properties of polyether-sulfone/halloysite nanocomposites prepared by melt compounding. *Polym. Degrad. Stab.* 98:1993–2004.
84. Ozkoc, G., Kemaloglu, S. 2009. Morphology, biodegradability, mechanical, and thermal properties of nanocomposite films based on PLA and plasticized PLA. *J. Appl. Polym. Sci.* 114:2481–2487.

85. E. Olewnik-Kruszkowska. 2015. Effect of UV irradiation on thermal properties of nanocomposites based on polylactide. *J. Therm. Anal. Calorim.* 119:219–228.
86. Chow, W.S., Teoh, E.L. 2015. Flexible and flame resistant poly(lactic acid)/organomontmorillonite nano-composites. *J. Appl. Polym. Sci.* 132:41253.
87. Tham, W.L., Poh, B.T., Ishak, Z.A.M., Chow, W.S. 2014. Thermal behaviors and mechanical properties of halloysite nanotube-reinforced poly(lactic acid) nanocomposites. *J. Therm. Anal. Calorim.* 118:1639–1647.
88. Liu, M., Pu, M., Ma, H. 2012. Preparation, structure and thermal properties of polylactide/sepiolite nano-composites with and without organic modifiers. *Comp. Sci. Technol.* 72:1508–1514.
89. Qi, K., Zhang, G. 2014. Effect of organoclay on the morphology, mechanical, and thermal properties of polyimide/organoclay nanocomposite foams. *Polym. Compos.* 35:2311–2317.
90. Dorigato, A., D'Amato, M., Pegoretti, A. 2012. Thermo-mechanical properties of high density polyethylene—Fumed silica nanocomposites: Effect of filler surface area and treatment. *J. Polym. Res.* 19:9889.
91. Chen, Z., Song, C., Bai, R., Wei, Z., Zhang, F. 2012. Effects of mesoporous SBA-15 contents on the properties of polystyrene composites via in-situ emulsion polymerization. *J. Polym. Res.* 19:9846.
92. Salami-Kalajahi, M., Haddadi-Asl, V., Rahimi-Razin, S., Behboodi-Sadabad, F., Najafi, M., Roghani-Mamaqani, H. 2012. A study on the properties of PMMA/silica nanocomposites prepared via RAFT polymerization. *J. Polym. Res.* 19:9793.
93. Yu, C.B., Wei, C., Lv, J., Liu, H.X., Meng, L.T. 2012. Preparation and thermal properties of mesoporous silica/phenolic resin nanocomposites via *in situ* polymerization. *Express Polym. Lett.* 6:783–793.
94. Feng, Y., Wang, B., Wang, F. et al. 2014. Thermal degradation mechanism and kinetics of polycarbonate/silica nanocomposites. *Polym. Degrad. Stab.* 107:129–138.
95. Xie, Z., Dao, B., Hodgkin, J., Hoang, M., Hill, A., Gray, S. 2011. Synthesis and characterization of hybrid organic–inorganic materials based on sulphonated polyamideimide and silica. *J. Polym. Res.* 18:965–973.
96. Lin, J., Liu, Y., Yang, W. et al. 2014. Structure and mechanical properties of the hybrid films of well dispersed $SiO_2$ nanoparticle in polyimide ($PI/SiO_2$) prepared by sol–gel process. *J. Polym. Res.* 21:531–539.
97. Al Arbash, A., Ahmad, Z., Al-Sagheer, F., Ali, A.A.M. 2006. Microstructure and thermomechanical properties of polyimide-silica nanocomposites. *J. Nanomater.* Article ID 58648, 1–9.
98. Ye, X., Wang, J., Xu, Y. et al. 2014. Mechanical properties and thermostability of polyimide/mesoporous silica nanocomposite via effectively using the pores. *J. Appl. Polym. Sci.* 131:41173.
99. Liu, M.Q., Duan, J.P., Shi, X.Z., Lu, J.J., Huang, W. 2015. Thermo-stabilized, porous polyimide microspheres prepared from nanosized $SiO_2$ templating via *in situ* polymerization. *Express Polym. Lett.* 9:14–22.
100. Mallakpour, S., Nikkhoo, E. 2013. Morphological and thermal properties of nanocomposites contain poly(amide-imide) reinforced with bioactive *N*-trimellitylimido-L-valine modified $TiO_2$ nanoparticles. *J. Polym. Res.* 20:78.
101. Kajohnchaiyagual, J., Jubsilp, C., Dueramae, I., Rimdusit, S. 2014. Thermal and mechanical properties enhancement obtained in highly filled alumina–polybenzoxazine composites. *Polym. Compos.* 35:2269–2279.
102. Rafiee, Z., Golriz, L. 2015. Polyimide nanocomposite films containing $\alpha$-$Fe_2O_3$ nanoparticles. *J. Polym. Res.* 22:630.
103. Van Krevelen, D.W., Hoftyzer, P.J. 1976. *Properties of Polymers,* 3rd ed. Amsterdam, Elsevier.
104. Pielichowska, K. 2012. The influence of molecular weight on the properties of polyacetal/hydroxyapatite nanocomposites. Part 2. *In vitro* assessment. *J. Polym. Res.* 19:9788.
105. Mallakpour, S., Barati, A. 2012. Preparation and characterization of optically active poly(amide-imide)/$TiO_2$ bionanocomposites containing N-trimellitylimido-L-isoleucine linkages: Using ionic liquid and ultrasonic irradiation. *J. Polym. Res.* 19:9802.
106. Abdolmaleki, A., Bazyar, Z. 2014. Synthesis and characterization of novel aromatic poly(benzimidazole-amide)/Ag and Cu nanocomposites through transition metal complexation. *Polym. Bull.* 71:1773–1795.
107. Rajeshwari, P., Dey, T.K. 2014. Structural and thermal properties of HDPE/n-AlN polymer nanocomposites. *J. Therm. Anal. Calorim.* 118:1513–1530.
108. Avcu, E., Çoban, O., Bora, M.Ö., Fidan, S., Sinmazçelik, T., Ersoy, O. 2014. Possible use of volcanic ash as a filler in polyphenylene sulfide composites: Thermal, mechanical, and erosive wear properties. *Polym. Compos.* 35:1826–1833.

109. Su, C.-H., Chiu, Y.-P., Teng, C.-C., Chiang, C.-L. 2010. Preparation, characterization and thermal properties of organic–inorganic composites involving epoxy and polyhedral oligomeric silsesquioxane (POSS). *J. Polym. Res.* 17:673–681.

110. Pan, M., Zhang, C., Liu, B., Mu, J. 2013. Dielectric and thermal properties of epoxy resin nanocomposites containing polyhedral oligomeric silsesquioxane. *J. Mater. Sci. Res.* 2:153–162.

111. Li, Z., Yang, R. 2014. Study of the synergistic effect of polyhedral oligomeric octadiphenylsulfonyl-silsesquioxane and 9,10-dihydro-9-oxa-10-phosphaphenanthrene-10-oxide on flame-retarded epoxy resins. *Polym. Degrad. Stab.* 109:233–239.

112. Kodal, M., Sirin, H., Ozkoc, G. 2014. Effects of reactive and nonreactive POSS types on the mechanical, thermal, and morphological properties of plasticized poly(lactic acid). *Polym. Eng. Sci.* 54:264–275.

113. Wang, R., Wang, S., Zhang, Y. 2009. Morphology, rheological behaviour, and thermal stability of PLA/PBS/POSS composites. *J. Appl. Polym. Sci.* 113:3095–3102.

114. Moniruzzaman, M., Karen, I.W. 2006. Polymer nanocomposites containing carbon nanotubes. *Macromolecules* 39:5194–5205.

115. Bredeau, S., Peeterbroeck, S., Bonduel, D., Alexandre, M., Dubois, P. 2008. From carbon nanotube coatings to high-performance polymer nanocomposites. *Polym. Int.* 57:547–553.

116. Breuer, O., Sundararaj, U. 2004. Big returns from small fibers: A review of polymer/carbon nanotube composites. *Polym. Compos.* 25:641–647.

117. Huxtable, S.T., Cahill, D.G., Shenogin, S. et al. 2003. P. Interfacial heat flow in carbon nanotube suspensions. *Nat. Mater.* 2:731–734.

118. Corcione, C.E., Frigione, M. 2012. Characterization of nanocomposites by thermal analysis. *Materials* 5:2960–2980.

119. Song, P., Yu, Y., Wu, Q., Fu, S. 2012. Facile fabrication of HDPE-g-MA/nanodiamond nanocomposites via one-step reactive blending. *Nanoscale Res. Lett.* 7: 355.

120. Gulotty R., Castellino, M., Jagdale, P., Tagliaferro, A., Balandin, A.A. 2013. Effects of functionalization on thermal properties of single-wall and multi-wall carbon nanotube–polymer nanocomposites. *ACS Nano* 7:5114–5121.

121. Mohsin, M.E.A., Arsad, A., Alothman, O.Y. 2014. Enhanced thermal, mechanical and morphological properties of CNT/HDPE nanocomposite using MMT as secondary filler. *World Acad. Sci., Eng. Technol. Int. J. Chem., Nucl., Metall. and Mater. Eng.* 8:127–130.

122. Park, S.J., Chae, S.W., Rhee, J.M., Kang, S.J. 2010. A study on electrical and thermal properties of polyimide/MWNT nanocomposites *Bull. Korean Chem. Soc.* 31:2279–2282.

123. Faghihi, K., Shabanian, M. 2011. Thermal and optical properties of silver–polyimide nanocomposite based on diphenyl sulphone moieties in the main chain. *J. Chil. Chem. Soc.* 56:665–667.

124. Zhang, H., Wang, C., Zhang, Y. 2015. Preparation and properties of styrene–butadiene rubber nanocomposites blended with carbon black–graphene hybrid filler. *J. Appl. Polym. Sci.* 132:41309.

125. Inuwa, I.M., Hassan, A., Samsudin, S.A., Kassim, M.H.M., Jawaid, M. 2014. Mechanical and thermal properties of exfoliated graphite nanoplatelets reinforced polyethylene terephthalate/polypropylene composites. *Polym. Compos.* 35, 2029–2035.

126. Sefadi, S.J., Luyt, A.S., Pionteck, J. 2015. Effect of surfactant on EG dispersion in EVA and thermal and mechanical properties of the system. *J. Appl. Polym. Sci.* 132:41352.

127. Long, Y., Pu, Z., Huang, X., Jia, K., Liu, X. 2014. Effect of CuPc@MWCNTs on rheological, thermal, mechanical and dielectric properties of polyarylene ether nitriles (PEN) terminated with phthalonitriles. *J. Polym. Res.* 21:525.

128. Kim, I.H, Jeong, Y.G. 2010. Polylactide/exfoliated graphite nanocomposites with enhanced thermal stability, mechanical modulus, and electrical conductivity. *J. Polym. Sci. B: Polym. Phys.* 48:850–858.

129. Li, X., Xiao, Y., Bergeret, A., Longerey, M., Che, J. 2014. Preparation of polylactide/graphene composites from liquid-phase exfoliated graphite sheets. *Polym. Compos.* 35:396–403.

130. Chieng, B.W., Ibrahim, N.A., Yunus, W.M.Z.W., Hussein, M.Z., Loo, Y.Y. 2014. Effect of graphene nanoplatelets as nanofiller in plasticized poly(lactic acid) nanocomposites. Thermal properties and mechanical properties. *J. Therm. Anal. Calorim.* 118:1551–1559.

131. Wang, X., Yang, H., Song, L., Hu, Y., Xing, W., Lu, H. 2011. Morphology, mechanical and thermal properties of graphene–reinforced poly(butylene succinate) nanocomposites. *Compos. Sci. Technol.* 72:1–6.

132. Zhang, J., Li, D., Fu, M., Zhang, Y., Zhang, L., Zhao, P. 2014. Synergistic effect of adipic acid pentaerythritol ester with calcium and zinc stearates on polyvinylchloride thermal stability. *J. Vinyl Addit. Technol.* doi: 10.1002/vnl.21436.

133. Kwon, J., Kim, J., Lee, J., Han, P., Park, D., Han, H. 2014. Fabrication of polyimide composite films based on carbon black for high-temperature resistance. *Polym. Compos.* 35:2214–2220.

134. Raza, R., Ali, S., Ahmed, F., Ullah, S., Grant, J.T. 2014. Thermal and thermomechanical behavior of amino functionalized and metal decorated MWCNTs/PMMA nanocomposite films. *Polym. Compos.* 35:1807–1817.
135. Selvi, M., Devaraju, S., Sethuraman, K., Alagar, M. 2014. Carbon black–polybenzoxazine nanocomposites for high K dielectric applications. *Polym. Compos.* 35:2121–2128.
136. Zhao, L., Guo, Z., Cao, Z., Zhang, T., Fang, Z., Peng, M. 2013. Thermal and thermo-oxidative degradation of high density polyethylene/fullerene composites *Polym. Degrad. Stab.* 98:1953–1962.
137. Radhakrishnan, V.K., Zagarola, S.W., Davis, E.W., Davis, V.A. 2011. Thermal properties of polypropylene nanocomposites: Effects of carbon nanomaterials and processing. *Polym. Eng. Sci.* 51:460–473.
138. Shahil, K.M.F., Balandin, A.A. 2012. Graphene—Multilayer graphene nanocomposites as highly efficient thermal interface materials. *Nano Lett.* 12:861–867.
139. Yu, A., Ramesh, P., Sun, X., Bekyarova, E., Itkis, M.E., Haddon, R.C. 2008. Enhanced thermal conductivity in a hybrid graphite nanoplatelet—Carbonnanotube filler for epoxy composites. *Adv. Mater.* 20:4740–4744.
140. Ma, W.S., Li, J., Zhao, X.S. 2013. Improving the thermal and mechanical properties of silicone polymer by incorporating functionalized graphene oxide. *J. Mater. Sci.* 48:5287–5294.
141. Cheng, H.K.F., Basu, T., Sahoo, N.G., Li, L., Chan, S.H. 2012. Current advances in the carbon nanotube/thermotropic main-chain liquid crystalline polymer nanocomposites and their blends. *Polymers* 4:889–912.
142. Hsiao, M.C., Ma, C.C.M., Chiang, J.C. et al. 2013. Thermally conductive and electrically insulating epoxy nanocomposites with thermally reduced grapheneoxide–silica hybrid nanosheets. *Nanoscale* 5:5863–5871.
143. Wu, C., Huang, X., Wu, X., Xie, L., Yang, K., Jiang, P. 2013. Graphene oxide-encapsulated carbon nanotube hybrids for high dielectric performance nanocomposites with enhanced energy storage density. *Nanoscale* 5:3847–55.
144. Ha,, H.W., Choudhury, A., Kamal, T., Kim, D.H., Park, S.Y. 2012. Effect of chemical modification of graphene on mechanical, electrical, and thermal properties of polyimide/graphene nanocomposites. *ACS Appl. Mater. Interfaces* 4:4623–30.
145. Huang, X., Iizuka, T., Jiang, P., Ohki, Y., Tanaka, T. 2012. Role of interface on the thermal conductivity of highly filled dielectric epoxy/AlN composites. *J. Phys. Chem. C* 116:13629–39.
146. Hu, K., Kulkarni, D.D., Choi, I., Tsukruk, V.V. 2014. Graphene–polymer nanocomposites for structural and functional applications. *Prog. Polym. Sci.* 39:1934–1972.
147. Kakroodi, A.R., Cheng, S., Sain, M., Asiri, A. 2014. Mechanical, thermal, and morphological properties of nano-composites based on polyvinyl alcohol and cellulose nanofiber from *Aloe vera* Rind, *J. Nanomater.*, 7 pages, Article ID 903498. doi: 10.1155/2014/903498.
148. Gavgani, J.N., Adelnia, H., Sadeghi, G.M.M., Zafari, F. 2014. Intumescent flame retardant poly-urethane/starch composites: Thermal, mechanical, and rheological properties *J. Appl. Polym. Sci.* 131:41158.
149. Chruściel, J.J., Leśniak, E. 2012. Chapter 9: Modification of thermoplastics with reactive silanes and siloxanes. In *Thermoplastic Elastomers*, ed. El-Sonbati, A.Z., 155–192. Rijeka, Croatia, InTech - Open Science.
150. Chruściel, J.J., Leśniak, E. 2015. Modification of epoxy resins with functional silanes, polysiloxanes, silsesquioxanes, silica and silicates. *Prog. Polym. Sci.* 41:67–121.

# 9 Optical Properties of Nanocomposites

*Enis S. Džunuzović and Jasna V. Džunuzović*

## CONTENTS

## 9.1 INTRODUCTION

The optical characteristics of inorganic nanoparticles (NP) have been well known for quite a long time and this knowledge has been applied for the creation of optically functional materials based on these NPs and different polymers used as a processable matrix. Remarkable optical properties of certain polymer nanocomposites (NCs), such as absorption (ultraviolet [UV] and visible), transparency, color, high refractive index, fluorescence, luminescence, different nonlinear optical properties, etc. are the reasons for such large interest of various researchers from different disciplines all over the world in this type of material. In these materials, NPs improve optical properties, while a polymer matrix is used to stabilize the size and growth of NPs (Beecroft and Ober 1997). Polymer NCs, which show such intriguing optical properties, include NCs prepared from different polymers and metal NPs, metal oxide NPs, carbon-based NPs (fullerenes, graphene, and carbon nanotubes) or semiconductor nanocrystals (quantum dots), as well as NCs based on conjugated polymer matrices and different inorganic NPs. Since polymer NCs are prepared by synergistic combination of polymers and NPs, optical properties of these specific materials can be tailored by adequate selection of the organic and inorganic building blocks with desirable properties and by simply changing the portion of NPs introduced in the polymer matrix. This led to the rapid growth of the area of polymer NCs' possible applications, such as for optical detectors, solar cells, in optical amplification, photovoltaics, biosensors, for waveguiding in integrated optical devices, as optical coatings, optical switches, high refractive index devices, for encapsulation of light emitting diodes (LEDs), as lens materials, etc.

Linear and nonlinear optical properties of polymer NCs depend on the optical properties of the polymer matrix and applied NPs, as well as on other different parameters such as type, size, shape, surface characteristics, concentration, and spatial distribution of NPs in the matrix. Therefore, in order to fully understand optical properties of polymer NCs, it is necessary first to briefly describe the most important optical properties of its constituents, that is, polymers and NPs.

## 9.2  OPTICAL PROPERTIES OF POLYMERS

### 9.2.1  LINEAR OPTICAL PROPERTIES OF POLYMERS

Absorption, reflection, refraction, and scattering belong to the most important linear optical properties of polymers (Figure 9.1). The proportion of each of these phenomena depends on the properties of the polymer. During the light absorption of sample, atoms or groups of atoms absorb electromagnetic energy, the absorbance of the light which passes through the sample increases, while the transmission of light decreases and if visible light is absorbed, color will appear. Reflection represents the portion of the light reflected from the surface after it strikes the polymer. The rest of the incident light will be transmitted into the polymer. Reflection of the light depends on the values of linear refractive index ($n_0$) of the polymer and air. Since polymer and air have different values of $n_0$, refraction, that is, change of the light direction occurs. Scattering of the light appears as a consequence of the optical (morphological) inhomogeneities (density fluctuations, voids, presence of crystals, etc.), which scatter the light in different directions. There are two types of scattering: Rayleigh scattering where scattering centers are smaller than one-tenth of the wavelength and Mie scattering where scattering centers have the same size as the applied wavelength. If the surface of the polymer is rough, then surface scattering will appear. However, if the surface of the polymer is smooth, then reflection from such a surface is called specular reflection. A polymer surface will be glossy if the relative proportion of specular reflection is higher than the proportion of surface scattering, but if specular reflection is close to zero, the surface of the polymer will be matte (Demir and Wegner 2012). Gloss can be defined as the measure of the surface ability to reflect light and it represents the portion of the incident light that is reflected at the same angle as the angle of incident light beam (usually 45°).

Due to the presence of absorption and scattering, the intensity of light that passes through the polymer decreases on its way ($I_t < I_0$, where $I_0$ and $I_t$ are intensity of the incident and transmitted light, respectively). This phenomenon is called attenuation. Furthermore, the portion of the incident light which passes through the polymer (transmittance) depends on the intensities of the absorption and scattering. The wavelength region of interest for the optical transmittance usually starts at 200 nm (~200–400 nm is UV range while ~400–800 nm represents the visible region) up to 1700 nm (~800–1700 nm is the near infrared [IR] region). If the light absorption is very small and scattering is zero, then the polymer will have large transmittance and it will be transparent. On the other hand, the polymer will appear opaque (milky) if it has large scattering power. Generally amorphous, homogenous polymers like polystyrene (PS) and poly(methyl methacrylate) (PMMA) are

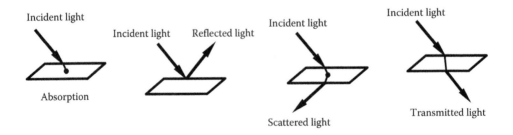

**FIGURE 9.1**  Schematic representation of some interactions of light with a polymer.

transparent, while highly crystalline polymers, such as high density polyethylene, are opaque. The increase of the crystallinity leads to the increase of the polymer density, which further reduces the speed of light that passes through the sample and increases the refractive index. If the size of crystals is higher than the wavelength of visible light, the light will be scattered, leading to the decrease of the polymer clarity (Shah 2007). Between transparency and opaqueness lies translucency (Demir and Wegner 2012). On the other hand, haze represents the percentage of the transmitted light that deviates more than 2.5° from an incident parallel beam. It is caused by light scattered from the optical inhomogeneities present in the polymer or from the surface imperfections and it depends on the thickness of the polymer (Demir and Wegner 2012).

Beside those previously mentioned, some undoped conjugated polymers also exhibit photoluminescence (PL) in the near IR and visible spectral range, that is, they are capable of reemitting visible light after the absorption of photons, due to the transition of electrons from an excited energy to the lower energy state. Depending on the delay time between absorption and reemission, different luminescence phenomena can occur. If reemission occurs in a time period lower than 1 s, then this phenomenon is called fluorescence (photons that are emitted have lower energy than absorbed ones), while in the case of longer times it is termed phosphorescence. On the other hand, electroluminescence (EL) appears when material is emitting light due to the presence of electric field or passage of electric current. Polymers that possess these optical properties are usually called light emitting polymers (LEP).

Refractive index is defined as the ratio between the speed of light in a vacuum and its speed in the examined medium, and it depends on the molecular structure and polarizability of molecules. For polymers $n_0$ is usually between 1.3 and 1.7 (only a few polymers have high $n_0$, such as poly(thiophene) with $n_0 = 2.12$) and due to their chemical structure it is difficult to prepare transparent polymers that have $n_0$ higher than 1.7. However, $n_0$ can be easily modified by the addition of adequate NPs. In Table 9.1 are given values of $n_0$ for some nonconjugated and conjugated polymers. Complex refractive index ($n$) depends on the wavelength and it can be expressed in the following manner:

$$n = n_0 + ik \tag{9.1}$$

where $k$ is the extinction coefficient defined as

$$k = \frac{\alpha_0 c}{2\omega} = \frac{\alpha_0 \lambda}{4\pi} \tag{9.2}$$

In Equation 9.2 $c$ is the speed of light, $\omega$ is the frequency of the optical field, $\lambda$ is the wavelength of light, while $\alpha_0$ represents the linear absorption coefficient, defined as the fraction of the energy absorbed during its passing through a unit of thickness of material.

**TABLE 9.1**

**Values of Linear Refractive Index ($n_0$) of Several Nonconjugated and Conjugated Polymers**

| Nonconjugated Polymers | $n_0$ | Conjugated Polymers | $n_0$ |
|---|---|---|---|
| Poly(methyl methacrylate) | 1.49 | Poly(N-vinylcarbazole) | 1.69 |
| Poly(vinyl alcohol) | 1.52–1.55 | Poly(p-phenylene vinylene) | 2.10 |
| Polystyrene | 1.59–1.60 | Polyaniline | 1.85 |

*Source:* Adapted from Mark, J. E. 1999. *Polymer Data Handbook.* New York: Oxford University Press.

With decreasing wavelength of the incident light, $n_0$ increases. The parameter used to describe the dependence of $n_0$ of a given material on wavelength, the so-called dispersion of $n_0$, is called Abbe number ($V$), and it reflects the slope of the $n_0$ dispersion curve. High values of Abbe number are connected with low dispersion and low chromatic aberration of transparent materials (Asai et al. 2013). Abbe number can be determined from the refractive index at three different wavelengths (e.g., D-line at 589 nm, F-line at 486 nm, and C-line at 656 nm) using the following equation (Tao et al. 2011):

$$V = \frac{n_D - 1}{n_F - n_C} \tag{9.3}$$

For anisotropic material, it is necessary to define the state of light polarization relative to a reference axis in the examined sample. In this case, usually two values of $n_0$ are quoted, as well as the maximum difference between the $n_0$ values, measured in two mutually perpendicular directions, that is, birefringence ($\Delta n$) (Brown 1999). In other words, birefringence can be defined as a measure of optical anisotropy of polymer, which can be induced by application of mechanical stress (e.g., during preparation of uniaxially drawn polymer films) (Matsuda and Ando 2003) or it can be photoinduced after addition of, for example, azo-dyes (Costanzo et al. 2014). For the application of polymers as optical (antireflective) coatings, in holographic recording systems, in optical filters, lenses, optical waveguides and switching, optoelectronic devices, etc. it is desirable that the material has high $n_0$ and low $\Delta n$ value, combined with high transparency and mostly long-term UV light stability and good thermal stability.

### 9.2.2 Nonlinear Optical Properties of Polymers

The interaction of electric fields, for example, oscillatory field, with a material induces oscillation of electrons of material at the same frequency and appearance of polarization ($P$). If the strength of the applied electric field ($E$) is relatively small in comparison to the interatomic columbic fields of the material ($E < 10^8$–$10^9$ V/m), then the polarization response of the material will be linear (Kanis et al. 1994). In that case, polarization can be expressed as a function of the first-order susceptibility ($\chi^{(1)}$):

$$P = \chi^{(1)}E \tag{9.4}$$

From the susceptibility $\chi^{(1)}$ (polarizability), linear optical properties, such as linear absorption coefficient, linear refractive index, etc. directly follow (Barford 2005). However, if the intensity of the applied electric field is much higher ($E > 10^9$ V/m) such as in the case of powerful optical lasers, the polarization response of the material will be nonlinear. For example, if a material composed of N polymers per volume unit is influenced by very high electric fields, $E(\omega_1)$, $E(\omega_2)$, $E(\omega_3)$, …, then the response of a material to an electric field at a frequency $\omega_\sigma = \omega_1 + \omega_2 + \omega_3 + \cdots$, measured by its polarization $P(\omega_\sigma)$ is

$$P(\omega_\sigma) = \chi^{(1)}(-\omega_\sigma; \omega_1)E(\omega_1) + \chi^{(2)}(-\omega_\sigma; \omega_1, \omega_2)E(\omega_1)E(\omega_2)$$
$$+ \chi^{(3)}(-\omega_\sigma; \omega_1, \omega_2, \omega_3)E(\omega_1)E(\omega_2)E(\omega_3) + \cdots \tag{9.5}$$

where $\chi^{(n)}$ is an $n + 1$ rank tensor and coefficients $\chi^{(2)}$ and $\chi^{(3)}$ are the second (hyperpolarizability) and third-order susceptibilities, respectively (Williams 1983; Kanis et al. 1994). Nonlinear susceptibilities ($\chi^{(2)}$ and $\chi^{(3)}$) of a material depend on its symmetry for even order $\chi$, the nature of the electronic environment of the medium and nature of the components of the interacting field (Williams 1983).

Noncentrosymmetric materials can show the second-order nonlinear optical properties ($\chi^{(2)}$ processes) and might be used in optoelectronic switching and for frequency doubling, since in this case the energy of photons is doubled after their interaction with media (Beecroft and Ober 1997). In a centrosymmetric media, the even order tensors ($\chi^{(2)}$) are zero, that is, the electric field contribution to the second-order nonlinear optical properties is forbidden by symmetry, since in this case under a reversal of the electric field the polarization must reverse sign. On the other hand, the odd tensors have no restrictions considering symmetry. Conjugated polymers are often centrosymmetric, and therefore the lowest nonzero nonlinear susceptibility is for these polymers $\chi^{(3)}$ (Barford 2005). $\chi^{(3)}$ response of conjugated polymers is excellent because they have large effective $\pi$-electron delocalization through the backbone of the polymer. Due to that, these polymers can be successfully used for the optical switching and signal processing applications (Williams 1983).

The third-order susceptibility is a complex number ($\chi^{(3)} = \mathrm{Re}\,\chi^{(3)} + i\,\mathrm{Im}\,\chi^{(3)}$), and it depends on the wavelength used for the measurements. Beside $\chi^{(3)}$ ($m^2/V^2$), the third-order nonlinear optical properties of polymer can also be quantitatively characterized by a nonlinear refractive index ($n_2$, $m^2/W$) and nonlinear absorption coefficient ($\beta$, $m/W$). When linear absorption is negligible, $n_2$ can be connected with the real part and $\beta$ with the imaginary part of $\chi^{(3)}$ if the international system (SI) of units is applied:

$$\mathrm{Re}\,\chi^{(3)} = \frac{4n_2 n_0^2 c \varepsilon_0}{3} \tag{9.6}$$

$$\mathrm{Im}\,\chi^{(3)} = \frac{\beta n_0^2 \varepsilon_0 c \lambda}{3\pi} \tag{9.7}$$

where $\varepsilon_0$ is the electric permittivity of free space ($8.85 \times 10^{-12}$ F/m) (Coso and Solis 2004). The value of parameter $K = \beta\lambda/n_2$ is applied to estimate the optical switching effectiveness of a material (for $K < 1$ material can be used in optical switching applications) (Sun et al. 2008).

## 9.3 OPTICAL PROPERTIES OF NANOPARTICLES

### 9.3.1 LINEAR OPTICAL PROPERTIES OF NANOPARTICLES

The most important linear optical properties of NPs are scattering, extinction (the sum of scattering and absorption), color, refractive index, energy band gap, etc. Linear optical properties of NPs depend on the chemical composition, size, shape, orientation and number of NPs, optical constants of the NPs and surrounding medium, the distance between NPs, the frequency and polarization state of the incident light, and can be quite different from their bulk properties. From all these, the size of NPs has probably the largest influence on their optical properties. By decreasing NPs size, the surface/volume ratio increases, leading to the enhancement of the importance of NPs surface properties. For example, when the size of NPs is 5 nm, they are composed of several thousands of atoms with around 40% of atoms located on their surface. On the other hand, when the size of particles is 0.1 μm, they have $10^7$ atoms from which only 1% is placed on their surface (Hanemann and Szabó 2010). One of the most interesting size-dependent properties is the appearance of resonant light absorption of gold (Au) and silver (Ag) NPs. Namely, when an external electromagnetic field is applied on Au and Ag in bulk, then the collective excitation of conductive electrons will result in a relaxator. However, in the case of Au and Ag NPs, a surface plasmon resonance (SPR) will appear as a consequence of the collective conduction band electron oscillation (Zhang 2009). When the resonant structures of Au and Ag NPs are in the extinction, then these particles will have characteristic color, that is, Au will be red and Ag will be yellow. With increasing NPs size, the peak of SPR shifts to longer wavelengths (red shift) and it becomes broader. Broadening of the SPR peak also occurs when the shape of noble metal NPs is changed from spherical to nonspherical. Furthermore, it has been shown that the presence of structural anisotropy of Ag NPs can induce more than one SPR band (Vodnik et al. 2013). SPR of metal NPs also depends on the chemical nature of the particles, particle–particle interactions, embedding environment, distance between NPs and surrounding media, etc. If the dielectric properties of surrounding medium are changed, then SPR peak of metal NPs will shift. For example, red shifts are observed when $n_0$ of the environment was higher than 1, as well as with increasing the dielectric constant of the surrounding media which is in agreement with Drude model (Kreibig and Vollmer 1995; Vodnik et al. 2009, 2010, 2012; Vukoje et al. 2014). The position of SPR peak is around 520 nm for Au and 400 nm for Ag in water (Zhang 2009). Generally speaking, optical properties of metal NPs in the visible spectral region strongly depend on the surface plasmon absorption.

Furthermore, the effect known as quantum size (or quantum confinement) effect, that is, the shift of the wavelength for photoluminescence and absorption to shorter wavelengths (blue shift), appears with decreasing semiconductor NPs (quantum dots [QDs]) size (usually below 10 nm). Simultaneously, with decreasing particle dimensions, the energy level spacing increases, as schematically shown in Figure 9.2. Using this effect, it is possible to control the wavelength for photoluminescence by controlling the size of the QDs (Quinten 2011). Generally speaking, optical properties of semiconductor NPs depend more on their size, while optical properties of metal NPs are more sensitive to their shape (Zhang 2009). For very small NPs, optical properties are described using quantum-mechanical concepts (Quinten 2011), while for larger particles Mie's theory of light scattering and absorption by a single, isolated spherical particle is applied (Mie 1908). For the nonspherical NPs, usually adequate approximations and numerical methods are used. These theoretical considerations are extensively covered in the literature (Zhang 2009; Quinten 2011).

As already mentioned, linear optical properties of NPs depend also on their chemical composition. In order to improve compatibility between NPs and polymer matrix by creating chemical bonds between them, it is often necessary to modify the surface of NPs with adequate compounds. The optical properties of surface modified NPs can be quite different than those of bare NPs. Janković et al. (2010) have reported that the onset of absorption of titanium dioxide ($TiO_2$) NPs surface modified with catecholate type ligands is red shifted in comparison to the unmodified NPs (Janković et al. 2010). These results were explained by the excitation of localized electrons from the modifier surface into the conduction band continuum states of $TiO_2$. Similar was obtained by Džunuzović et al. (2007, 2010) for $TiO_2$ NPs surface modified with 6-palmitate ascorbic acid, as well as for the $TiO_2$ NPs surface modified with gallates (Džunuzović et al. 2012, 2013; Radoman et al. 2013). As an example, in Figure 9.3 are given absorption spectra of $TiO_2$ colloid in water and $TiO_2$ NPs surface modified with 6-palmitate ascorbic acid in toluene (left) and $TiO_2$ colloid and $TiO_2$ NPs surface modified with lauryl gallate in chloroform (right). On the other hand, the modification of $TiO_2$ NPs with chlorine atoms led to the blue shift in the UV absorption spectrum (Park et al. 2009).

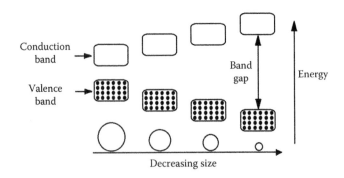

**FIGURE 9.2** Schematic illustration of the quantum confinement effect.

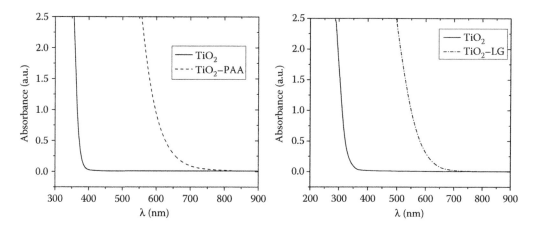

**FIGURE 9.3** Absorption spectra of $TiO_2$ colloid in water and $TiO_2$ NPs surface modified with 6-palmitate ascorbic acid ($TiO_2$–PAA) in toluene (left) and $TiO_2$ colloid and $TiO_2$ NPs surface modified with lauryl gallate ($TiO_2$–LG) in chloroform (right). (Data reproduced from (left) Džunuzović, E. S. et al. *Hemijska Industrija* 64:473–89 and (right) Radoman, T. S. et al. *Hemijska Industrija* 67:923–32 with the permission of Association of Chemical Engineers of Serbia.)

In contrast to the polymers, $n_0$ of inorganic NPs can have values lower than 1 and higher than 4 (Table 9.2). For example, $n_0$ of Au NPs is far below 1 in a broad range of wavelengths, $n_0$ of $TiO_2$ is 2.54 for anatase or 2.75 for rutile form, while lead sulfide (PbS) has $n_0$ of 4.0–4.3 (for diameters of PbS > 20 nm, at 633 nm) (Caseri 2009). That is why NPs can be applied as effective additives for modification of $n_0$ values of polymers by preparation of polymer NCs. Furthermore, oxide NPs, such as $TiO_2$, silica ($SiO_2$), zinc oxide (ZnO), alumina ($Al_2O_3$), etc. are especially interesting for the preparation of transparent polymer NCs. Namely, metal oxides such as ZnO and $TiO_2$ do not absorb light in the visible and near IR spectral region (their absorption edge is around 400 nm and band gap energy higher than ~3 eV) and are therefore often colorless or white, while metal oxides such as hematite ($Fe_2O_3$) and magnetite ($Fe_3O_4$) absorb light in the visible spectral region, have lower band gap, and are therefore considered semiconductors (Table 9.3) (Zhang 2009). Regardless of these distinctions, certain metal oxides such as $TiO_2$ are in practice often considered semiconductors. The band gap represents the energy difference between the top of the valence band (ground state) and the bottom of the higher energy band, that is, conduction band. The band gap energy is the energy that the electron needs to have to be able to jump from the valence band to the conduction band. So,

**TABLE 9.2**

**Values of Linear Refractive Index ($n_0$) of Several Inorganic Materials**

| Inorganic Material | $n_0$ | Wavelength (nm) | Inorganic Material | $n_0$ | Wavelength (nm) |
|---|---|---|---|---|---|
| Ag | 0.17 | 413 | $TiO_2$ (anatase) | 2.54 | 550 |
| Au | 1.64 | 413 | $TiO_2$ (rutile) | 2.75 | 550 |
| ZnO | 2.0 | 370 | ZnS | 2.36 | 388 |
| CdS | 2.35 | 512 | PbS | 4.0–4.3 | 633 |

*Source:* Adapted from Caseri, W. 2009. *Chemical Engineering Communications* 196:549–72; Hanemann, T. and D. V. Szabó. 2010. *Materials* 3:3468–517; Matras-Postolek, K. and D. Bogdal. 2010. *Advances in Polymer Science* 230:221–82.

**TABLE 9.3**

**Values of the Band Gap Energy ($E_g$) of Several Inorganic, UV Absorbing Materials**

| Inorganic Material | $E_g$ (eV) | Wavelength (nm) | Inorganic Material | $E_g$ (eV) | Wavelength (nm) |
|---|---|---|---|---|---|
| TiO$_2$ (anatase) | 3.20 | 550 | ZnS | 3.68 | 388 |
| TiO$_2$ (rutile) | 3.03 | 550 | CeO$_2$ | 3.15 | 393 |
| ZnO | 3.20 | 388 | CdS | 2.50 | 512 |

*Source:* Adapted from Hanemann, T. and D. V. Szabó. 2010. *Materials* 3:3468–517; Matras-Postolek, K. and D. Bogdal. 2010. *Advances in Polymer Science* 230:221–82.

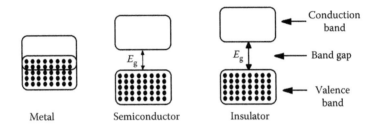

**FIGURE 9.4** Schematic representation of different band gap energies ($E_g$) of metal, semiconductor, and insulator.

material will be more electrically conductive if it has a lower band gap. The comparison of band gap energy of metal, semiconductor, and insulator is presented in Figure 9.4.

### 9.3.2 NONLINEAR OPTICAL PROPERTIES OF NANOPARTICLES

At high excitation intensities, NPs can show several interesting nonlinear optical properties, such as the second- and third-order nonlinear optical properties, luminescence up-conversion (LUC), and saturable absorption. LUC represents the nonlinear optical property where the emitted light has a higher frequency or shorter wavelength than the incident light (Zhang 2009). On the other hand, saturable absorption (absorption saturation or light induced transparency) can be defined as a decrease of the absorption of light with increasing light intensity, since in that case the upper energy states fill and block further absorption (Hasan et al. 2009). The absorption coefficient of saturable absorber ($\alpha$) is given by the following equation:

$$\alpha = \frac{\alpha_0}{1 + I_0/I_{sat}} \tag{9.8}$$

where $\alpha_0$ is the linear absorption coefficient at the wavelength of excitation and $I_{sat}$ is the saturation intensity, that is, intensity necessary to reduce $\alpha$ to half of the initial value (Hasan et al. 2009; Murali et al. 2013). NPs can also show reverse saturable absorption, which is the opposite phenomenon than the saturable absorption, and it appears when excited states formed by optical "pumping" (optical absorption) of the ground state have a higher absorption cross section than the ground state (Wang et al. 2004). The above-mentioned nonlinear optical properties of NPs can be applied in optical switching, as well as for optical limiting application, which represents the ability of a material to attenuate dangerous intense laser beams, allowing at the same time high transmittance for low-intensity ambient light (Hasan et al. 2009). Because of that, optical limiting materials are used to protect human eyes and sensitive optical instruments.

## 9.4 OPTICAL PROPERTIES OF POLYMER NANOCOMPOSITES

### 9.4.1 LINEAR OPTICAL PROPERTIES OF POLYMER NANOCOMPOSITES

#### 9.4.1.1 Transparency of Polymer Nanocomposites

Two of the most desirable optical properties of the polymer NC materials are transparency and high refractive index value, which can be obtained by dispersing NPs with high refractive index, a steep absorption in the near UV region (below 400 nm), and no absorption in the visible region (TiO$_2$, ZnO, SiO$_2$, Al$_2$O$_3$, zirconia [ZrO$_2$], zinc sulfide [ZnS], etc.) into transparent polymers (PMMA, PS, poly(vinyl alcohol) [PVA], polyimides [PI], etc.). This type of material can be used as transparent UV filters or coatings for materials sensitive to UV light, especially if TiO$_2$ and ZnO are used as nanofillers. In contrast to the conventional composites, where after the addition of particles into the polymer matrix a loss of transparency occurs due to the light scattering from the microsized particles or their agglomerates, in the case of polymer NCs optical transparency of polymers can be retained because light scattering by nanosized particles which are smaller than the wavelength of visible light is not significant (Althues et al. 2007). Namely, when particles are placed in front of the electromagnetic radiation, then the extinction of the incident beam will occur, due to the scattering by the particles and absorption in the particles (assuming that the surrounding medium is nonabsorbing) (Bohren and Huffman 1983). In the case of polymer composites, it is usually assumed that the scattering losses are dominant. If the size of particles is similar or larger than the wavelength of light, then Mie theory is applied for the description of the light scattering by particles (Figure 9.5a). On the other hand, the scattering of light by particles where the dimensions are smaller than the wavelength of light (less than one-tenth of the wavelength of visible light) is called Rayleigh scattering (Figure 9.5b).

According to the Rayleigh scattering theory, if the intensity of unpolarized incident light is $I_0$, then the intensity of scattered light ($I_s$) will be

$$I_s = \frac{8\pi^4 N a^6}{\lambda^4 r^2} \left(\frac{n^2-1}{n^2+2}\right)^2 (1+\cos^2\theta)I_0 \tag{9.9}$$

where $N$ is the number of scattering particles, $a$ is the radius of the particle, $\theta$ is the scattering angle, $r$ is the distance of the detector from the particle, and $n$ ($= n_p/n_m$) is the ratio of refractive index of the particle ($n_p$) and polymer matrix ($n_m$) (Bohren and Huffman 1983). When the scattering particle is absorptive, a complex relative refractive index is used in the above equation (Kerker 1969). From Equation 9.9 it can be concluded that with increasing radius of particle, the intensity of scattered

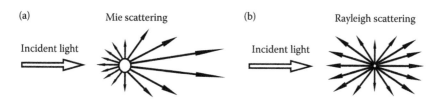

**FIGURE 9.5** Schematic representation of the (a) Mie and (b) Rayleigh scattering by particles.

light will also increase, which can lead to the appearance of the turbidity in the polymer NCs. Using the Rayleigh scattering theory, the loss of the light intensity due to the scattering can be expressed in the following manner:

$$T = \frac{I_t}{I_0} = \exp\left(-\frac{3\varphi_p x a^3}{4\lambda^4}\left(\frac{n_p}{n_m}-1\right)\right) \tag{9.10}$$

where $T$ is the transmittance (or transmission) of polymer NCs, $\varphi_p$ is the volume fraction of NPs, and $x$ is the optical path length (i.e., thickness of polymer NC sample) (Novak 1993). Equation 9.10 is valid for the unpolarized incident light, nonabsorbing, small, isotropically scattering spherical particles, and for low particle content. If $n_p < n_m$, then Equation 9.11 must be used (Caseri 2009):

$$T = \frac{I}{I_0} = \exp\left(-\frac{32\varphi_p x \pi^4 a^3 n_m^4}{\lambda^4}\left|\frac{(n_p/n_m)^2-1}{(n_p/n_m)^2+2}\right|^2\right) \tag{9.11}$$

According to Equations 9.10 and 9.11, polymer NCs will be completely transparent if $n_p = n_m$ (the so-called index matching), regardless of the NPs size. Li et al. prepared core/shell NPs (diameter ~59 nm) with different shell thickness composed from $SiO_2$ core ($n_0 = 1.44$) and $TiO_2$ shell ($n_0 = 2.50$) and obtained that the addition of 1 wt.% of these NPs, with shell weight content of 36.5%, into transparent epoxy matrix ($n_0 = 1.54$) led to the formation of transparent polymer NC (thickness ~4 mm), since at this shell thickness the index matching between matrix and filler was accomplished (Li et al. 2008). Tan et al. obtained that PS/cerium fluoride: ytterbium–erbium ($CeF_3$:Yb–Er) NCs have better transparency than PMMA/$CeF_3$:Yb–Er NCs, even at high NPs loading ($\geq$10 vol.%), due to the smaller index mismatch $\Delta n \approx 0.03$ between $CeF_3$ and PS than $\Delta n \approx 0.12$ between $CeF_3$ and PMMA (Tan et al. 2010). Virtually transparent polymer NCs can be obtained when the diameter of the applied NPs is lower than ~40 nm, since in that case the optical scattering can be significantly lowered even if there is a large difference between $n_0$ values of the polymer and NPs. This is of course valid only when the thickness of the polymer NC is relatively low (a few micrometers) and when the dispersion of NPs in the polymer matrix is good. However, for certain applications, such as transfer of information, it is necessary that NPs imbedded in polymer matrix have a diameter lower than 10 nm. Polymer NCs prepared with such small NPs will be transparent even at high NPs portion and large refractive index differences between polymer and NPs. For example, Nussbaumer et al. (2003) synthesized PVA/$TiO_2$ films of thickness between 40 and 80 μm, using $TiO_2$ NPs (diameter ~2.5 nm), which contained only rutile crystal modification, and obtained that NC films with concentration of $TiO_2$ NPs below 25 wt.% absorb UV but not visible light (Nussbaumer et al. 2003). The authors concluded that these NCs can be used as efficient optically transparent UV filters. The thickness of polymer NCs also plays a significant role in obtaining transparent material, since with increasing thickness of NC sample, the path length (Equations 9.10 and 9.11) and scattering volume increase as well, and the importance of multiple scattering effect becomes larger, which eventually can lead to the formation of opaque material (Liao et al. 2014). On the other hand, when the diameter of NPs is higher (>100 nm), it is also necessary to take care that the refractive index of NPs is similar to that of the polymer matrix (the difference between $n$ of NPs and polymer matrix in this case must be less than 0.02) in order to avoid scattering and opaque appearance of polymer NCs. The transparency loss and multiple scattering effects also occur with increasing fraction of the incorporated NPs, since with high NPs loadings agglomerates can appear inside the matrix and consequently Rayleigh or even Mie scattering, which must be avoided by ensuring uniform dispersion of NPs in polymer. The appearance of larger aggregates in PMMA/indium oxide ($In_2O_3$)

NC films (thickness ~100 µm) with increasing NP content was the reason for the red shift of UV light absorption start of NCs and for the lowering of $T$, as observed by Singhal et al. (2013). Furthermore, Sarwar et al. (2009) obtained that transmittance of NC films prepared from polyamide and $Al_2O_3$ increases with increasing NP content up to 10 wt.%, but with further increase of $Al_2O_3$ content, $T$ decreases due to the lowering of the homogeneity of the NCs by the presence of agglomerates (Sarwar et al. 2009). Uniform dispersion of NPs in polymer matrix can be obtained by increasing the compatibility between NPs and polymer, that is, by increasing solubility of NPs in organic media and enhancing its miscibility with polymer. In order to do that, it is necessary to have chemical bonds between NPs and polymer matrix, which can be achieved usually by surface functionalization of the applied NPs or by preparation of NPs using *in situ* reactions in an organic matrix. For example, Li et al. (2007) have used an *in situ* sol–gel polymerization technique to prepare transparent PMMA/ZnO NCs, which showed significant UV shielding effect over the full UV range, even at very low ZnO content of 0.017 wt.% (Li et al. 2007). The diameter of the applied ZnO NPs was $2.8 \pm 0.4$ nm. Tsai et al. (2014) prepared transparent polymer NCs using *in situ* polymerization and anatase $TiO_2$ NPs surface modified with phenyl acetic and hexanoic acid (compatible with poly(ethoxylated (6) bisphenol A dimethacrylate)) and with acetic acid (compatible with poly(4-vinyl benzyl alcohol)) (Tsai et al. 2014). These authors obtained that the transparency of the synthesized NCs increased with increasing content of $TiO_2$ NPs and their dispersity. Furthermore, Imai et al. (2009) have shown that dispersibility of $TiO_2$ (7 nm diameter) and $ZrO_2$ (5 nm diameter) NPs in poly(bisphenol A carbonate) can be significantly improved by surface modification of NPs with phosphoric acid 2-ethylhexyl esters and simultaneous incorporation of sulfonic acid moiety into the polycarbonate chain (Imai et al. 2009). The obtained NCs were highly transparent up to the content of 42 wt.% of $TiO_2$ and 50 wt.% of $ZrO_2$. Demir et al. (2007) performed surface modification of ZnO NPs with *tert*-butylphosphonic acid and used so prepared NPs (22 nm in diameter) to synthesize PMMA/ZnO NC films of ~2.0 µm with relatively high $n_0$ and high $T$ of $91.3 \pm 0.7$% up to the 2.37 vol.% (11 wt.%), that is, NCs that were almost transparent as pure PMMA ($T_{PMMA} = 92$%) (Demir et al. 2007). Yen et al. (2013) prepared transparent polyimide/$TiO_2$ films that contained relatively high NP content (50 wt.%) and thickness (15 µm) using poly(*o*-hydroxy-imide) as matrix (Yen et al. 2013).

### 9.4.1.2   Refractive Index of Polymer Nanocomposites

Beside high optical transparency, optically functional polymer NCs are often required to have high refractive index value for successful applications in optical filters, waveguides, lenses, solar cells, optical adhesives, reflectors, antireflection films, etc. This requirement can be efficiently accomplished by combination of NPs with high $n_0$ values and low absorption coefficients ($TiO_2$, ZnO, $ZrO_2$, ZnS, and PbS) with polymers. The refractive index of polymer NCs, $n_{NC}$, can be approximately calculated using the mixing rule, as shown below (Liu et al. 2008)

$$n_{NC} = \varphi_p n_p + \varphi_m n_m = \frac{\rho_p n_m - w_p(\rho_p n_m - \rho_m n_p)}{\rho_p - w_p(\rho_p - \rho_m)} \tag{9.12}$$

where $\varphi_p$ and $\varphi_m$ are volume fractions of NPs and polymer, respectively, $\rho_p$ and $\rho_m$ are densities of NPs and polymer matrix, respectively, and $w_p$ is the weight fraction of NPs. If $\varphi_p$ is small and densities of polymer matrix and NPs are similar, then Equation 9.12 can be written as follows:

$$n_{NC} = w_p n_p + n_m \tag{9.13}$$

According to Equations 9.12 and 9.13, with increasing volume or weight fraction of NPs, the $n_{NC}$ value also increases. As already mentioned, higher fractions of NPs homogenously dispersed in polymer matrix usually can be introduced using *in situ* preparation techniques, that is, *in situ* polymerization in the presence of surface modified NPs or *in situ* preparation of NPs in polymer matrix. The change of the $n_{NC}$ with increasing NPs loading can also be theoretically predicted using Maxwell-Garnett effective medium theory (Maxwell-Garnett 1904):

$$\frac{\varepsilon_c - \varepsilon_m}{\varepsilon_c + 2\varepsilon_m} = \varphi_p \left( \frac{\varepsilon_p - \varepsilon_m}{\varepsilon_p + 2\varepsilon_m} \right) \tag{9.14}$$

where $\varepsilon_c$, $\varepsilon_m$, and $\varepsilon_p$ are dielectric constants of the NC, matrix, and NP, respectively. The refractive index can be calculated using Equation 9.14 and the relation: $n^2 = \varepsilon$. This theory is valid only for the spherical particles with sizes smaller than the wavelength of the incident light and in the case when there is no interaction of neighbored particles, that is, for relatively low particle loading.

Complex refractive index of polymer NCs can be calculated from reflectance (or reflectivity, $R$) and extinction coefficient ($k$, Equation 9.2) using Equation 9.15 (Saini et al. 2013):

$$R = \frac{(n-1)^2 + k^2}{(n+1)^2 + k^2} \tag{9.15}$$

An example of the dependence of reflectance of polymer NC based on PMMA and TiO$_2$ NPs surface modified with 6-palmitate ascorbic acid on wavelength, which was used as confirmation of the presence of NPs in polymer matrix, is given in Figure 9.6.

The change (dispersion) of the $n_{NC}$ with wavelength can be described by the Cauchy model (Equation 9.16), in which $A$, $B$, and $C$ are constants (Demir et al. 2007). If transparent NC films are prepared, then constant $C$ is close to zero and can be ignored (Ruiterkamp et al. 2011). The dispersion of refractive index of polymer NCs is especially important to know for designing devices that will be applied for spectral dispersion and in optical communications.

$$n(\lambda) = A + \frac{B}{\lambda^2} + \frac{C}{\lambda^4} \tag{9.16}$$

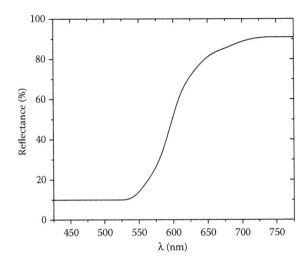

**FIGURE 9.6** The dependence of reflectance of NC based on PMMA and TiO$_2$ NPs surface modified with 6-palmitate ascorbic acid (PMMA/TiO$_2$-PAA-R (3)) on wavelength. (Data reproduced from Džunuzović, E. S. et al. 2010. *Hemijska Industrija* 64:473–89 with the permission of Association of Chemical Engineers of Serbia.)

One of the most applied nanofillers to increase the refractive index of polymer NCs is $TiO_2$. Liu et al. (2008) prepared homogenous and transparent polyimide/$TiO_2$ NC film (0.16 μm thickness) using sulfur containing PI with $n_{m, 633nm} = 1.68$ and relatively low birefringence and 45 wt.% of anatase $TiO_2$ NPs modified with silica ($n_{p, 589nm} = 2.0$), and obtained that $n_{NC}$ of the synthesized NC is 1.81 at 633 nm (Liu et al. 2008). Chau et al. (2009) have shown that $n_{NC}$ of epoxy/ $TiO_2$ NC films (1 μm thickness), synthesized using surface modified anatase $TiO_2$ NPs, increases with increasing weight percentage of NPs, and the highest increase was observed for the NC prepared with hexylamine/acetic acid-capped $TiO_2$ NPs (Chau et al. 2009). The increase of $n_{NC}$ with increasing weight percentage of NPs at 486, 589, and 656 nm was also observed by Imai et al. (2009), and obtained results were in a good agreement with the Maxwell-Garnett theory for the lower NPs loading (Imai et al. 2009). Tsai et al. (2014) prepared 50 μm thick transparent polyacrylate/$TiO_2$ and poly(vinyl benzyl alcohol)/$TiO_2$ NC films by *in situ* polymerization, starting from hexanoic acid and phenyl acetic acid surface modified $TiO_2$ NPs and obtained that $n_{NC}$ increases with increasing $TiO_2$ content and aromatic structure in polymer matrix (Tsai et al. 2014). According to these authors, at 633 nm poly(4-vinyl benzyl alcohol)/$TiO_2$ NC has $n_{NC} = 1.73$ at 60 wt.% of $TiO_2$. On the other hand, Liou et al. (2010) synthesized polyimide/$TiO_2$ hybrid films of thickness between 20 and 30 μm, starting from organosoluble PI with hydroxyl groups and titanium butoxide, and obtained that at 50 wt.% titania content, the value of $n_{NC}$ was 1.83 (Liou et al. 2010). Tao et al. (2011) observed that $n_{NC}$ of transparent polymer NCs, prepared by grafting poly(glycidyl methacrylate) onto anatase $TiO_2$ NPs surface modified with phosphate-azide ligand, linearly increases from 1.5 to 1.8 at 500 nm by increasing NP loading to 30 vol.% (60 wt.%), indicating homogenous dispersion of NPs in the polymer matrix (Tao et al. 2011). Furthermore, they have also obtained that Abbe number decreases with increasing NP loading and that dispersion curves of NCs (400–800 nm) become steeper at higher NP loading. Chang et al. (2006) reported that $n_{NC}$ of polyimide/$TiO_2$ linearly increases at 633 nm from 1.69 to 1.82 between approximately 8 and 36 vol.% (Chang et al. 2006). The linear increase of $n_{NC}$ with increasing volume fraction of NPs at 586 nm was also observed for transparent NCs, formed using poly(benzyl acrylate) as matrix and rutile $TiO_2$ NPs surface modified with alkanephosphonic acid, and measured $n_{NC}$ values coincide with Maxwell-Garnett effective medium theory (Ruiterkamp et al. 2011). Similar was obtained by Nussbaumer et al. (2003) for PVA/$TiO_2$ films prepared with rutile $TiO_2$ NPs (Nussbaumer et al. 2003). After extrapolation of the linear fit to 100%, a lower value of refractive index (2.30) was obtained than was expected for the bulk rutile $TiO_2$, which was explained by nonlinear dependence of refractive index at higher NP content. The linear dependence of $n_{NC}$ of PMMA/$TiO_2$ NCs on weight fraction of $TiO_2$ NPs surface modified with oleic and phosphonic acid (8–32 wt.%) at 1500 nm was observed by Convertino et al. (2007). The $n_{NC}$ weight fraction relation of these NCs prepared with different capping ligand had slightly different slope, which was explained by different morphology of the prepared NC films. On the other hand, Liu et al. (2011) prepared hybrid films (thickness 90–140 nm) from anatase $TiO_2$ NPs, titanium alkoxide and epoxy resin using *ex situ* synthesis method, and obtained that $n_{NC}$ increases with increasing weight fraction of $TiO_2$ NPs to a certain value and then decreases with further increase of $TiO_2$ NPs content (Liu et al. 2011). At 633 nm, the maximum $n_{NC}$ of 1.972 was obtained for 90 wt.% of $TiO_2$ NPs, which was higher than the value obtained when the matrix was either titanium alkoxide or epoxy resin. These results were explained by the ability of epoxy resin to fill porous regions of the hybrid film, so when the content of epoxy resin is not high enough, then it cannot fill the spaces between the $TiO_2$ NPs and thus $n_{NC}$ decreases.

Beside $TiO_2$, $ZrO_2$ NPs have been also applied to increase $n_{NC}$ of polymer NCs. For example, Xu et al. (2009) have shown that refractive index of UV-curable poly(urethane acrylate)s can be increased from 1.475 for pure polymer matrix to 1.625 for NCs by the addition of 20 wt.% of $ZrO_2$ NPs (Xu et al. 2009). In another report, Tao et al. (2013) determined that the refractive index of transparent and thick (1 mm) epoxy resin/$ZrO_2$ NCs linearly increases with increasing volume fraction of NPs to 1.65 for 20 vol.% ($n_0$ of pure epoxy resin was 1.51) (Tao et al. 2013).

TABLE 9.4

Tuning the Refractive Index of Polymer Nanocomposites by Addition of Different
Nanoparticles in Polymer Matrix

| Polymer Matrix | Nanoparticles and Their Content in NCs | Refractive Index Increase | References |
|---|---|---|---|
| Poly(vinyl alcohol) | TiO$_2$ NPs (10.5 vol.%) | 0.088 (at 589 nm) | Nussbaumer et al. (2003) |
| Poly(4-vinyl benzyl alcohol) | TiO$_2$ NPs surface modified with acetic acid (60 wt.%) | 0.120 (at 633 nm) | Tsai et al. (2014) |
| Sulfur containing polyimide | TiO$_2$ NPs modified with silica (45 wt.%) | 0.130 (at 633 nm) | Liu et al. (2008) |
| Polyimide | TiO$_2$ NPs (36 vol.%) | 0.160 (at 633 nm) | Chang et al. (2006) |
| Epoxy resin | Hexylamine/acetic acid-capped TiO$_2$ NPs (>70 wt.%) | 0.165 | Chau et al. (2009) |
| Poly(glycidyl methacrylate) | TiO$_2$ NPs surface modified with phosphate-azide ligand (60 wt.%) | 0.300 (at 500 nm) | Tao et al. (2011) |
| Epoxy resin | ZrO$_2$ NPs modified with (3-glycidyloxypropyl) trimethoxysilane ligand (50 wt.%) | 0.140 (at 600 nm) | Tao et al. (2013) |
| Poly(urethane acrylate) | ZrO$_2$ NPs (20 wt.%) | 0.150 (at 633 nm) | Xu et al. (2009) |
| Polythiourethane | PbS NPs (67 wt.%) | 0.090 (at 633 nm) | Lü et al. (2005) |
| Poly(vinlypyrrolidone) | Mercaptoethanol capped ZnS NPs (80 wt.%,) | 0.246 (at 590 nm) | Zhang et al. (2013a) |

Semiconducting NPs with high refractive index, such as PbS and ZnS, are also used for the synthesis of polymer NCs with high refractive index. Weibel et al. (1991) prepared poly(ethylene oxide)/PbS NCs, consisting of 88 wt.% of PbS, using coprecipitation of PbS and polymer matrix, and obtained extremely high $n_{NC}$, which was between 2.9 and 3.4 at wavelengths 632.8 and 1295 nm (Weibel et al. 1991). Furthermore, Lü et al. (2005) used an *in situ* method and lead-containing precursor (0–67 wt.%) with hydroxyl groups to incorporate PbS NPs into polythiourethane matrix in order to obtain NCs with $n_{NC}$ between 1.574 and 1.665 (Lü et al. 2005). Transparent NCs of poly-(dimethyl)-block-(phenyl)siloxane matrix ($n_0 = 1.54$) and ZnS (1–5 nm) with $n_{NC}$ of 1.68 at ZnS content of 4.6 vol.% were prepared by Sergienko et al. (2012). Zhang et al. (2013a) have used mercaptoethanol capped ZnS (ME–ZnS) NPs (~3 nm) to prepare poly(vinlypyrrolidone)/ZnS NC optically transparent films by simple blending, and obtained that with increasing content of ME–ZnS NPs from 0 to 80 wt.%, $n_{NC}$ and Abbe number changes from 1.506 to 1.752 and from 55.6 to 20.4, respectively (Zhang et al. 2013a). The change of the refractive index of different polymer/NPs systems after addition of NPs is summarized in Table 9.4.

Increase of the $n_{NC}$ value of polymer NCs can also be accomplished using graphene NPs (Zhang et al. 2012), carbon black (Xue et al. 2012), silicon NPs (Zhang et al. 2013b), in some cases silver nanoparticles (Chahal et al. 2012), etc. On the other hand, if polymer NC is prepared using NPs with $n_p$ lower than $n_m$, then $n_{NC}$ will be reduced. Usually gold NPs are used for that purpose, since $n_0$ of Au is smaller than 1 in a broad wavelength range. An example of such NCs are gelatin/Au NC films, which $n_{NC}$ linearly decreased to around 1.0 at 632.8 nm with increasing Au content from 0.7 up to 48 vol.% (9.5–92.9 wt.%) (Zimmermann et al. 1993).

### 9.4.1.3   UV Absorption in Polymer Nanocomposites

The effect of incorporation of NPs in polymer matrix, the possible aggregates formation, appearance of quantum confinement effect, etc. can be easily investigated by UV–visible (UV–vis) absorption spectroscopy, which is probably one of the simplest and most useful techniques for the investigation of optical properties of polymer NCs. Using the UV–vis spectroscopy, the light absorption due to the electronic transition in the sample is measured, and the results are presented as dependence of absorption coefficient versus wavelength. For example, it has been shown that the presence of 30 nm size $TiO_2$ NPs in PS matrix leads to the strong increase of absorbance compared to the pure PS, due to the UV absorption of $TiO_2$ NPs, as observed by Jaleh et al. (2011). These authors have also investigated the influence of the UV irradiation on the optical properties of prepared NCs and they observed that absorbance of NCs increases in the range of 230–600 nm after exposing the samples to UV. This was explained by the production of photogenerated electrons and holes under UV irradiation due to the presence of $TiO_2$ NPs, which degrade PS and form cavities in the sample, causing in this manner scattering of the light and decrease of the transmittance. Also, Pucci et al. (2006) observed change of the color of PVA/Au film from yellow (absorption at around 300 nm) to purple after only 5 min of UV irradiation, due to the conversion of $Au^{3+}$ ions dissolved in PVA into Au(0) NPs and the appearance of surface plasmon absorption band at around 550 nm (Pucci et al. 2006). Furthermore, Chahal et al. (2012) have proved the presence of Ag NPs in PVA by the appearance of SPR peak in the absorption spectra of the prepared NCs, and they have also observed that the absorption intensity of NCs increases with increasing applied γ irradiation dose (Chahal et al. 2012). On the other hand, Gasaymeh et al. (2010) observed blue shift and the higher intensity of the absorption band for the poly(vinyl pyrolidone)/cadmium sulfide (CdS) NCs prepared using $CdSO_4$ (9 wt.%) as NPs precursor with increasing γ-dose from 10 to 50 kGy, indicating formation of monodisperse CdS NPs of smaller size (Gasaymeh et al. 2010). However, with increasing content of $CdSO_4$ from 3 to 16 wt.% a red shift and the higher absorption intensity occurred, which is assigned to the formation of larger size nanocrystallites. Ramesh and Radhakrishnan (2011) prepared PVA/ Ag NC film by a facile *in situ* preparation method, and observed that in the presence of mercury the absorption peak of NC was blue shifted, indicating that prepared NC can be used as a sensor for Hg (Ramesh and Radhakrishnan 2011). Furthermore, Dirix et al. (1999a,b) obtained that the absorption spectrum of uniaxially oriented high density polyethylene/Ag nanocomposite films, prepared by drawing techniques, strongly depend on the polarization direction of the incident light (Dirix et al. 1999a,b). The color of the prepared NCs was changed from bright yellow to red by changing the vibration direction of linearly polarized light from perpendicular to parallel to the drawing axis. Similar was obtained for polyethylene/Au NCs (Dirix et al. 1999c).

The transitions that occur in NCs can be examined from the optical band gap energy values ($E_g$) near the absorption edge, which can be determined using the UV–vis measurements and the following relation (Tauc et al. 1966; Hemalatha et al. 2014):

$$(\alpha_0 h\nu)^{1/n} = B(h\nu - E_g) \tag{9.17}$$

where $\alpha_0 = 2.303A/x$ from the Beer–Lambert's relation, $A$ is absorbance, $x$ is thickness of the polymer NC sample, $B$ is a factor that depends on the structure of the specimen and inter band transition probability, and in certain frequency range it can be assumed that it is constant, $h$ is Planck's constant, and $\nu$ is the frequency, while $h\nu$ is the incident photon energy. Index $n$ characterizes the type of electronic transition which causes the optical absorption and its values can be 1/2 for allowed direct, 3/2 for forbidden direct, 2 for allowed indirect, and 3 for forbidden indirect transitions. For example, from the dependence of $(\alpha h\nu)^{1/2}$ versus $h\nu$, indirect band gap energy can be calculated by extrapolating the linear part of the graph to $h\nu$ axis, while by plotting $(\alpha h\nu)^2$ against $h\nu$ the direct band gap energy is determined. As already mentioned, when the particle size is reduced (lower than 10 nm), then due to the quantum confinement effect a blue shift of the absorption edge and an increase of the band gap energy occur in comparison to the bulk material. This phenomenon was observed for the PS-*co*-maleic acid/CdS NCs, for which absorption onset (490 nm) is blue shifted in comparison to the bulk CdS (512 nm) (Nair et al. 2005). Furthermore, Porel et al. (2005) prepared PVA/Au NCs by *in situ* formation of polygonal Au nanoplates in PVA films and obtained that with decreasing size of Au NPs from pentagons to hexagons to triangles, the blue shift of absorption maxima occurred (Porel et al. 2005). The blue shift was also observed for the PMMA/TiO$_2$ NCs prepared from the methyl methacrylate, 3-(trimethoxysilyl)propyl methacrylate and TiO$_2$ precursor titanium-isopropoxide (Ti-iP) by *in situ* polymerization and sol–gel process (Yuwono et al. 2003). The band gap energy of the prepared NCs (3.91–4.72 eV) was higher than the bulk value (3.20 eV), indicating that the size of the TiO$_2$ NPs in PMMA is very small and most likely lower than 10 nm. On the other hand, Sun et al. (2010) applied a similar procedure to prepare PMMA/TiO$_2$ NCs using titanium butoxide (Ti(OBu)$_4$) and acetylacetone as surfactant, and obtained that with increasing content of Ti(OBu)$_4$ from 20 to 60 wt.%, the absorption onset of NCs is red shifted and the band gap energies lowered (Sun et al. 2010). The modest increase and red shift of the absorbance, and lowering of the gap energies from 2.89 (pure polymer) to 2.63 eV (35 wt.% of NPs) of poly(9,9′-di-n-octylfluorenyl-2.7-diyl)(PFO)/TiO$_2$ NCs on addition of TiO$_2$ NPs was observed by Jumali et al. (2012). Hemalatha et al. (2014) prepared ZnO NPs and PVA/ZnO NCs by a solution casting technique using different content of ZnO NPs and observed that the absorption spectrum of ZnO NPs (371 nm; 3.318 eV) is blue shifted compared to the bulk ZnO (380 nm; 3.268 eV) at room temperature, while the absorption spectrum of PVA/ZnO NCs showed a red shift (from 437 to 528 nm) in comparison to the pure PVA (241 nm; 5.15 eV) and ZnO NPs with increasing content of ZnO from 5 to 20 mol% (Hemalatha et al. 2014). At the same time, indirect band gap energy of the synthesized NCs decreased from 4.76 to 2.38 eV, which was explained by the development of microstrain in PVA/ZnO NCs on addition of ZnO NPs and simultaneous formation of defects in the polymer matrix. The authors concluded that NCs showing such optical properties can have possible application in optoelectronic devices. On the other hand, Chandrakala et al. (2014) reported that absorption spectra of NCs prepared from PVA and ZnO–cerium oxide (Ce$_2$O$_3$) NPs show no red or blue shift, but only variations in the absorbance intensity with increasing NPs content because of the uneven particle distribution (Chandrakala et al. 2014). However, they have also observed that $E_g$ decreases from 6.98 eV for pure PVA to 6.45 eV for NCs with 1.0 wt.% NPs, due to the interactions of the NPs with polar groups of polymer matrix, leading to the complex formation. Furthermore, Džunuzović et al. (2012, 2013) obtained that absorption edge of TiO$_2$ NPs surface modified with gallates, PMMA/TiO$_2$ and PS/TiO$_2$ NCs prepared with these NPs are red shifted compared to the pure PMMA, PS, and unmodified TiO$_2$ NPs (Džunuzović et al. 2012, 2013). Similar was obtained by Convertino et al. (2007), who observed increased absorption and red shift of the absorption edge (shift to lower energy) with increasing content of TiO$_2$ NPs surface modified with oleic and phosphonic acid in PMMA, indicating the presence of NPs aggregates (Convertino et al. 2007).

Furthermore, Deepa et al. (2011) reported that the presence of Ag and Au NPs in conjugated polymer poly(3,4-ethylenedioxypyrolle) films increases the absorption coefficient and decreases direct band gap energy of polymer matrix, increasing in this manner its electrochromic switching ability (Deepa et al. 2011). Mahendia et al. (2011) observed that the intensity of SPR peak of PVA/Ag NCs increases and its position moves to the higher wavelengths, while the optical band gap values decrease with increasing NPs concentration, due to the formation of charge transfer complexes (CTCs) (Mahendia et al. 2011). The formed CTCs act as trap levels between the highest occupied molecular orbital (HOMO) and the lowest unoccupied molecular orbital (LUMO) PVA bands. Similar results were obtained for the PVA/Ag NCs prepared by Chahal et al. (2012).

### 9.4.1.4  Photo- and Electroluminescence of Polymer Nanocomposites

In order to use polymer NCs in applications such as organic light emitting diodes (OLEDs) with tunable emission colors, solid state lightning, *in vitro* cell imaging, dye-sensitized solar cells, field-emission devices, chemical sensors, etc., different luminescent NPs can be applied as fillers. By changing the size of semiconductor NPs, emission of color can vary due to the quantum confinement effect. Photoluminescence of polymer NCs is quite different than for the pure polymer; it depends on the type of polymer and NPs, as well as on the NPs size and content. Using PL spectroscopy, a PL spectrum is recorded, which represents a plot of PL intensity versus wavelength for a fixed excitation wavelength. Pucci et al. (2004) investigated optical properties of ultrahigh molecular weight polyethylene (UHMWPE) films loaded with gold NPs coupled with a thiol-bearing terthiophene and observed that they exhibit strongly enhanced photoluminescence compared to the nonchromophoric control samples (Pucci et al. 2004). Guo et al. (2007) reported that PL spectra of vinyl ester resin/ZnO depend on the content of NPs, that is, that pure polymer showed no luminescence, while investigated NCs exhibited significant luminescence at 1 wt.% of ZnO (Guo et al. 2007). With further increase of the ZnO content, the intensity increased, but there was no influence on the emission maximum. Jumali et al. (2012) obtained that intensity of fluorescence spectra at 355 nm of PFO/TiO$_2$ NCs were enhanced with increasing NPs content (Jumali et al. 2012). The obtained increase originated from the polymer matrix, since TiO$_2$ NPs showed no PL emission. However, TiO$_2$ NPs served for trapping electrons produced during polymer excitation, due to their strong electron affinity. As a consequence, more holes were formed and due to the recombination of excitons at the NC surface, the PL intensity increased. The increase of the PL intensity with increasing NPs content was also observed for the polyaniline (PANI)/Ag NCs at an excitation wavelength of 330 nm (Gupta et al. 2010). According to the results obtained by Hemalatha et al. (2014), the PL emission spectrum of ZnO NPs and PVA/ZnO NCs have luminescence emission in the blue region and the emission intensity varies with NPs content with the optimum intensity at 10 mol.% of ZnO (Hemalatha et al. 2014). Similar was obtained by Chandrakala et al. (2014) for PVA/ZnO–Ce$_2$O$_3$ NCs. On the other hand, luminescence peak of poly(3-hexylthiophene-2,5-diyl)/silicon nanowire NCs decreased with increasing content of NPs due to the agglomeration of NPs at higher content (Braik et al. 2014). Hussain et al. (2014) reported that PL intensity of NCs prepared from plasma polymerized aniline (PPani)/TiO$_2$ has significantly decreased with increasing time of UV illumination (Hussain et al. 2014). Observed PL quenching was explained by the increase of the donor/acceptor (PPani)/TiO$_2$ interfacial area and phase interpenetration. Similar results were obtained by Han et al. (2008) for the poly(N-vinylcarbazole)/TiO$_2$ NCs. This was explained by the dissociation of the photogenerated excitons before luminescence could occur. Kuila et al. (2007) observed that in solution, luminescence intensity of poly(3-hexyl thiophene)/hexadecyl amine capped Ag NCs continuously decreased with increasing NPs content, while NC thin films had higher fluorescence intensity compared to the pure polymer matrix (Kuila et al. 2007). These authors have also observed that in solution there is no shift of the PL peak because NPs had no effect on the excited state of the polymer, while blue shift of the PL band of NC thin films appeared with increasing Ag content, which was explained by the coupling of the plasmon vibration with the electronic levels of the excitons of polymer. Yang and Yoon (2004) have also observed that with increasing NPs content, the PL spectra of poly(p-phenylene vinylene)/TiO$_2$ NCs was blue shifted. However, for the polybutanediolmonoacrylate/ZnO NCs it has been observed that excitation and emission wavelengths increase with increasing NPs size (Althues et al. 2009).

The application of NCs, obtained by copolymerization of zinc methacrylate with styrene, followed by H$_2$S treatment, in electroluminescence devices was investigated by Yang et al. (1997). The prepared NCs showed electroluminescence with an emission peak at 440 nm. Al-Asbahi et al. (2013) prepared PFO/TiO$_2$ NCs, and used them as an emissive layer in OLED devices (Al-Asbahi et al. 2013). Up to 25 wt.% of TiO$_2$, all devices showed higher EL intensity than pure PFO, which decreased with increasing NPs content, that is, a device with 5 wt.% TiO$_2$ had the highest EL intensity due to the best electron–holes recombination.

### 9.4.2 NONLINEAR OPTICAL PROPERTIES OF POLYMER NANOCOMPOSITES

For the application of polymer NCs in nonlinear optical applications such as signal processing, as optical switches, for optical pulse compression and limiting, etc. it is important to examine their nonlinear optical properties as well. Such material is required to have large nonlinear refractive index, minimal one- or multiphoton absorption losses, and fast response time (Sezar et al. 2009). Nonlinear optical properties of polymer NCs are usually related to NCs with metallic and semiconductive NPs.

A widely used and quite simple technique to characterize and measure nonlinear absorption, scattering, and refraction, for example, nonlinear absorption coefficient and nonlinear refractive index, is $z$-scan. The open aperture $z$-scan is used to measure the total transmittance through the investigated sample versus the incident laser intensity, while the sample is moved along the z-axis through the focus of the lens (Wang et al. 2011). Schematic representation of the open aperture $z$-scan apparatus is given in Figure 9.7. The open aperture $z$-scan is applied to find $\beta$ and Im $\chi^{(3)}$, while from the closed aperture $z$-scan data, divided by that of an open aperture $z$-scan data, $n_2$ and Re $\chi^{(3)}$ can be determined. From the shape of a $z$-scan curve, information considering the nature of the nonlinearity is determined. Yuwono et al. (2003) obtained that PMMA/TiO$_2$ NCs show positive nonlinearity, that is, the normalized transmission has a prefocal transmission minimum (valley) and then postfocal transmission maximum (peak) (Yuwono et al. 2003). Furthermore, nonlinear optical properties of PMMA/TiO$_2$ NCs strongly depend on the content of used titanium-isopropoxide precursor, that is, $\beta$ and $n_2$ increase with increasing content of Ti-iP up to 60 wt.%. Similar results were obtained for PMMA/TiO$_2$ NCs prepared using Ti(OBu)$_4$ precursor (Sun et al. 2010). On the other hand, PMMA-$co$-MA/TiO$_2$ nanorods NCs prepared by Sciancalepore et al. (2008) exhibited negative nonlinearity with $n_2 = -6 \times 10^{-15}$ cm$^2$/W at all excitation intensities, indicating pure third-order origin (Sciancalepore et al. 2008). Sezar et al. (2009) determined from the closed and open aperture $z$-scan that PANI/Ag NC films of 130 nm thickness have negative refractive nonlinearity ($n_2 = -(0.57 \pm 0.03) \times 10^{-14}$ m$^2$/W) and negative nonlinear absorption coefficient ($\beta = -(2.74 \pm 0.1) \times 10^{-7}$ m/W), as well as negative values of the Im $\chi^{(3)}$ and Re $\chi^{(3)}$ (Sezar et al. 2009). Negative nonlinearity was also observed for the sulfonated PS/CdS NCs (Du et al. 2002). By assuming that only third-order nonlinear mechanisms exist, Du et al. (2002) obtained that negative nonlinear refractive index increased linearly with increasing input irradiance and with decreasing CdS content. Deng et al. (2008) obtained negative $n_2$ of PMMA/Ag NC films and that enhanced nonlinear optical properties appeared because of the SPR of Ag NPs (Deng et al. 2008). Also, according to Lu et al. (2012), a large third-order nonlinear susceptibility of poly(2-methoxy-5-(2-ethylhexyloxy)-1,4-phenylenevinylene)/Ag NCs is associated with strong SPR of Ag NPs (Lu et al. 2012). The appearance of the large nonlinear optical properties of NCs in the last two cases is based on the specific resonance conditions, when an energetic transition frequency such as plasmon frequency, coincides with the exciting laser frequency. Due to that, the excitation is localized in the highly polarizable NPs which strongly amplify the local field formed by the induced irradiation

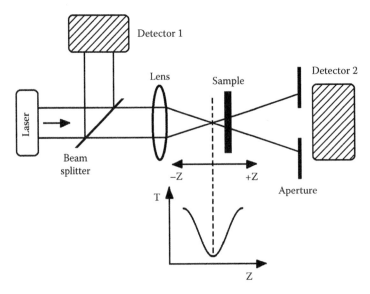

**FIGURE 9.7**   Schematic representation of the open aperture z-scan apparatus.

(Pomogailo and Kestelman 2005). Sun et al. (2008) investigated the influence of the Ag content on the nonlinear optical properties of PMMA/Ag NC films and obtained that $\chi^{(3)}$ increases with increasing Ag content and that NCs show positive nonlinearity at low Ag content, while at 2.4 wt.% of Ag, PMMA/Ag NCs show negative nonlinearity (Sun et al. 2008). Since for the PMMA/Ag NC film with 0.8 wt.% Ag parameter $K < 1$, this NC can be applied for the optical switching application at 532 nm. Liu et al. (2012) prepared NCs by the addition of the colloidal cadmium selenide (CdSe) QDs into a polymerizable ionic liquid monomer with one methacryloyl group, and obtained by $z$-scan technique that NC films at CdSe content as high as 3.6 wt.% exhibit saturable absorption, negative third-order and positive fifth-order nonlinear refraction, and therefore possible application in nonlinear photonics applications (Liu et al. 2012).

Sezar et al. (2009) investigated nonlinear optical properties of PANI/Ag NC films and observed the appearance of saturable and reverse saturable absorption using a 532 nm picosecond laser (Sezar et al. 2009). The change from the saturable to reverse saturable absorption occurred with increasing laser excitation levels, as observed from the open aperture $z$-scan of NCs at different laser intensities. Furthermore, Sezar et al. also observed that the transmission of light steadily drops after an initial rise with increasing laser intensity. The threshold of limiting of PANI/Ag NCs, which represents the incident irradiance for which $T$ falls to half of the initial value, is 20.8 GW/cm². The authors concluded that prepared PANI/Ag NCs can be applied for optical pulse compression and limiting. It has been shown that conjugated polymers are good candidates for the optical limiting applications, since they possess significant two-photon absorption cross section (Wang et al. 2011). Excellent optical limiting effect for ns laser pulses at 532 nm was also observed for the PFO/carbon nanotube (CNT) NCs and poly(m-phenylenevinylene-co-2,5-dioctoxy-p-phenylenevinylene)/CNT (O'Flaherty et al. 2003a,b). The outstanding optical limiting properties of graphene and CNT polymer composites were widely investigated and reviewed by Wang et al. (2011).

Kulyk et al. observed that highly transparent NC films of PMMA/ZnO with low ZnO content have higher values of the second- and third-order nonlinear susceptibilities than bulk ZnO (Kulyk et al. 2009). The second-order nonlinear optical properties were investigated by Gu et al. for the NCs prepared from poly(trimethylolpropane trimethylacrylate) and rare earth NPs obtained by lanthanide doping method (Gu et al. 2012). Within the excitation range of 750–850 nm, the prepared NCs show tunable second-order nonlinear optical properties, which are applicable for photonic devices. Pronounced second-order optical effect in potassium titanyl phosphate (KTiOPO₄) nanocrystallites (5.6 wt.%) introduced into the PMMA, at 1320 nm and at liquid helium temperature, was observed by Faugeroux et al. (2008). Interestingly, by substitution of K with Rb ions to obtain RbTiOPO₄ nanocrystallites or by replacing PMMA with poly(N-vinylcarbazole), the second-order susceptibility decreases from the maximum value of 2.2 pm/V. On the other hand, photoinduced frequency doubled second harmonic generation of 3.23 pm/V at laser wavelength of 1064 nm, under treatment with polarized UV light at 391 K was obtained for PMMA/KTiOPO₄ NCs at 7–8 wt.% of nanocrystyllites by Galceran et al. (2009).

## 9.5 CONCLUSION

In conclusion, an overview of the recent findings and advances on the investigation of linear and nonlinear optical properties of polymer NCs is summarized in this chapter. The importance of investigation of factors that influence the optical properties of polymer NCs and determine their application, such as type, size, shape, surface characteristics, concentration, and spatial distribution of NPs in the matrix, and optical properties of polymer matrix and applied NPs is highlighted. The research results presented in the literature thus far revealed several interesting facts:

1. The uniform dispersion of NPs with high refractive index, a steep absorption in the near UV region (below 400 nm) and no absorption in the visible region, such as $TiO_2$, ZnO, $SiO_2$, $Al_2O_3$, $ZrO_2$, and ZnS, into transparent polymers (PMMA, PS, PVA, PI, etc.) leads

to the production of transparent, high refractive index nanocomposite materials, which can be applied as optically transparent UV filters or coatings for materials sensitive to UV light. The transparency of polymer NCs can be accomplished if index-matching between polymer and NPs exists, regardless of the NPs size, and when diameter of the applied NPs is lower than ~10 nm, even at high NPs portion and large refractive index differences between polymer and NPs. Furthermore, in order to avoid optical scattering and opaque appearance of polymer NCs, uniform dispersion of NPs in polymer matrix must be achieved by increasing the compatibility between polymer and NPs.

2. Optically functional polymer NCs, which can be applied as optical filters, waveguides, lenses, solar cells, optical adhesives, reflectors, antireflection films, etc., can be prepared by combination of NPs with high $n_0$ values and low absorption coefficients ($TiO_2$, ZnO, $ZrO_2$, ZnS, and PbS) with polymers. The increase of the NP loading by simultaneously maintaining homogenous dispersion of NPs in the polymer matrix has proved to be an efficient method to increase the refractive index of nanocomposite materials.

3. The examination of UV absorption of NPs embedded in a polymer matrix can be used as a simple way to prove the presence of NPs in polymer, to investigate the change of NPs size or the formation of aggregates, to determine the band gap energy, to investigate the ability of polymer NCs to be used as sensors, etc.

4. When polymer NCs are prepared using different luminescent NPs, then they can find their application as organic light emitting diodes with tunable emission colors, in solid state lightning, *in vitro* cell imaging, dye-sensitized solar cells, field-emission devices, chemical sensors, etc.

5. Nonlinear optical properties of polymer NCs, such as large nonlinear refractive index, minimal one- or multiphoton absorption losses, high values of the second- and third-order nonlinear susceptibilities and fast response time can be obtained by introduction of metallic and semiconductive NPs into polymer matrix, in particular, into conjugated polymers. Polymer NCs that show such properties can be applied for photonic devices, in signal processing, as optical switches, for optical pulse compression and limiting, etc.

According to the numerous published papers dealing with the optical properties of polymer NCs, it can be concluded that research in this field of science still has a lot to offer to the scientific audience. The reason for that lies in rapid development of these advanced materials and nanotechnology, intended to improve properties of the materials with simultaneous decrease of the final product price. From all these it is evident that polymer NCs represent suitable candidates for various different linear and nonlinear optical applications.

## ACKNOWLEDGMENTS

This work was financially supported by the Ministry of Education, Science and Technological Development of the Republic of Serbia (research project number: 172062).

## REFERENCES

Al-Asbahi, B. A., Jumali, M. H. H., Yap, C. C., and M. M. Salleh. 2013. Influence of $TiO_2$ nanoparticles on enhancement of optoelectronic properties of PFO-based light emitting diode. *Journal of Nanomaterials* 2013: Article ID 561534 (7 pp).
Althues, H., Henle, J., and S. Kaskel. 2007. Functional inorganic nanofillers for transparent polymers. *Chemical Society Review* 36:1454–65.
Althues, H., Pötschke, P., Kim, G. M., and S. Kaskel. 2009. Structure and mechanical properties of transparent ZnO/PBDMA nanocomposites. *Journal of Nanoscience and Nanotechnology* 9:2739–45.
Asai, T., Sakamoto, W., and T. Yogo. 2013. Synthesis of patterned and transparent $TiO_2$ nanoparticle/polymer hybrid films. *Materials Letters* 107:235–8.

Barford, W. 2005. *Electronic and Optical Properties of Conjugated Polymers*. Oxford: Oxford University Press.

Beecroft, L. L. and C. K. Ober. 1997. Nanocomposite materials for optical applications. *Chemistry of Materials* 9:1302–17.

Bohren, C. F. and D. R. Huffman. 1983. *Absorption and Scattering of Light by Small Particles*. New York: A Wiley-Interscience Publication.

Braik, M., Dridi, C., Rybak, A., Davenas, J., and D. Cornu. 2014. Correlation between nanostructural, optical, and photoelectrical properties of P3HT:DiNW nanocomposites for solar-cell application. *Physica Status Solidi B: Basic Solid State Physics* 3:670–6.

Brown, R. 1999. Optical properties. In *Handbook of Polymer Testing: Physical Methods*, ed. R. Brown, 647–57. New York, Basel: Marcel Dekker.

Caseri, W. 2009. Inorganic nanoparticles as optically effective additives for polymers. *Chemical Engineering Communications* 196:549–72.

Chahal, R. P., Mahendia, S., Tomar, A. K., and S. Kumar. 2012. $\gamma$-Irradiated PVA/Ag nanocomposite films: Materials for optical applications. *Journal of Alloys and Compounds* 538:212–9.

Chandrakala, H. N., Ramaraj, B., Shivakumaraiah, and Siddaramaiah. 2014. Optical properties and structural characteristics of zinc oxide–cerium oxide doped polyvinyl alcohol films. *Journal of Alloys and Compounds* 586:333–42.

Chang, C.-M., Chang, C.-L., and C.-C. Chang. 2006. Synthesis and optical properties of soluble polyimide/titania hybrid thin films. *Macromolecular Materials and Engineering* 291:1521–8.

Chau, J. L. H., Liu, H.-W., and W.-F. Su. 2009. Fabrication of hybrid surface-modified titania-epoxy nanocomposite films. *Journal of Physics and Chemistry of Solids* 70:1385–9.

Convertino, A., Leo, G., Tamborra, M. et al. 2007. TiO$_2$ colloidal nanocrystals functionalization of PMMA: A tailoring of optical properties and chemical adsorption. *Sensors and Actuators B* 126:138–43.

Coso, R. D. and J. Solis. 2004. Relation between nonlinear refractive index and third-order susceptibility in absorbing media. *Journal of the Optical Society of America B: Optical Physics* 21:640–4.

Costanzo, G. D., Ribba, L., Goyanes, S., and S. Ledesma. 2014. Enhancement of the optical response in a biodegradable polymer/azo-dye film by the addition of carbon nanotubes. *Journal of Physics D: Applied Physics* 47:135103 (8pp).

Deepa, M., Kharkwal, A., Joshi, A. G., and A. K. Srivastava. 2011. Charge transport and electrochemical response of poly(3,4-ethylenedioxypyrrole) films improved by noble-metal nanoparticles. *The Journal of Physical Chemistry B* 115:7321–31.

Demir, M. M. and G. Wegner. 2012. Challenges in the preparation of optical polymer composites with nanosized pigment particles: A review on recent efforts. *Macromolecular Materials and Engineering* 297:838–63.

Demir, M. M., Koynov, K., Akbey, Ü. et al. 2007. Optical properties of PMMA and surface-modified zincite nanoparticles. *Macromolecules* 40:1089–100.

Deng, Y., Sun, Y., Wang, P., Zhang, D., Ming, H., and Q. Zhang. 2008. *In situ* synthesis and nonlinear optical properties of Ag nanocomposite polymer film. *Physica E* 40:911–4.

Dirix, Y., Bastiaansen, C., Caseri, W., and P. Smith. 1999a. Preparation, structure and properties of uniaxially oriented polyethylene–silver nanocomposites. *Journal of Materials Science* 34:3859–66.

Dirix, Y., Bastiaansen, C., Caseri, W., and P. Smith. 1999b. Oriented pearl-necklace arrays of metallic nanoparticles in polymers: A new route toward polarization-dependent color filters. *Advanced Materials* 11:223–7.

Dirix, Y., Darribère, C., Heffels, W., Bastiaansen, C., Caseri, W., and P. Smith. 1999c. Optically anisotropic polyethylene–gold nanocomposites. *Applied Optics* 38:6581–6.

Du, H., Xu, G. Q., and W. S. Chin. 2002. Synthesis, characterization and nonlinear optical properties of hybridized CdS–polystyrene nanocomposites. *Chemistry of Materials* 14:4473–9.

Džunuzović, E., Jeremić, K., and J. M. Nedeljković. 2007. In situ radical polymerization of methyl methacrylate in a solution of surface modified TiO$_2$ and nanoparticles. *European Polymer Journal* 43:3719–26.

Džunuzović, E. S., Džunuzović, J. V., Marinković, A. D., Marinović-Cincović, M. T., Jeremić, K. B., and J. M. Nedeljković. 2012. Influence of surface modified TiO$_2$ nanoparticles by gallates on the properties of PMMA/TiO$_2$ nanocomposites. *European Polymer Journal* 48:1385–93.

Džunuzović, E. S., Džunuzović, J. V., Radoman, T. S. et al. 2013. Characterization of *in situ* prepared nanocomposites of PS and TiO$_2$ nanoparticles surface modified with alkyl gallates: Effect of alkyl chain length. *Polymer Composites* 34:399–407.

Džunuzović, E. S., Marinović-Cincović, M. T., Džunuzović, J. V., Jeremić, K. B., and J. M. Nedeljković. 2010. Influence of the way of synthesis of poly(methyl methacrylate) in the presence of surface modified TiO$_2$ nanoparticles on the properties of obtained nanocomposites. *Hemijska Industrija* 64:473–89.

Faugeroux, O., Majchrowski, A., Rutkowski, J., Klosowicz, S., Caussanel, M., and S. Tkaczyk. 2008. Manifestation of second-order nonlinear optical effects in KTP and RTP nanocrystallites incorporated into polymer matrices. *Physica E* 41:6–8.

Galceran, M., Pujol, M. C., Carvajal, J. J. et al. 2009. Synthesis and characterization of $KTiOPO_4$ nanocrystals and their PMMA nanocomposites. *Nanotechnology* 20:035705 (10pp).

Gasaymeh, S. S., Radiman, S., Heng, L. Y., and E. Saion. 2010. Gamma irradiation synthesis and influence the optical and thermal properties of cadmium sulphide (CdS)/poly(vinyl pyrolidone) nanocomposites. *American Journal of Applied Sciences* 7:500–8.

Gu, J., Yan, Y., Zhao, Y. S., and J. Yao. 2012. Controlled synthesis of bulk polymer nanocomposites with tunable second order nonlinear optical properties. *Advanced Materials* 24:2249–53.

Guo, Z.H., Wei, S. Y., Shedd, B., Scaffaro, R., Pereira, T., and H. T. Hahn. 2007. Particle surface engineering effect on the mechanical, optical and photoluminescent properties of ZnO/vinyl-ester resin nanocomposites. *Journal of Materials Chemistry* 17:806–13.

Gupta, K., Jana, P. C., and A. K. Meikap. 2010. Optical and electrical transport properties of polyaniline–silver nanocomposite. *Synthetic Metals* 160:1566–73.

Han, Y., Wu, G., Chen, H., and M. Wang. 2008. Preparation and optoelectronic properties of a novel poly(N-vinylcarbazole) with covalently bonded titanium dioxide. *Journal of Applied Polymer Science* 109:883–8.

Hanemann, T. and D. V. Szabó. 2010. Polymer–nanoparticle composites: From synthesis to modern application. *Materials* 3:3468–517.

Hasan, T., Sun, Z., Wang, F. et al. 2009. Nanotube–polymer composites for ultrafast photonics. *Advanced Materials* 21:3874–99.

Hemalatha, K. S., Rukmani, K., Suriyamurthy, N., and B. M. Nagabhushana. 2014. Synthesis, characterization and optical properties of hybrid PVA–ZnO nanocomposite: A composition dependent study. *Materials Research Bulletin* 51:438–46.

Hussain, A. A., Pal, A. R., and D. S. Patil. 2014. High photosensitivity with enhanced photoelectrical contribution in hybrid nanocomposite flexible UV photodetector. *Organic Electronics* 15:2107–15.

Imai, Y., Terehara, A., Hakuta, Y., Matsui, K., Hayashi, H., and N. Ueno. 2009. Transparent poly(bisphenol A carbonate)-based nanocomposites with high refractive index nanoparticles. *European Polymer Journal* 45:630–8.

Jaleh, B., Madad, M. S., Tabrizi, M. F., Habibi, S., Goldbedaghi, R., and M. R. Keymanesh. 2011. UV-degradation effect on optical and surface properties of polystyrene–$TiO_2$ nanocomposite film. *Journal of the Iranian Chemical Society* 8:S161–8.

Janković, I. A., Šaponjić, Z. V., Džunuzović, E. S., and J. M. Nedeljković. 2010. New hybrid properties of $TiO_2$ nanoparticles surface modified with catecholate type ligands. *Nanoscale Research Letters* 5:81–8.

Jumali, M. H. H., Al-Asbahi, B. A., Yap, C. C., Salleh, M. M., and M. S. Alsalhi. 2012. Optoelectronic property enhancement of conjugated polymer in poly(9,9′-di-n-octylfluorenyl-2.7-diyl)/titania nanocomposites. *Thin Solid Films* 524:257–62.

Kanis, D. R., Ratner, M. A., and T. J. Marks. 1994. Design and construction of molecular assemblies with large second-order optical nonlinearities. Quantum chemical aspects. *Chemical Reviews* 94:195–242.

Kerker, M. 1969. *The Scattering of Light and Other Electromagnetic Radiation.* New York: Academic Press, Inc.

Kreibig, U. and M. Vollmer. 1995. *Optical Properties of Metal Clusters.* Berlin: Springer-Verlag.

Kuila, B. K., Garai, A., and A. K. Nandi. 2007. Synthesis, optical, and electrical characterization of organically soluble silver NPs and their poly(3-hexyl thiophene) nanocomposites: Enhanced luminescence property in the nanocomposite thin films. *Chemistry of Materials* 19:5443–52.

Kulyk, B., Sahraoui, B., Krupka, O. et al. 2009. Linear and nonlinear optical properties of ZnO/PMMA nanocomposite films. *Journal of Applied Physics* 106:093102.

Li, S., Toprak, M. S., Jo, Y. S., Dobson, J., Kim, D. K., and M. Muhammer. 2007. Bulk synthesis of transparent and homogeneous polymeric hybrid materials with ZnO quantum dots and PMMA. *Advanced Materials* 19:4347–52.

Li, Y.-Q., Fu, S.-Y., Yang, Y., and Y.-W. Mai. 2008. Facile synthesis of highly transparent polymer nanocomposites by introduction of core–shell structured nanoparticles. *Chemistry of Materials* 20:2637–43.

Liao, C., Wu, Q., Su, T., Zhang. D., Wu, Q., and Q. Wang. 2014. Nanocomposite gels via *in situ* photoinitiation and disassembly of $TiO_2$–clay composites with polymers applied as UV protective films. *ACS Applied Materials & Interfaces* 6:1356–60.

Liou, G.-S., Lin, P.-H., Yen, H.-J., Yu, Y.-Y., Tsai, T.-W., and W.-C. Chen. 2010. Highly flexible and optical transparent 6F-PI/$TiO_2$ optical hybrid films with tunable refractive index and excellent thermal stability. *Journal of Materials Chemistry* 20:531–6.

Liu, B.-T., Tang, S.-J., Yu, Y.-Y., and S.-H. Lin. 2011. High-refractive-index polymer/inorganic hybrid films containing high $TiO_2$ contents. *Colloids and Surfaces A: Physicochemical and Engineering Aspects* 377:138–43.

Liu, J.-G., Nakamura, Y., Ogura, T., Shibasaki, Y., Ando, S., and M. Ueda. 2008. Optically transparent sulphur-containing polyimide–$TiO_2$ nanocomposite films with high refractive index and negative pattern formation from poly(amic acid)–$TiO_2$ nanocomposite film. *Chemistry of Materials* 20:273–81.

Liu, X., Adachi, Y., Tomita, Y., Oshima, J., Nakashima, T., and T. Kawai. 2012. High-order nonlinear optical response of a polymer nanocomposite film incorporating semiconductor CdSe quantum dots. *Optics Express* 20:13457–69.

Lu, C, Hu, X., Zhang, Y., Li, Z., Yang, H., and Q. Gong. 2012. Large nonlinearity enhancement of Ag/MEH–PPV nanocomposite by surface plasmon resonance at 1550 nm. *Plasmonics* 7:159–65.

Lü, C., Guan, C., Liu, Y., Cheng, Y., and B. Yang. 2005. PbS/polymer nanocomposite optical materials with high refractive index. *Chemistry of Materials* 17:2448–54.

Mahendia, S., Tomar, A. K., and S. Kumar. 2011. Nano-Ag doping induced changes in optical and electrical behaviour of PVA films. *Materials Science and Engineering B* 176:530–4.

Mark, J. E. 1999. *Polymer Data Handbook*. New York: Oxford University Press.

Matras-Postolek, K. and D. Bogdal. 2010. Polymer nanocomposites for electro-optics: Perspectives on processing technologies, material characterization, and future application. *Advances in Polymer Science* 230:221–82.

Matsuda, S. and S. Ando. 2003. Anisotropy in optical transmittance and molecular chain orientation of silver-dispersed uniaxially drawn polyimide films. *Polymers for Advanced Technologies* 14:458–70.

Maxwell-Garnett, J. C. 1904. Colours in metal glasses and in metallic films. *Philosophical Transactions of the Royal Society of London A* 203:385–420.

Mie, G. 1908. Beiträge zur Optik trüber Medien, speziell kolloidaler Metallösungen (Contributions to the optics of turbid media, especially colloidal metal suspensions). *Annals of Physics (Leipzig)* 25:377–445.

Murali, M. G., Dalimba, U., Yadav, V., Srivastava, R., and K. Safakath. 2013. Thiophene-based donor–acceptor conjugated polymer as potential optoelectronic and photonic material. *Journal of Chemical Sciences* 125:247–57.

Nair, P. S., Radhakrishnan, T., Revaprasadu, N., Kolawole, G. A., Luyt, A. S., and V. Djoković. 2005. Polystyrene-*co*-maleic acid/CdS nanocomposites: Preparation and properties. *Journal of Physics and Chemistry of Solids* 66:1302–6.

Novak, B. M. 1993. Hybrid nanocomposite materials – Between inorganic glasses and organic polymers. *Advanced Materials* 5:422–433.

Nussbaumer, R. J., Caseri, W. R., and P. Smith. 2003. Polymer–$TiO_2$ nanocomposites: A route towards visually transparent broadband UV filters and high refractive index materials. *Macromolecular Materials and Engineering* 288:44–9.

O'Flaherty, S. M., Hold, S. V., Brennan, M. E. et al. 2003b. Nonlinear optical response of multiwalled carbon-nanotube dispersions. *Journal of the Optical Society of America B* 20:49–58.

O'Flaherty, S. M., Murphy, R., Hold, S. V., Cadek, M., Coleman, J. N., and W. J. Blau. 2003a. Material investigation and optical limiting properties of carbon nanotube and nanoparticle dispersions. *The Journal of Physical Chemistry B* 107:958–64.

Park, J. T., Koh, J. H., Koh, J. K., and J. H. Kim. 2009. Surface-initiated atom transfer radical polymerization from $TiO_2$ nanoparticles. *Applied Surface Science* 255:3739–44.

Pomogailo, A.D. and V. N. Kestelman. 2005. *Metallopolymer Nanocomposites*. Heidelberg: Springer-Verlag.

Porel, S., Singh, S., and T. P. Radhakrishnan. 2005. Polygonal gold nanoplates in a polymer matrix. *Chemical Communications* 2387–9.

Pucci, A., Bernabó, M., Elvati, P. et al. 2006. Photoinduced formation of gold nanoparticles into vinyl alcohol based polymers. *Journal of Materials Chemistry* 16:1058–66.

Pucci, A., Tirelli, N., Willneff, E. A., Schroeder, S. L. M., Galembeck, F., and G. Ruggeri. 2004. Evidence and use of metal–chromophore interactions: Luminescence dichroism of terthiophene-coated gold nanoparticles in polyethylene oriented films. *Journal of Materials Chemistry* 14:3495–502.

Quinten, M. 2011. *Optical Properties of Nanoparticle System, Mie and Beyond*. Weinheim, Germany: Wiley-VCH Verlag & Co. KgaA.

Radoman, T. S., Džunuzović, J. V., Jeremić, K. B. et al. 2013. The influence of the size and surface modification of $TiO_2$ nanoparticles on the rheological properties of alkyd resin. *Hemijska Industrija* 67:923–32.

Ramesh, G. V. and T. P. Radhakrishnan. 2011. A universal sensor for mercury (Hg, Hg[I], Hg[II]) based on silver nanoparticle-embedded polymer thin film. *ACS Applied Materials and Interfaces* 3:988–94.

Ruiterkamp, G. J., Hempenius, M. A., Wormeester, H., and G. J. Vancso. 2011. Surface functionalization of titanium dioxide nanoparticles with alkanephosphonic acids for transparent nanocomposites. *Journal of Nanoparticle Research* 13:2779–90.

Saini, I., Rozra, J., Chandak, N., Aggarwal, S., Sharma, P. K., and A. Sharma. 2013. Tailoring of electrical, optical and structural properties of PVA by addition of Ag nanoparticles. *Materials Chemistry and Physics* 139:802–10.

Sarwar, M. I., Zulfiqar, S., and Z. Ahmad. 2009. Investigating the property profile of polyamide–alumina nanocomposite materials. *Scripta Materialia* 60:988–91.

Sciancalepore, C., Cassano, T., Curri, M. L. et al. 2008. TiO$_2$ nanorods/PMMA copolymer-based nanocomposites: Highly homogeneous linear and nonlinear optical material. *Nanotechnology* 19:205705 (8pp).

Sergienko, N., Godovsky, D., Zavin, B., Lee, M., and M. Ko. 2012. Nanocomposites of ZnS and poly-(dimethyl)-block-(phenyl)siloxane as a new high-refractive-index polymer media. *Nanoscale Research Letters* 7:181–7.

Sezar, A., Gurudas, U., Collins, B., Mckinlay, A., and D. M. Bubb. 2009. Nonlinear optical properties of conducting polyaniline and polyaniline–Ag composite thin films. *Chemical Physics Letters* 477:164–8.

Shah, V. 2007. *Handbook of Plastics Testing and Failure Analysis*. Hoboken, NJ: John Wiley & Sons, Inc.

Singhal, A., Dubey, K. A., Bhardwaj, Y. K., Jain, D., Choudhury, S., and A. K. Tyagi. 2013. UV-shielding transparent PMMA/In$_2$O$_3$ nanocomposite films based on In$_2$O$_3$ nanoparticles. *RSC Advances* 3:20913–21.

Sun, X., Chen, X., Fan, G., and S. Qu. 2010. Preparation and the optical nonlinearity of surface chemistry improved titania nanoparticles in poly(methyl methacrylate)-titania hybrid thin films. *Applied Surface Science* 256:2620–5.

Sun, Y., Liu, Y., Zhao, G., Zhou, X., Zhang, Q., and Y. Deng. 2008. Controlled formation of Ag/poly(methyl-methacrylate) thin films by RAFT technique for optical switcher. *Materials Chemistry and Physics* 111:301–4.

Tan, M. C., Patil, S. D., and R. E. Riman. 2010. Transparent infrared-emitting CeF$_3$:Yb–Er polymer nanocomposites for optical application. *ACS Applied Materials and Interfaces* 2:1884–91.

Tao, P., Li, Y., Rungta, A. et al. 2011. TiO$_2$ nanocomposites with high refractive index and transparency. *Journal of Materials Chemistry* 21:18623–9.

Tao, P., Li, Y., Siegel, R. W., and L. S. Schadler. 2013. Transparent dispensible high-refractive index ZrO$_2$/epoxy nanocomposites for LED encapsulation. *Journal of Applied Polymer Science* 130:3785–93.

Tauc, J., Grigorovici, R., and A. Vancu. 1966. Optical properties and electronic structure of amorphous germanium. *Physica Status Solidi B: Basic Solid State Physics* 15:627–37.

Tsai C.-M., Hsu, S.-H., Ho, C.-C. et al. 2014. High refractive index transparent nanocomposites prepared by *in situ* polymerization. *Journal of Materials Chemistry C* 2:2251–8.

Yang, B. D. and K.-H. Yoon. 2004. Effect of nanoparticles on the conjugated polymer in the PPV/TiO$_2$ nanocomposites. *Synthetic Metals* 142:21–4.

Yang, Y., Huang, J., Liu S., and J. Shen. 1997. Preparation, characterization and electroluminescence of ZnS nanocrystals in a polymer matrix. *Journal of Materials Chemistry* 7:131–3.

Yen, H.-J., Tsai, C.-L., Wang, P.-H., Lin, J.-J., and G.-S. Liou. 2013. Flexible, optically transparent, high refractive, and thermally stable polyimide–TiO$_2$ hybrids for antireflection coating. *RCS Advances* 3:17048–56.

Yuwono, A. H., Xue, J., Wang, J. et al. 2003. Transparent nanohybrids of nanocrystalline TiO$_2$ in PMMA with unique nonlinear optical behaviour. *Journal of Materials Chemistry* 13:1475–9.

Vodnik, V. V., Božanić, D. K., Džunuzović, E., Vuković, J., and J. M. Nedeljković. 2010. Thermal and optical properties of silver-poly(methyl methacrylate) nanocomposites prepared by *in-situ* radical polymerization. *European Polymer Journal* 46:137–44.

Vodnik, V. V., Božanić, D. K., Džunuzović, J. V., Vukoje, I., and J. M. Nedeljković. 2012. Silver/polystyrene nanocomposites: Optical and thermal properties. *Polymer Composites* 33:782–8.

Vodnik, V. V., Šaponjić, Z., Džunuzović, J. V., Bogdanović, U., Mitrić, M., and J. M. Nedeljković. 2013. Anisotropic silver nanoparticles as filler for the formation of hybrid nanocomposites. *Materials Research Bulletin* 48:52–7.

Vodnik, V. V., Vuković, J. V., and J. M. Nedeljković. 2009. Synthesis and characterization of silver-poly(methyl methacrylate) nanocomposites. *Colloid Polymer Science* 287:847–51.

Vukoje, I. D., Vodnik, V. V., Džunuzović, J. V., Džunuzović, E. S., Marinović-Cincović, M. T., Jeremić, K., and J. M. Nedeljković. 2014. Characterization of silver/polystyrene nanocomposites prepared by *in situ* bulk radical polymerization. *Materials Research Bulletin* 49:434–9.

Wang, C., Guo, Z.-X., Fu, S., Wu, W., and D. Zhu. 2004. Polymers containing fullerene or carbon nanotube structures. *Progress in Polymer Science* 29:1079–141.

Wang, J., Chen, Y., Li, R. et al. 2011. Graphene and carbon nanotube polymer composites for laser protection. *Journal of Inorganic and Organometallic Polymers and Materials* 21:736–46.

Xu, K., Zhou, S., and L. Wu. 2009. Effect of highly dispersible zirconia nanoparticles on the properties of UV-curable poly(urethane-acrylate) coatings. *Journal of Materials Science* 44:1613–21.

Weibel, M., Caseri, W., Suter, U. W., Kiess, H., and E. Wehrli. 1991. Preparation of polymer nanocomposites with "ultrahigh" refractive index. *Polymers for Advanced Technologies* 2:75–80.

Williams, D. J. 1983. *Nonlinear Optical Properties of Organic and Polymeric Materials*. Washington, D.C.: ACS Symposium Series 233, American Chemical Society.

Xue, P.-F., Wang, J.-B., Bao, Y.-B., Li, Q.-Y., and C.-F. Wu. 2012. Synthesis and characterization of functionalized carbon black/poly(vinyl alcohol) high refractive index nanocomposites. *Chinese Journal of Polymer Science* 30:652–63.

Zhang, G., Zhang, H., Wei, H. et al. 2013b. Creation of transparent nanocomposite films with a refractive index of 2.3 using polymerizable silicon nanoparticles. *Particle and Particle System Characterization* 30:653–7.

Zhang, G., Zhang, H., Zhang, X. et al. 2012. Embedding graphene nanoparticles into poly(*N,N'*-dimethylacrylamine) to prepare transparent nanocomposite films with high refractive index. *Journal of Materials Chemistry* 22:21218–24.

Zhang, J. Z. 2009. *Optical Properties and Spectroscopy of Nanomaterials*. Singapore: World Scientific Publishing Co. Pte. Ltd.

Zhang, Q., Goh, E. S. M., Beuerman, R. et al. 2013a. Development of optically transparent ZnS/poly(vinylpyrrolidone) nanocomposite films with high refractive indices and high Abbe numbers. *Journal of Applied Polymer Science* 129:1793–8.

Zimmermann, L., Weibel, M., Caseri, W., Suter, U. W., and P. Walther. 1993. Polymer nanocomposites with "ultralow" refractive index. *Polymers for Advanced Technologies* 4:1–7.

# 10 Rheological Behavior of Nanocomposites

*Nicole R. Demarquette and Danilo J. Carastan*

## CONTENTS

## 10.1 INTRODUCTION

### 10.1.1 NANOCOMPOSITES

For the last few decades, interest has arisen toward the development of nanocomposites consisting of a polymer matrix in which a nanoparticle is evenly distributed. Several types of nanofillers, with their three, two, or one dimensions in the nanoscale, such as exfoliated clay,[1–3] carbon derivatives such as carbon nanotubes, graphene,[4] ceramic oxides, nitrides, and carbides[5–7] have been dispersed into polymers. The resultant nanocomposites were shown to exhibit improvements in several properties, such as mechanical,[8,9] conductive, and dielectric,[10] gas barrier, UV,[11] and flame resistance,[12] among others when compared to pure polymers, and hence can be used in packaging, automotive, and aerospace industries.

There are three main methods to prepare these nanocomposites: in situ polymerization, solution casting, and melt mixing.[13] In situ polymerization consists of dispersing the nanoparticles in the monomer to be subsequently polymerized. During this process, it is expected that the nanoparticles form a well-dispersed structure within the polymeric matrix that is being polymerized. The proper selection of reagents and polymerization routes has led to the successful formation of highly dispersed structures; however, these routes are not always easily reproducible in industrial scale. Solution casting can be used when there exists a solvent that both dissolves the desired polymer and disperses the nanoparticles, forming a stable suspension. The nanocomposite can be obtained after solvent evaporation, as long as the dispersed nanoparticles are stable enough not to collapse back into large agglomerates during the process. This route is usually more adequate to prepare nanocomposites of water-soluble polymers, as it will not require the use of large amounts of organic solvents. Melt mixing has become the most popular and promising method to prepare nanocomposites, as it is more environmentally friendly and it gives more freedom of polymer selection, as no dissolution or polymerization steps are involved. It also makes use of common polymer processing techniques, which is interesting from the industrial point of view. Also, the shear and extension flows provided during melt mixing are essential for the ultimate state of nanoparticle dispersion.

A true nanocomposite is obviously expected to have its particles dispersed in the nanoscale. The classification of a nanocomposite may change depending on the nanoparticle used, but the most famous morphologies are the ones found in materials containing layered silicates (nanoclays): (1) microcomposite (or immiscible, not truly a nanocomposite), when there is not enough affinity between the nanoparticles and the polymer; (2) intercalated, when the polymer molecules penetrate the clay interlayer spaces resulting in particles (tactoids) with alternating nanometer thick polymer/silicate layers; and (3) exfoliated, when the clay layers are individually separated and well distributed in the polymeric matrix. In practice, the nanocomposites often present a combination of different morphologies, as illustrated in Figure 10.1.

The morphology of nanocomposites is normally accessed by x-ray diffraction or scattering techniques and microscopy at different scales, especially transmission electron microscopy (TEM). While these techniques are efficient tools, they normally probe a small area unless a statistical analysis involving observations at many different scales is carried out.[14–16] Melt rheology, on the other hand, can provide a quantification of the overall degree of intercalation, exfoliation, distribution, and dispersion of the nanoparticles. It can even be used to calculate the percolation threshold of nanoparticles using some physical models.

Rheology not only is a useful characterization tool to evaluate if there is good nanoparticle dispersion, but also to be able to see if the morphology stays stable with time, temperature, and stress during processing, for example. Rheology will also reveal the stress state in the polymer and its effects during processing. Low viscosity polymers, for example, will provide poor shear stress

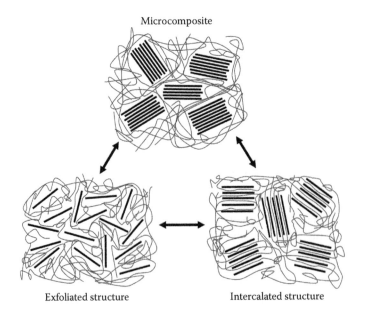

**FIGURE 10.1**  Typical morphologies of clay-containing nanocomposites.

transfer to the nanoparticles when compared to more viscous, high molecular weight materials, which will eventually affect the formation of the nanocomposite during melt processing.[17]

## 10.1.2 RHEOLOGY

Rheology is the science that studies the flow of matter. When studying the rheological behavior of a fluid, a deformation is normally applied to this fluid and the resultant stress is measured; alternatively, a stress can be applied and the resultant deformation measured. If a constant stress is applied to a polymer, it will initially deform and continue deforming with time, as it is a viscoelastic fluid with a rheological behavior intermediate to the one of a Newtonian fluid and a Hookean solid. If small or slow deformations are applied, the resultant stress will vary linearly with the amplitude of deformation; alternatively, if a stress of small amplitude is applied, the resultant deformation will vary linearly with the magnitude of the stress. This situation corresponds to linear viscoelasticity, which can be used to study the chain characteristics of the polymer[18,19] or to characterize the morphology in the case of multiphase polymeric materials such as blends and nanocomposites, for example.[20–22] If the deformations are large and rapid, the molecules will be disturbed from their equilibrium and nonlinear responses will be observed. These large deformations correspond to the ones that polymers normally undergo during processing and yield to nonlinear viscoelasticity. Although the study of nonlinear viscoelasticity may be mathematically more difficult, its study is of great interest to facilitate an effective processing.

Studying the rheological behavior of nanocomposites is therefore very important as it enables getting a picture of the state of dispersion of the nanoparticles within the matrix and provides information on how to process these materials as rheological properties govern the flow of the polymers during processing.

Polymeric materials are normally tested using two types of deformation: shear and extensional flow.[23] Figure 10.2 illustrates schematically shear and uniaxial extensional flows. Both types of flow are present in polymer processing and the knowledge of rheological behavior for both types of flow is necessary if one wants to get a clear picture of the processing of nanocomposites. Extensional flows normally result in larger deformations and in a nonlinear viscoelastic behavior of the material.

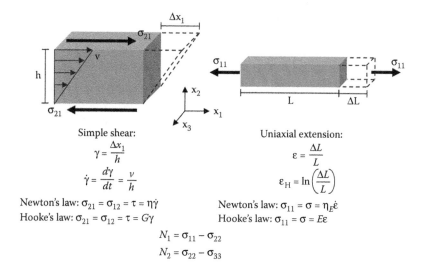

Simple shear:
$$\gamma = \frac{\Delta x_1}{h}$$
$$\dot{\gamma} = \frac{d\gamma}{dt} = \frac{v}{h}$$

Uniaxial extension:
$$\varepsilon = \frac{\Delta L}{L}$$
$$\varepsilon_H = \ln\left(\frac{\Delta L}{L}\right)$$

Newton's law: $\sigma_{21} = \sigma_{12} = \tau = \eta\dot{\gamma}$
Hooke's law: $\sigma_{21} = \sigma_{12} = \tau = G\gamma$

Newton's law: $\sigma_{11} = \sigma = \eta_E\dot{\varepsilon}$
Hooke's law: $\sigma_{11} = \sigma = E\varepsilon$

$$N_1 = \sigma_{11} - \sigma_{22}$$
$$N_2 = \sigma_{22} - \sigma_{33}$$

**FIGURE 10.2** Simple shear and uniaxial extensional flows and some basic definitions: $\sigma_{ij}$ is the ij component of the stress tensor, $\sigma$ is the normal stress, $\tau$ is the shear stress, $\varepsilon$ is the normal strain, $\varepsilon_H$ is the Hencky (true) strain, $\eta$ is the shear viscosity, $G$ is the shear modulus, $\eta_E$ is the extensional viscosity, $E$ is Young's modulus, and $N_1$ and $N_2$ are the first and the second normal stress differences, respectively.

Although rheological measurements in extensional flows are normally more difficult to carry out, they are of extreme importance for processing techniques such as fiber spinning[24] and film blowing,[25] where polymer stretching occurs.

### 10.1.3 PLAN OF THIS CHAPTER

In this chapter, the rheological behavior of polymer nanocomposites will be reviewed. Major emphasis will be given on materials containing nanoclays, but occasionally the effect of adding other nanoparticles, such as carbon nanotubes or silica nanospheres on the rheological properties of the nanocomposites will also be discussed. In a first part, it will be shown how the rheological characterization in the viscoelastic regime can be used as a tool to characterize the state of dispersion of nanoparticles within thermoplastics. The second part will present the rheological behavior of nanocomposites in the nonlinear viscoelastic regime. In particular, it will be shown how the morphology of nanocomposites evolves under shear and elongational flows.

## 10.2 RHEOLOGICAL BEHAVIOR OF POLYMER NANOCOMPOSITES IN THE LINEAR VISCOELASTIC REGIME

### 10.2.1 CLAY CONTAINING NANOCOMPOSITES

As mentioned above, the linear viscoelastic regime corresponds to small and slow deformations that do not disturb the polymer chains from their equilibrium state. The study of linear viscoelastic behavior of nanocomposites can therefore be used to get information on the microstructure of these materials.

One of the tests most commonly used when studying the linear viscoelastic behavior of a molten polymer is the small amplitude oscillatory shear (SAOS), which consists of applying a sinusoidal deformation to the polymer given by

$$\gamma(t) = \gamma_o \sin \omega t \tag{10.1}$$

and evaluating the resultant stress which can be written as

$$\sigma(t) = \sigma_o \sin(\omega t + \delta) \tag{10.2}$$

where $\gamma_0$ and $\sigma_0$ are the strain and stress amplitudes, respectively, $\omega$ is the frequency, $t$ is the time, and $\delta$ is the phase shift, called loss angle, due to the viscoelastic behavior of the polymer.

After some trigonometrical rearrangements, two physical quantities of rheological importance can be inferred from Equation 10.2: the storage, $G'$, and loss, $G''$, moduli that are given by Equations 10.3 and 10.4:

$$G'(\omega) = \frac{\sigma_o}{\gamma_o} \cos(\delta) \tag{10.3}$$

$$G''(\omega) = \frac{\sigma_o}{\gamma_o} \sin(\delta) \tag{10.4}$$

The storage modulus corresponds to the part in phase with the deformation and therefore to the elastic part of the response of the material, whereas the loss modulus corresponds to the part 90° out of phase of the deformation and therefore to the viscous part of the response of the material.

More details about the tests that can be carried out to study the linear viscoelastic behavior of molten polymers can be found in Dealy[23] and Barnes.[26]

Small amplitude oscillatory shear is normally carried out using a rotational rheometer (either stress- or strain-controlled) using either cone-plate or parallel plate geometries. Samples obtained normally by compression or injection molding are inserted in the rheometer. The plates are approximated making a "sandwich" with the sample, which is then heated. The excess sample, on the edge of the plate, is then removed. The shear or stress history that was programmed to test the sample can then start. However, prior to any test, especially with multiphase materials, it is important that the samples be annealed in place at the test temperature for a certain time to remove the effect of loading history[27,28] and to be tested while having an equilibrium morphology, although this annealing treatment may not be sufficient as the morphology of clay containing nanocomposites has been shown to evolve with time.[29–31] More details on the evolution of morphology with time will be presented later.

When carrying out small amplitude oscillatory shear, it is also very important to verify that the magnitude of the strain amplitude $\gamma_0$ (when using a controlled strain rheometer) or stress amplitude $\sigma_0$ (when using a controlled stress rheometer) corresponds to the linear viscoelastic range. Strain or stress sweeps at different frequencies are then normally carried out and the magnitude of the storage modulus evaluated as a function of time. If the strain or stress corresponds to linear viscoelastic behavior, the storage modulus will then be independent of the magnitude of strain or stress.

It has been observed in the literature that the onset of nonlinearity for nanocomposites is lower than the one for pure polymers and it decreases with filler content. As an example, Figure 10.3 shows the normalized stress modulus (where $G'_0$ is the storage modulus at lower strains) as a function of strain $\gamma_0$ for PET nanocomposites at a frequency of 6.28 rad/s obtained by Ghanbari et al.[32] The authors defined the dashed line as the transition from the linear to nonlinear viscoelastic behavior with $G'/G'_0 < 0.9$, in other words when $G'$ had decreased by more than 10% from its original value. It can be seen that the onset of nonlinearity, which is shown in the inset of Figure 10.3, decreases as the concentration of clay increases and displays a power law dependency on the volume fraction of clay. Such a behavior, reported in many studies,[33–37] was shown to depend on the degree of clay dispersion[38] and concentration of clay[39] and was attributed to the shear thinning of the clay-containing nanocomposites.

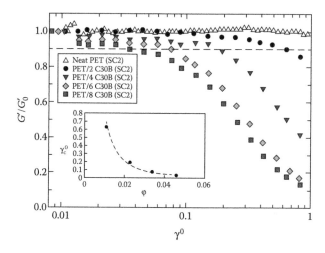

**FIGURE 10.3    (See color insert.)** Normalized storage modulus versus strain amplitude for neat PET and PET/C30B nanocomposites at 6.28 rad/s. The inset shows the maximum strain amplitude for the linear viscoelastic behavior as a function of the clay volume fraction. (With kind permission from Springer Science + Business Media: *Rheologica Acta,* Morphological and rheological properties of PET/clay nanocomposites, 52, 2013, 59–74, Ghanbari, A. et al.)

Once the strain/stress range corresponding to linear viscoelastic regime is determined, the rheological behavior of polymeric materials can be evaluated by small amplitude oscillatory shear. It is then important that the material be probed at small frequencies as the signature of nanocomposite morphology is observed for these frequency values. Typically, tests should be performed between 0.01 and 300 rad/s. However, depending on the number of points that is measured by the rheometer per frequency decade, the tests could last around two hours during which the morphology of the nanocomposites is likely to suffer evolution[29–31] and one should take this into account when analyzing the data.

Figure 10.4 shows the typical viscoelastic behavior of a molten polymer when subjected to small amplitude oscillatory shear. At low frequencies the storage modulus is proportional to $\omega$,[2] loss modulus to $\omega$, and complex viscosity independent of $\omega$. This is called the terminal behavior.

On addition of nanoparticles, the storage modulus often increases at low frequencies (sometimes by several orders of magnitude) and its frequency dependence decreases. In some cases, a plateau of storage modulus is even observed at low frequencies, indicating a large deviation from terminal behavior, sometimes called a "pseudo-solid-like" behavior. Krishnamoorti et al.[40,41] were among the first who observed this effect in polymer/clay nanocomposites, when studying the dynamic behavior of polyamide 6 and polycaprolactone nanocomposites. This nonterminal behavior has been observed since for clay-containing nanocomposites of many different polymers, such as polyethylene,[42] polypropylene,[30,37,43–48] polyamide 6,[17,49–51] polystyrene,[52–57] polylactic acid,[58–60] polycaprolactone,[61–63] polyethylene terephthalate,[32] polycarbonate,[64] poly(butylene succinate-*co*-adipate),[65] acrylonitrile-butadiene-styrene (ABS),[66] and poly(ethylene-*co*-vinyl acetate) (EVA),[67] to name a few. Similar effects have also been observed for polymer blends to which clay was added[68] and for block copolymer/clay nanocomposites,[69] although in this case, if the copolymer is in the ordered state, the rheological behavior will be a result of the interaction between the morphology of the copolymer and the clay nanoparticles.[70–73]

The nonterminal effect appears at different concentrations or degrees of nanoparticle dispersion depending on the system studied. The solid-like behavior observed at low frequencies is usually attributed to the formation of nanoparticle percolating networks, such as clay platelets stacked like a house of cards.[74] Similar behavior has been observed for other common fillers such as talc[75] but at much higher loadings. The Newtonian zero shear complex viscosity plateau also tends to disappear

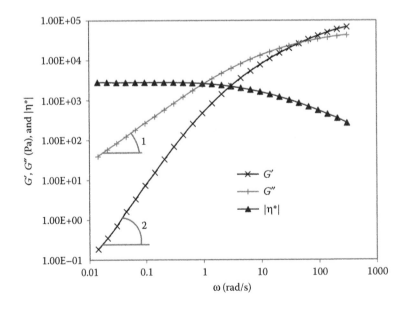

**FIGURE 10.4**  Dynamic data at a temperature of 200°C for a typical sample of polystyrene.

as another consequence of the same phenomenon, and the sample exhibits strong shear thinning behavior instead.[17,54,55,64,73,76]

The linear viscoelastic response of layered silicate nanocomposites was shown to be dependent on clay concentration,[36,46,69] on the state of clay dispersion,[77,78] on the affinity between the polymer and the nanoparticles, the type of surfactant used for the modification of the clay,[33,78,79] the size of clay platelets,[73] and the chemical modification of the polymer.[78] All these variables affect the linear viscoelastic behavior of the nanocomposites essentially because they may give rise to a network structure due to particle/particle and/or polymer/particle interactions.[40,69,80,81]

The nonterminal solid-like behavior usually occurs when clay nanoparticles are in the exfoliated state, but sometimes intercalated structures also exhibit this behavior.[43,47,56,69] Carastan et al. prepared polystyrene and block copolymer nanocomposites by incorporating organoclays using three different techniques: solution casting, melt mixing, and a hybrid master batch process combining solution and melt mixing.[16,70,71,77] The degree of clay dispersion was strongly dependent on the processing technique used, as evidenced by the x-ray diffraction and transmission electron microscopy analyses. Pure melt mixing led to the least dispersed samples, followed by the master batch process, and then solution casting. These different degrees of clay dispersion were sensed by the rheological measurements. The composite prepared by solution casting alone exhibited a much higher value of $G'$ at low frequencies than the other composites. Besides having a more dispersed structure, these studies showed that solution casting promotes the formation of a well-structured percolated network, as the clay particles arrange themselves in a "house of cards" structure as the solvent is being evaporated. The melt mixing step promoted the alignment of clay tactoids, decreasing the overall solid-like behavior.

In a very systematic study, Mitchell and Krishnamoorti[73] evaluated the effect of the size of clay nanoplatelets on the linear viscoelastic behavior of a polystyrene–polyisoprene block copolymer. In order to study solely the effect of clay morphology, part of the tests was carried out when the block copolymer was in the disordered state, and the three types of clay tested, namely fluorohectorite ($F$), montmorillonite ($M$), and laponite ($L$), with respective average platelet diameters of about 10 μm, 1 μm, and 30 nm, were organically modified in such a way that "the surface coverage of the silicate layer and conformation of salt at their surface be equivalent."[73] The nanocomposites containing the larger clay platelets presented an intercalated structure, whereas the laponite nanocomposites presented an exfoliated morphology. They observed that all composites were shear thinning and presented similar rheological behavior independently of clay size at high frequencies. However, at low frequencies, both nanocomposites with a larger size clay platelet presented a reduction in frequency dependence of $G'$ and increase of $\eta^*$. Such a phenomenon could not be observed for the smaller clay platelets in spite of the exfoliated structure observed. The authors attributed this behavior to the fact that although the structure of the nanocomposites was exfoliated, the size of the clay platelet was too small for the platelets to form a percolating network at this concentration of 5 wt%.

Analyzing experimental data for polystyrene/clay nanocomposites, Zhao et al.[57] summarized graphically the effect of clay concentration and dispersion on the small amplitude oscillatory shear behavior of clay-containing nanocomposites (see Figure 10.5). For very low concentrations of clay, the nanocomposites show a terminal behavior with $G' \propto \omega^2$ and $G'' \propto \omega$. When the clay concentration increases but is below the percolation threshold (see Figure 10.5b), $G'$ increases and is proportional to $\omega$. Once the clay concentration is above the percolation threshold, $G'$ becomes higher than $G''$ at low frequencies and $G'' \propto \omega^0$. Further increase of clay concentration results in a storage modulus that is larger than the loss modulus at all frequencies. The crossover point between $G'$ and $G''$ can also be used to analyze the effect of nanoparticles on the rheological behavior of molten polymers. As this point changes its position whether a polymer has a more elastic (increase in $G'$) or a more fluid behavior (increase in $G''$), it is often used to correlate the effects of polymer structure, such as the molecular weight, with rheological properties. Likewise, the addition of dispersed clay nanoparticles increases the elasticity of the polymer, affecting the position of the crossover point.

Another test that can be used to evaluate the linear viscoelastic behavior of molten polymer is the stress relaxation test. During a stress relaxation experiment, a sudden deformation, $\gamma_o$, is imposed

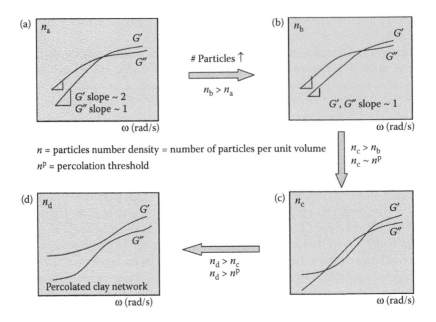

**FIGURE 10.5** Schematic representation of the speculated rheological response to the increase in the number of particles per unit volume. (Reprinted from *Polymer,* 46, Zhao, J.; Morgan, A. B.; Harris, J. D., Rheological characterization of polystyrene–clay nanocomposites to compare the degree of exfoliation and dispersion, 8641–60, Copyright (2005), with permission from Elsevier.)

to the material and the resultant stress, $\sigma(t)$ is evaluated as a function of time. If the amplitude of deformation is small enough, the stress relaxation modulus defined as

$$G(t) = \frac{\sigma(t)}{\gamma_o} \tag{10.5}$$

will decay exponentially as a function of time due to the viscoelastic behavior of the material, not depending on the amplitude of deformation.

This stress relaxation modulus can be expressed as a sum of exponential decays, known as the generalized Maxwell Model given by

$$G(t) = \sum_{i=1}^{n} G_i \exp\left(-\frac{t}{\lambda_i}\right) \tag{10.6}$$

where $G_i$ and $\lambda_i$ are the moduli and relaxation times corresponding to each Maxwell element of the discrete relaxation spectrum, with the largest $\lambda_i$ called the terminal relaxation time.

If a material tested in stress relaxation presents a small molar mass or is slightly entangled, the relaxation will be fast and its terminal relaxation time will be smaller than the one for a high molar mass, very entangled polymer.

Figure 10.6 presents the relaxation modulus in the linear viscoelastic regime for nanocomposites of styrene-isoprene diblock copolymers with three different concentrations of clay obtained by Ren and Krishnamoorti.[82] The solid lines represent the best fit to Equation 10.6. It can be seen that as the concentration of clay is increasing, the relaxation is much slower exhibiting a solid-like behavior. Similar behavior was obtained by Bandyopadhyay et al.[65] for PLA/clay nanocomposites.

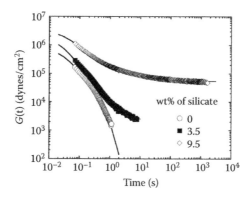

**FIGURE 10.6**    Time dependence of the linear stress relaxation modulus $G(t)$ at 85°C for the unfilled PSPI copolymer and hybrids with 3.5 and 9.5 wt% clay nanocomposites. The solid lines represent the best fits of the data to the empirical Equation 10.6. (Reprinted with permission from Ren, J.; Krishnamoorti, R.; Nonlinear viscoelastic properties of layered-silicate-based intercalated nanocomposites. *Macromolecules* 36, 2003, 4443–51. Copyright 2003 American Chemical Society.)

**FIGURE 10.7**    SEM image of the clay network from a 5 wt% PP nanocomposites after the polymer matrix had been burned away. (Reprinted with permission from Vermant, J. et al. Quantifying dispersion of layered nanocomposites via melt rheology. *Journal of Rheology* 51, 2007: 429–50 Copyright 2007, American Institute of Physics.)

Ren et al.[69] attributed the long relaxation times (larger than 100 s) and the nonterminal behavior observed during small amplitude oscillatory shear tests to a possible formation of a network with stacks of clay percolating. This mesostructure results in "physical jamming" of the dispersed clay preventing the composite to relax at long times; above a certain volume fraction, the clay platelets are unable to rotate and relax even at long times. The presence of the clay network was proven by Vermant et al.,[27] who observed the morphology of a sample of 5% clay polypropylene nanocomposite (see Figure 10.7) treated at 900°F in vacuum. This treatment was meant to burn the polymer without affecting the clay structure.

## 10.2.2    Effect of Other Nanoparticles

The nonterminal solid-like behavior observed in small amplitude oscillatory shear tests is certainly not exclusive of clay-containing nanocomposites. Most other nanoparticles cause similar effects on polymers, such as carbon nanotubes,[83–86] graphene,[87–90] silica,[91,92] cellulose nanocrystals,[93,94] and polyhedral oligomeric silsesquioxane (POSS),[95] among others.

Hassanabadi et al.[96,97] studied the effect of two different particles geometries (nanospheres of $CaCO_3$ and clay platelets) on the linear and nonlinear viscoelastic behavior (see below) of ethylene vinyl acetate nanocomposites both in shear and extension (see below). The effect of the addition of clay on the modulus at low frequencies during small amplitude oscillatory shear experiments was shown to be much larger than the one of addition of spherical particles; in the case of clay platelets, the probability of network formation is much higher and the relaxation time of the material increases.

The shape of the nanoparticles therefore plays a very important role in the rheological properties of nanocomposites. Nanoparticles with high aspect ratios, such as nanoclays, carbon nanotubes, nanofibers, and graphene tend to have a low percolation threshold, as opposed to more isotropic nanoparticles such as silica and $CaCO_3$ spherical nanoparticles and the Laponite synthetic clay. All layered nanoparticles, such as nanoclays and graphene,[87–89] have essentially the same behavior, as by increasing the exfoliation the nonterminal solid-like behavior becomes more pronounced. Carbon nanotubes behave in a somewhat similar manner, forming percolating networks usually by entanglements and the formation of bundles.[83,84] Cipiriano et al.[85] studied the effect of the aspect ratio of multiwall carbon nanotubes (MWNTs) on the linear viscoelastic properties of polystyrene nanocomposites, and they observed an increase in the nonterminal behavior for the samples containing higher aspect ratio nanotubes.

More isotropic nanoparticles, on the other hand, behave somewhat differently. Fumed silica, for example, is a hydrophilic type of nanoparticle that usually forms rather strong aggregates of the small spherical individual nanoparticles. The aggregates tend to easily form percolating networks when mixed with polymers, exhibiting the nonterminal behavior. Bartholome et al.[92] compared polystyrene nanocomposites containing untreated fumed silica and polystyrene-grafted silica nanoparticles. The linear viscoelastic behavior of the samples tested is presented in Figure 10.8. It is very clear that the treated samples have a weaker solid-like behavior than the nanocomposite containing unmodified silica. The authors verified that the PS-treated silica was much better dispersed in the matrix. This example shows that not always when increasing the nanoparticle dispersion the material will have a more solid-like behavior. Isotropic nanoparticles have somehow the opposite behavior of anisotropic particles like nanoclays.

This effect can also be understood in terms of the formation of a percolated network. The untreated nanoparticles are naturally more prone to forming networks, as there are strong particle–particle interactions. When some treatment improves the dispersion of these nanoparticles into

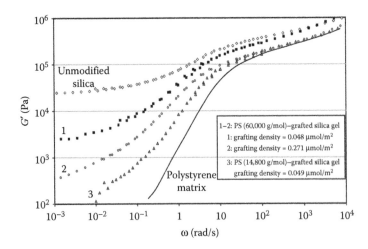

**FIGURE 10.8**  Storage modulus of PS nanocomposites filled with 5 vol% of silica and PS–grated silica master curve at T = 160°C. (Reprinted from *Polymer*, 49, Cassagnau, P., Melt rheology of organoclay and fumed silica nanocomposites, 2183–96, Copyright 2008, with permission from Elsevier.)

almost individual spheres, they increase their interaction with the polymer matrix, but the perco-lated network does not form so easily anymore, except at very high loadings, as the percolation threshold of spheres is much higher than the one of plates or fibers.[80,81] This behavior is the same observed for the $CaCO_3$ nanospheres and Laponite nanoclays discussed before.

## 10.2.3 EFFECT OF MATRIX ANISOTROPY

The formation of a percolated network by the nanoparticles governs the linear viscoelastic behavior of a molten polymer when the matrix is isotropic. In the case of structured, anisotropic polymer systems, the rheological properties are usually dominated by the ordered microstructure, as is the case of block copolymers in the ordered state. When disordered, molten block copolymers behave essentially like isotropic homopolymers, and the nanoparticles have a strong effect on their rheo-logical properties, causing the change from a terminal to a nonterminal behavior, as previously dis-cussed.[69,71,73] However, block copolymers are known for their ability of forming regularly ordered structures in the nanoscale due to phase separation when their blocks are not miscible in each other. Depending on the composition, affinity between the blocks and temperature, they tend to form ordered domains with regular shapes, such as lamellae, cylinders, or spheres.[98]

These ordered domains have a strong influence on the rheological properties of the molten copolymers. Whereas a homopolymer or a disordered block copolymer has a terminal viscoelastic behavior, with a $G'$ slope of around 2 at low frequencies in a double log plot, the ordered states usually exhibit a nonterminal behavior. This effect occurs because of a restriction for molecular relaxation due to the phase separation of chemically connected blocks. A block copolymer with a randomly oriented lamellar structure presents a $G'$ slope of around 0.5, and a similarly random, yet more ordered cylindrical structure renders a slightly lower slope, of about 0.3. A copolymer with a morphology composed of spheres arranged in a body-centered cubic structure has a complete tridimensional restriction for molecular relaxation, and these materials tend to have a nonterminal plateau of $G'$ with a slope of around zero.[99] Figure 10.9 shows the typical rheological behavior for different ordered block copolymers when subjected to small amplitude oscillatory shear.

When clay or other nanoparticles are added to copolymers with these ordered nanostructures, the viscoelastic response is dominated by the structure of the domains. The nanoparticles have therefore

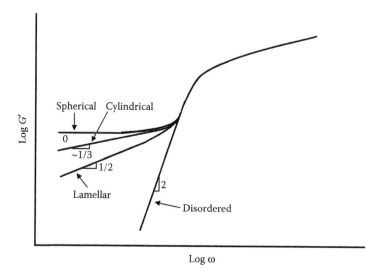

**FIGURE 10.9** Schematic small amplitude oscillatory shear curves showing the effect of different block copolymer structures in the low frequency region. (Adapted from Kossuth, M. B.; Morse, D. C.; Bates, F. S., *Journal of Rheology* 43, 1999, 167–96.)

a marginal effect on the rheological behavior.[73] Although it is very difficult to observe a footprint of the presence of nanoparticules on the rheological behavior of the nanocomposites, as it is masked by the response of the ordered nanostructure of the block copolymer, the formation of a percolated network can restrict even more the relaxation of the copolymer molecules and a further decrease in the slope of $G'$ in the low frequency region in the case of lamellar or cylindrical structures can be observed. However, in the case of a highly ordered spherical structure, presenting a truly horizontal $G'$ plateau, the clay particles can act as defects, so that they might end up increasing the slope of the plateau,[71] that is, increasing the possibility of molecular relaxation due to the presence of defects.

### 10.2.4 Phenomenological and Mathematical Models

Layered nanoparticles, such as nanoclays, tend to increase the nonterminal behavior with an increasing degree of exfoliation. Using a simple volume filling calculation, Ren et al.[69] assumed that the tactoids could be modeled as hydrodynamic spheres, as shown in Figure 10.10. The hydrodynamic radius of the spheres indicates the onset of percolation, below which incomplete rotation/relaxation of the tactoids takes place, forming the percolated network. The authors formulated a relation that relates the number of silicate layers, $n_{per}$ per tactoids, as a function of weight fraction at percolation $\Phi_{per}$.

$$n_{per} = \frac{4}{3\Phi_{per}} \left[ \frac{w_{silper}\rho_{org}}{w_{silper}\rho_{org} + (1 - w_{silper})\rho_{sil}} \right] \frac{R_h}{h_{sil}} \qquad (10.7)$$

where $\Phi_{per}$ is the volume fraction for percolation of random spheres, that is, 0.3, $R_h$ is the radius of the hydrodynamic sphere, $h_{sil}$ is the thickness of the silicate layer, $\rho_{org}$ and $\rho_{sil}$ are the respective densities of the polymer and the silicate and $w_{silper}$ is the weight fraction of silicate at rheological percolation.

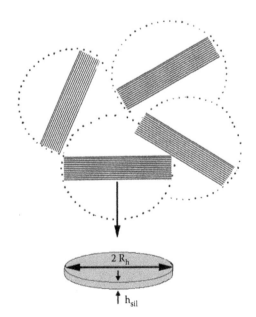

**FIGURE 10.10** Schematic representation of the tactoids of layered silicates and their interaction with each other, resulting in incomplete relaxation of the hybrids. (Reprinted with permission from Ren, J.; Silva, A. S.; Krishnamoorti, R. Linear viscoelasticity of disordered polystyrene–polyisoprene block copolymer-based layered–silicate nanocomposites. *Macromolecules* 33, 2000: 3739–46. Copyright 2000 American Chemical Society.)

Using this simple model, the authors were able to estimate from rheological measurements the thickness of the clay tactoids, which corroborated the values obtained using x-ray diffraction.

Several other scaling relations were suggested to relate the rheological behavior evaluated during small amplitude oscillatory shear tests to the state of dispersion of the clay. It has been shown in the literature that the storage modulus at low frequencies scales with the clay concentration as Reference 27

$$G' \sim (\varphi - \varphi_{per})^{\nu} \tag{10.8}$$

or

Reference 66

$$G'_c - G'_m = G_o \varphi^n \tag{10.9}$$

where $\nu$ is a power law exponent, $\varphi$ and $\varphi_{per}$ are the clay and percolation volume concentrations, respectively, $G'_c$ and $G'_m$ are the storage moduli at low frequencies for the composite and matrix, respectively, and $G_o$ is a constant. The exponent $n$ can be considered as a measure of fractal dimension of the clay 3D network being of greater magnitude for a more percolated network.

In order to determine the percolation threshold, Vermant et al.[27] suggested to draw a plot of the storage modulus at low frequencies as a function of clay concentration $\varphi$. Above a certain concentration $G'$ starts to increase linearly with $\varphi$, the intercept of the straight line with the $y = 0$ axis provides the percolation concentration. Figure 10.11 shows this analysis for two different PP/clay nanocomposites, labeled $C$ (higher molecular weight) and $D$ (lower molecular weight). For each system, the curves show a clear transition where the low frequency $G'$ (measured at 0.03 rad/s) starts to linearly increase with clay concentration, which is an indication of the percolation threshold.

Using the analysis of Vermant et al.[27] and Equation 10.7 of Ren et al.,[69] it is then possible to characterize the morphology of the nanocomposites and infer the aspect ratio for the nanoparticles.

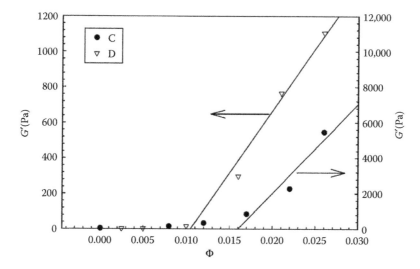

**FIGURE 10.11** Plot of the storage modulus at low frequency (0.03 rad/s) versus volume fraction for two different PP clay nanocomposites (C is a system based on a higher molecular weight PP, and D on a lower molecular weight PP). (Reprinted with permission from Vermant, J. et al. Quantifying dispersion of layered nanocomposites via melt rheology. *Journal of Rheology* 51, 2007: 429–50. Copyright 2007, American Institute of Physics.)

For concentrations above percolation, Vermant et al.[27] suggested use of the relations of Shih et al.[100] given by

$$G'_p \sim \varphi^{(3+x)/(3-d_f)} \tag{10.10}$$

$$\gamma_{crit} \sim \varphi^{-(1+x)/(3-d_f)} \tag{10.11}$$

where $G'_p$ is the plateau modulus, $\gamma_{crit}$ is the critical strain for linearity and $d_f$ is the fractal dimension of the aggregate network, and $x$ is the exponent that connects the particle volume fraction with aggregate size.[27] The values of $d_f$ and $x$ can be used to get a picture of the network with the larger $d_f$ and the smallest $x$ leading to a more open fractal structure.

Another approach to infer information about clay dispersion from linear viscoelastic data consists of using the dependence of the storage modulus and viscosity at low frequencies with frequency. Several authors suggested that Equations 10.12 and 10.13 can be used as criteria for state of clay dispersion,[39] the magnitude of $n$ representing the state of exfoliation (small values of $n$ corresponding to more exfoliated states):

$$G' = A\omega^n \tag{10.12}$$

$$\eta^* = B\omega^n \tag{10.13}$$

However, according to Vergnes et al.,[38,101,102] these criteria depend on the range of frequencies used to fit the data and can be very subjective. Therefore, they showed that the melt yield stress, $\sigma_0$, that can be inferred from the description of the complex viscosity of the nanocomposites as a function of frequency, based on Carreau–Yasuda model[103] (Equation 10.14), is a much more reliable quantitative parameter to evaluate the state of dispersion of clay.

$$\eta^*(\omega) = \frac{\sigma_o}{\omega} + \eta_o[1 + \lambda\omega^a]^{m-1/1} \tag{10.14}$$

where $\eta_o$ is the zero shear viscosity, $\lambda$ is the time constant, $a$ is the Yasuda parameter, and $m$ is a dimensionless power law index.

In particular, Vergnes et al.[101] used this equation to evaluate the state of clay dispersion in PP/PP-MA blends as a function of maleic anhydride content and also the morphological evolution along the screw in the extruder. Figure 10.12 shows the fitting of rheological data of PP/clay nanocomposites obtained at 220°C to Equation 10.14. The data for pure polypropylene are compared to the ones of the $X/Y/Z$ blend where $X$, $Y$, and $Z$ correspond to the concentration of PP, PP–MA, and clay, respectively. The line corresponds to the fit to Equation 10.14, which is excellent. The authors observed that the fitting parameters $m$ and $\lambda$ did not depend much on the state of exfoliation, whereas the melt yield stress, $\sigma_0$ depended greatly on the state of dispersion of the clay. Figure 10.13 shows $\sigma_0$ as well as the interlayer spacing as a function of PP–MA loading for PP/PP–MA/Cloisite 20A nanocomposites. For low concentration of PP–MA, the interlayer spacing increased but no yield stress was observed as most of the clay tactoids were intercalated by polymers. As the content of PP–MA increased, the number of exfoliated clay platelets increased leading to an increase of $\sigma_0$. The authors attributed the leveling off of $\sigma_0$ to a maximum of exfoliation.

All these quantitative criteria show that rheology can be used as a tool to get some qualitative and even quantitative information on the state of dispersion of clay within the polymer. However, care should be taken as the morphology of clay containing nanocomposites was shown to evolve as a function of time[29,30,31] and proper experimental protocols should be followed.

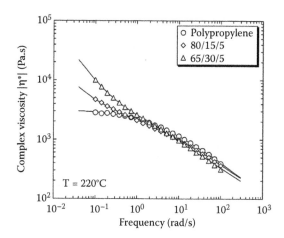

**FIGURE 10.12** Comparison of the complex viscosity $|\eta^*|$ of PP/PP-g-MA/Cloisite 20A composites prepared with different amounts of PP-g-MA. (Reprinted from *Polymer*, 46, Lertwimolnun, W.; Vergnes, B., Influence of compatibilizer and processing conditions on the dispersion of nanoclay in a polypropylene matrix, 3462–71, Copyright 2005, with permission from Elsevier.)

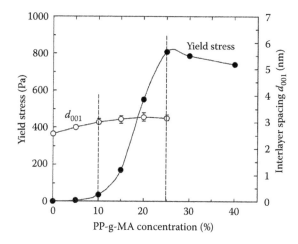

**FIGURE 10.13** Comparison of interlayer spacing and melt yield stress at 220°C as a function of PP-g-MA loading for the PP/PP-g-MA Cloisite 20A composites. (Reprinted from *Polymer*, 46, Lertwimolnun, W.; Vergnes, B., Influence of compatibilizer and processing conditions on the dispersion of nanoclay in a polypropylene matrix, 3462–71, Copyright 2005, with permission from Elsevier.)

Zouari et al.[31] evaluated dynamic data of PP/clay nanocomposites using successive frequency sweeps. Fitting their data to Equation 10.14, they evaluated the yield stress of the nanocomposites as a function of time. It was shown to follow a kinetics in two steps (see Equations 10.15 and 10.16), which have been attributed by the authors to the disorientation of the clay platelets and then to the aggregation of these platelets in a 3D network by van der Waals forces. This effect can be seen in Figure 10.14, which shows that the time evolution of the nanocomposites is also temperature dependant. Similar behavior had been observed by Treece et al.[29,30] except that in this case they fitted $G'$ to $t^\beta$ and they had two values of beta.

$$\sigma_o = C_1\sqrt{t} \quad \text{for} \quad t < t_{crit} \tag{10.15}$$

$$\sigma_o = C_1\sqrt{t_{crit}} + C_2(t - t_{crit}) \quad \text{for} \quad t > t_{crit} \tag{10.16}$$

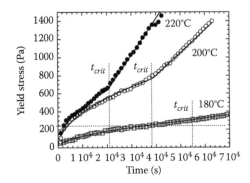

**FIGURE 10.14** Time evolution of the melt yield stress $\sigma_0$ of a PP-clay nanocomposite for different temperatures (180°C, 200°C, and 220°C). Vertical lines indicate the critical times associated with the change in temporal kinetics for the different temperatures. The horizontal line indicates the times associated with a same yield stress for different temperatures. (Reprinted with permission from Zouari, R. et al. Time evolution of the structure of organoclay/polypropylene nanocomposites and application of the time-temperature superposition principle. *Journal of Rheology* 56, 2012: 725–42. Copyright 2012, American Institute of Physics.)

This time dependence will have a large influence on the applicability of time-temperature superposition (TTS) for nanocomposites. The TTS principle implies that rheological curves obtained at different temperatures can be shifted horizontally and vertically (to take into account changes of density) to obtain a master curve at a reference temperature but for larger frequency or time ranges. More details can be found in Ferry.[104] According to Zouari et al.,[31] TTS is not expected to be valid for clay-containing nanocomposites because of the time evolution undergone by the viscoelastic data at low frequencies that is temperature dependant as shown in Figure 10.14. However, when a proper annealing of the samples was performed so that the samples presented similar yield stress prior to testing, TTS was shown to work.

## 10.3 RHEOLOGICAL BEHAVIOR POLYMER NANOCOMPOSITES IN THE NONLINEAR VISCOELASTIC REGIME

### 10.3.1 Introduction

As mentioned in the introduction, the nonlinear viscoelastic regime corresponds to fast and strong deformations. The rheological response depends then on the size, rate, and kinematics of the deformation.[23] The study of nonlinear viscoelasticity of molten polymers is much more complex than the study of linear viscoelasticity as there are no constitutive equations that adequately predict the rheological behavior of polymers in this regime. However, its study is of extreme importance as the flows that molten polymers undergo during processing correspond to the nonlinear viscoelastic regime. To understand the nonlinear viscoelastic behavior of molten polymers, experiments should be carried out either in simple shear or uniaxial extension as both these types of deformations are encountered in polymer processing. In shear, rheological quantities such as viscosity and first normal stress difference will be obtained as a function of shear rate. In order to perform these studies, well-defined transient shear flows should be used, such as step shear, which is similar to the flows used to study stress relaxation in the linear regime but with larger values of $\gamma_0$; startup of steady shear, which consists of shearing a sample at a constant shear rate; large amplitude oscillatory shear or any other defined shear history implying large and/or fast deformations. In extensional flows, the tensile stress growth coefficient of the material as a function of time or quantities such as melt strength and breaking stretching ratio can be obtained (more details in Section 3.4).

Much fewer studies of the rheological behavior of polymer nanocomposites in the nonlinear viscoelastic regime have been carried out in the literature. However, it enables the understanding of the evolution of the morphology under flow. Whereas linear viscoelastic behavior is dominated by the 3D network structure (particle–particle interactions), nonlinear viscoelastic behavior is dominated by the orientation of the nanoparticles, including the disruption of their network structure and an increasing importance of polymer–nanoparticle interactions.

In the following, some of the main findings in the literature for the rheological behavior of polymer nanocomposites when subjected to large step shear, startup of steady shear, and other well-defined shear histories will be reported. Then, the viscosity curves in shear and the behavior in extension, both of fundamental importance for the understanding of melt processing of those materials, will be discussed. Since nonlinear viscoelastic behavior is governed by the evolution of morphology, attention will be given to the studies that report the morphological changes of nanocomposites under flow. At last the influence of nanoparticle addition on technological processing issues such as die swell and sharkskin effect, two phenomena that are encountered in extrusion, will be addressed.

## 10.3.2 Behavior of Polymer Nanocomposites under Well-Defined Transient Shear Flows

Very few studies reported the behavior of clay containing nanocomposites in step shear.[47,82] Ren and Krishnamoorti[82] performed stress relaxation experiments of nanocomposites of styrene-isoprene diblock copolymers to which montmorillonite clay was added in concentrations ranging from 2.1 to 9.5 wt% in the linear and nonlinear viscoelastic regimes. For low strains (in the linear viscoelastic regime) they observed that the relaxation modulus data were independent of strain and that for higher strains, the stress relaxation modulus $G(t, \gamma)$ obeys the time strain separation and can be written as

$$G(t,\gamma) = G(t)h(\gamma) \quad \text{for} \quad t \geq \tau \tag{10.17}$$

where $G(t)$ is the relaxation modulus as evaluated in the linear viscoelastic regime, $h(\gamma)$ is the damping function which accounts for the shear thinning of the material, and $\tau$ is a characteristic time before which the time strain separation does not apply.

They observed that the onset of strain for shear thinning (for $G$ to depend on $\gamma$) decreased when the clay loading increased, as can be seen in Figure 10.15, and the damping function could be fitted to

$$h(\gamma) = \frac{1}{1 + a\gamma^2} \tag{10.18}$$

for the different nanocomposites obtained.

They also observed that time-strain separability was valid for much smaller ranges of strains as the concentration of clay increased. The premature shear thinning observed was attributed by the authors to a rotation of the anisotropic distribution of the clays when submitted to external flows for low clay loadings and to a disruption of the mesostructure for the higher clay loadings. These effects of nanoparticle rotation and disruption of the percolating network are probably the main effects that dominate the nonlinear viscoelastic behavior of nanocomposites, as discussed below for startup of steady shear tests.

The stress monitored for a molten polymer subjected to startup of steady shear (provided the shear rate is high enough) as a function of time normally presents at short times a stress overshoot[105] and then it levels off. The magnitude of this overshoot has been shown to be even greater for clay-containing nanocomposites[36,64,65,96,97,106,107] and to depend on the nanoparticle concentration, nanoparticle shape,[96,97] and orientation of nanoplatelets.

Figure 10.16 shows the results obtained by Hassanabadi et al.,[96] who evaluated the effect of particle geometries on the rheological behavior of ethylene-*co*-vinyl acetate (EVA) composites when subjected to steady shear. Two particle geometries were evaluated, $CaCO_3$ nanospheres with a diameter of around 70 nm and clay platelets with a diameter of around 300 nm. It can be seen that the presence of clay results in larger stress overshoot. In other words, clay increases the extent of nonlinearity. This behavior was attributed to the clay network that can occur at lower nanoparticle concentrations. The authors also showed that $\sigma_{max} \sim \dot{\gamma}^n$ with $n$ decreasing with increasing clay concentration. The presence of spherical $CaCO_3$ nanoparticles had only a minor influence on the nonlinear viscoelastic behavior of the polymer, as the formation of a percolating network occurred only at the highest concentration (15 wt%). For the clay nanocomposites, the network could be sensed at a concentration as low as 5 wt%.

In order to understand the physical meaning of the stress overshoot, Solomon et al.[46] studied the effect of a rest time posterior to the application of startup steady shear at a shear rate of 0.01 s$^{-1}$ for 300 s in one direction, and on a subsequent startup of steady shear in the reverse direction at the same shear rate (0.01 s$^{-1}$). Figure 10.17 shows the stress response for a 4.8 wt% PP—montmorillonite

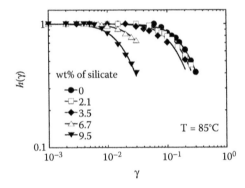

**FIGURE 10.15**  Silicate content dependence of the damping function for block copolymer–clay nanocomposites. The solid curves represent the empirical fitting to Equation 10.18. (Reprinted with permission from Ren, J.; Krishnamoorti, R. Nonlinear viscoelastic properties of layered-silicate-based intercalated nanocomposites. *Macromolecules* 36, 2003: 4443–51. Copyright 2003 American Chemical Society.)

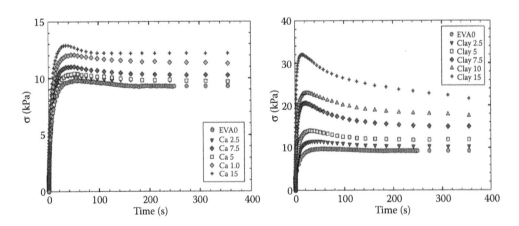

**FIGURE 10.16**  Stress as a function of time for $CaCO_3$ (left) and clay (right) nanocomposites at a rate of 0.1 s$^{-1}$ and 110°C. (With kind permission from Springer Science + Business Media: *Rheologica Acta,* Relationships between linear and nonlinear shear response of polymer nano-composites, 51, 2012, 991–1005, Hassanabadi, H. M.; Rodrigue, D.)

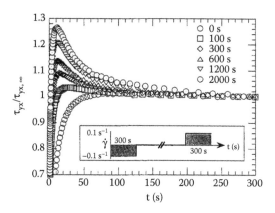

**FIGURE 10.17**  Results of flow reversal studies of a 4.80 wt% clay PP nanocomposites. After an initial episode of steady shear ($\dot{\gamma} = 0.1\,s^{-1}$) followed by a rest time of varying duration, the stress response on the startup of steady shear flow ($\dot{\gamma} = 0.1\,s^{-1}$) in the reverse direction was monitored for 300 s. (Reprinted with permission from Solomon, M. J. et al. Rheology of polypropylene/clay hybrid materials. *Macromolecules* 34, 2001: 1864–72. Copyright 2001 American Chemical Society.)

nanocomposite to the second startup of steady shear. In the inset of the figure, the shear history undergone by the samples is summarized. It can be seen that the magnitude of the stress overshoot is larger for the samples that were left at rest for longer times. The changes observed as a function of resting time were attributed to the evolution of the clay structure during rest time. These results suggest that the clay network sensed during small amplitude oscillatory shear is easily disrupted by deformation. However, on cessation of flow reconstruction of the network is observed a phenomenon attributed by the authors to the attractive interactions that exist between the platelet domains.

The reconstruction of the clay network is expected to happen by the disorientation of the clay platelets initially aligned by shear. If this effect were to happen due only to Brownian motion, the rather large clay platelets would take a very long time to change their orientation. To analyze this effect, Lele et al.[47] used in situ x-ray diffraction to evaluate the orientation of clay particles during rheological experiments on PP nanocomposites. After the cessation of steady shear flow, Hermans' function $S$[108] was used to evaluate the orientation of the nanoparticles with time, as shown in Figure 10.18. Two samples were studied, one uncompatibilized and another one containing polypropylene modified with maleic anhydride (PP–MA) as a compatibilizer. Initially, the clay nanoparticles in the compatibilized sample were highly aligned, but then quickly decreased their orientation. The uncompatibilized sample, on the other hand, was not so strongly oriented since the beginning, and its structure did not change considerably with time. The authors argue that if attractive interactions between clay particles were to be the cause for particle disorientation, both samples should have a similar behavior. As only the compatibilized composite showed strong disorientation, one possible explanation is that the relaxation of polymer molecules, being more strongly attached to the clay in the compatibilized case, might have accelerated the formation of a network, unlike the uncompatibilized sample.

Therefore, it is still not completely clear if nonlinear viscoelastic response of polymer nanocomposites is due to particle–particle or polymer–particle interactions. It is more likely that both interactions are important. The stress overshoot is clearly affected by the presence of a percolated network, but transient data allows the estimation of the longest relaxation time, whereas this is not possible using linear data, when the nonterminal behavior occurs.[96] This means that during transient tests, such as startup of steady shear, the terminal relaxation of the polymer chains is allowed to occur, although it is affected by the presence of the nanoparticles (polymer–particle interactions), whereas for the linear behavior there is usually no terminal relaxation, due to a stable formation of a percolated network (particle–particle interactions). The influence of polymer–particle interactions

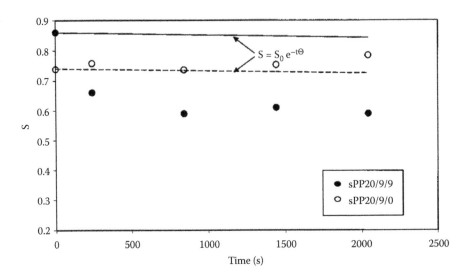

**FIGURE 10.18** Orientation function S as a function of time for PP nanocomposites with (closed symbols) and without (open symbols) PP–MA compatibilizer after cessation of steady shear at 190°C. (Reprinted with permission from Lele, A. et al. In situ rheo-x-ray investigation of flow-induced orientation in layered silicate–syndiotactic polypropylene nanocomposite melt. *Journal of Rheology* 46, 2002: 1091–110. Copyright 2002, American Institute of Physics.)

on the nonlinear behavior seems however perhaps to be more important, affecting especially the phenomenon of nanoparticle orientation.

The stress overshoots were successfully predicted by Rajabian et al.,[109] who modeled clay nanoparticles as thin oblate spheroid particles. Their model predicted that during a startup of steady shear experiment, the particles initially have a 3D random orientation and are oriented by shear. Their predictions showed that the magnitude of stress overshoot increased on clay concentration increase and that the steady shear decreased with increasing shear rate. This stress overshoot, only observed for higher shear rates, was attributed to the rotation of nanoparticles. Similar conclusions had been reached by Letwilmolnum et al.,[110] who attributed the origin of the overshoot to orientation of clay platelets and not to the disruption of clay network.

Nevertheless, at high nanoparticle loadings, the percolated network apparently starts to dominate the nonlinear behavior. The results of Hassanabadi et al.[96,97] were compared to a modified version of Doi-Edwards theory,[111] which could not predict the rheological properties of the clay-containing nanocomposites above a certain concentration due to the presence of the percolating network, which is not taken into account by models based on chain dynamics.

### 10.3.3 SHEAR VISCOSITY AS A FUNCTION OF SHEAR RATE

At low frequencies or shear rates, the viscosity for the nanocomposites is higher than the one of the unfilled systems. However, at high shear rates the viscosity follows the one of the unfilled system.[112] In some cases, it has even been shown that there is more shear thinning for the clay-containing nanocomposites than for the pure resin,[36] most likely due to an orientation of clay platelets that occurs during shearing at high shear rates.

This change of morphology can explain the failure of the empirical Cox-Merz rule, which relates dynamic data to steady shear data (see Equation 10.19) that has been observed for clay-containing nanocomposites.[17,36,40,66,69,113,114]

$$\eta^*(\omega) = \eta(\dot{\gamma}) \quad \text{for} \quad \omega = \dot{\gamma} \tag{10.19}$$

When subjected to small amplitude oscillatory shear, the nanocomposites present a rather stable 3D network, but this mesoscale structure is disrupted in steady shear or capillary rheometry.

Figure 10.19 shows data of complex viscosity versus frequency and steady shear viscosity versus shear rate for polyamide nanocomposites with three different molecular weights (HMW: high molecular weight; MMW: medium molecular weight; LMW: low molecular weight) obtained by Fornes et al.[17] The nanocomposites were prepared by melt mixing and contained 3 wt% MMT. Due to higher shear stresses generated by the higher molecular weight (therefore, higher viscosity) polyamide during processing, the HMW and MMW composites presented an exfoliated structure whereas the LMW presented a blend of exfoliated and intercalated structure. It can be seen from Figure 10.19 that for the three polyamides, the viscosity increased on addition of clay. It can also be

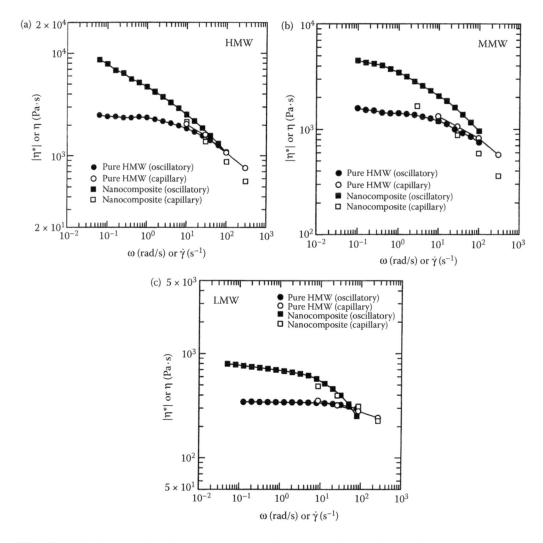

**FIGURE 10.19** Complex viscosity versus frequency from a dynamic parallel plate rheometer (solid points) and steady shear viscosity versus shear rate from a capillary rheometer (open points) at 240°C for (a) pure high molecular weight polyamide and its clay nanocomposites, (b) pure medium molecular weight polyamide and its clay nanocomposites, and (c) low molecular weight polyamide and its clay nanocomposites. The composites contain 3 wt% clay. (Reprinted from *Polymer*, 42, Fornes, T. D. et al. Nylon 6 nanocomposites: The effect of matrix molecular weight, 9929–40, Copyright 2001, with permission from Elsevier.)

seen that both exfoliated structures presented a non-Newtonian behavior at low frequencies whereas the LMW PA clay-containing nanocomposite presented a smaller divergence from the Newtonian behavior. Also, it can be observed that although the Cox-Merz rule was valid for the three polyamides, it failed for the three nanocomposites.

When submitted to shear flows, molten polymers also exhibit normal shear stresses, due to their viscoelastic nature. Very few reports of $N_1$ for nanocomposites have been made in the literature. While Gupta et al.[67] observed a decrease of $N_1$ at high shear rates for increasing clay content in EVA nanocomposites, Krishnamoorti et al.[115] did not observe any influence of the addition of clay on PS–PI clay nanocomposites. Gupta et al. attributed this difference to the fact that the morphology that they obtained was exfoliated, whereas the one observed by Krishnamoorti et al. was intercalated.

### 10.3.4  RHEOLOGICAL BEHAVIOR IN EXTENSIONAL FLOW

Many polymer processing operations, such as fiber spinning, film blowing, blow molding, and thermoforming are governed by the elongational rheological properties of the molten material. Therefore, it is important to understand how the presence of nanoparticles affects the rheological properties of a polymer under elongational flows. However, mainly due to experimental difficulties, relatively few studies have been conducted on the subject. Okamoto et al.[74] were among the first to study the extensional rheological properties of polymer–clay nanocomposites. They observed that the addition of clays results in an increase of transient elongational viscosity of polypropylene. The nanocomposites also exhibited strain hardening, an effect not observed for the pure PP. This strain-induced hardening, which is a time-dependent thickening (rheopectic behavior) occurred during extension, is usually a desired feature for a polymer system, as it results in higher melt strength during processing. Normally, polymers containing long-chain branches or polarity may exhibit this response under extensional flows, which is not the case for PP.

In order to understand the data they obtained, Okamoto et al.[74] made some morphological observations of the samples recovered after flow. Figure 10.20 presents typical morphologies that they obtained for different stretching rates. Figure 10.20a corresponds to larger extension rates and Figure 10.20b corresponds to small extension rates. It can be seen that stretching at high extension rates results in an orientation of clay perpendicular to the stretching direction. This unusual alignment of clay particles and the strain hardening effect possibly occurred due to strong interactions between the nanoparticles and PP–MA molecules used as a compatibilizer. Lee et al.[114] studied the elongational viscosity of PP nanocomposites with and without PP–MA compatibilizer, and the results showed that only the compatibilized samples exhibited strain hardening. This behavior has been observed for other systems based on PP.[34,107,116] Apparently, when a polymer does not normally exhibit strain hardening in extension, the presence of exfoliated nanoparticles with good affinity for the matrix induces the emergence of this behavior. Similar results have been observed in poly(methyl methacrylate) (PMMA) nanocomposites.[117,118] In another study, however, an intercalated PS nanocomposite did not display the strain hardening, and the clay had almost no effect on the extensional viscosity of the polymer.[116]

There are other polymers that normally exhibit strain-induced hardening in extension, such as low density polyethylene (LDPE) or EVA. In such systems, even though the presence of clay can increase the overall transient extensional viscosity of the polymers, it tends to decrease the strain hardening effect observed for the pure polymer. This effect has been generally observed in EVA[36,67,97,112,119,120] and LDPE[112] systems.

By conducting poststretching transmission electron microscopy observations, Gupta et al.[67] observed that on stretching the microstructure of EVA–bentonite nanocomposites with a well-dispersed morphology changed from a dispersed state of clay nanoplatelets to a less dispersed state. They attributed that to the fact that uniaxial extensional flow results in a contraction in the two other directions which results in an approximation of clay platelets as shown schematically in Figure 10.21.

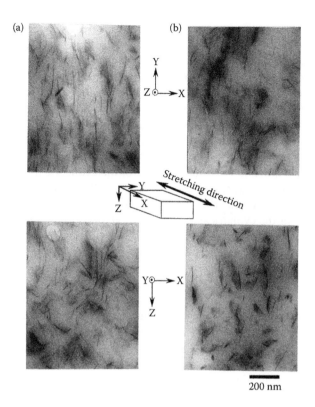

**FIGURE 10.20** TEM micrographs showing PP/clay elongated at 150° with (a) $\dot{\varepsilon}_o = 1s^{-1}$ up to $\varepsilon = 1.3$ ($\lambda = 3.7$) and (b) $\dot{\varepsilon}_o = 0.01s^{-1}$ up to $\varepsilon = 0.5$ ($\lambda = 1.7$). Upper pictures are in the x–y plane and lower ones are in the x–z plane along the stretching direction. (Reprinted with permission from Okamoto, M. et al. A house of cards structure in polypropylene/clay nanocomposites under elongational flow. *Nano Letters* 1, 2001: 295–98. Copyright 2001 American Chemical Society.)

It has been observed as well that Trouton's rule,

$$3\eta_o(\dot{\gamma}_o;t) \cong \eta_E(\dot{\varepsilon}_o;t) \tag{10.20}$$

where $\eta_0$ is the zero shear viscosity and $\eta_E$ is the elongational viscosity, usually does not hold for clay-containing nanocomposites. The rule may be valid for systems with low nanoparticle content or a poor dispersion, but the deviation increases as the filler content increases.[67] The formation of a

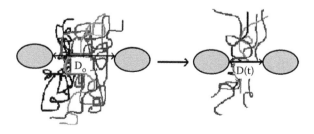

**FIGURE 10.21** Idealized illustration of the decrease in distance (D) between clay–clay layers (represented by ellipses) after stretching. (Reprinted from *Journal of Non-Newtonian Fluid Mechanics*, 128, Gupta, R. K.; Pasanovic-Zujo, V.; Bhattacharya, S. N., Shear and extensional rheology of EVA/layered silicate-nanocomposites, 116–25, Copyright 2005, with permission from Elsevier.)

percolated network and strong interactions between the nanoparticles and the matrix are probably the cause for the deviation from Trouton's rule.

In anisotropic systems, such as ordered block copolymers, clay nanoparticles tend to align to the flow direction, as well as the copolymer domains. Carastan et al. prepared nanocomposites of styrene-*b*-(ethylene/butylene)-*b*-styrene (SEBS) triblock copolymers by extrusion.[121] The copolymer tested had a cylindrical morphology, and the cylinders and clay nanoparticles were aligned during processing in the direction of extrusion. The samples were tested in uniaxial elongational in the directions parallel and perpendicular to the extrusion direction.[122] In the parallel direction, the copolymer behaved according to Trouton's rule, and the clay nanoparticles did not affect the viscosity, as they were aligned in the same direction of the cylinders. In the perpendicular direction, the copolymer exhibited strong extension softening followed by hardening, as displayed in Figure 10.22. Small angle x-ray scattering (SAXS) results showed that the strain softening occurred because of the rotation of the cylindrical domains to align in the direction of elongation. The clay nanoparticles increased the transient extensional viscosity because they also rotated during the test. The study showed that the clay nanoparticles affect the orientation of the cylindrical domains of the copolymer depending on the degree of clay dispersion and affinity between the clay and the copolymer blocks, especially for high elongation rates.

The effect of other nanoparticles on the extensional viscosity of polymers has been studied even less than clay. Apparently the effects are similar: if there is not enough affinity between the nanoparticles and the matrix, little effect of the nanoparticles on the elongational viscosity is observed; if the interactions are strong, strain hardening occurs. Studies of the effect of carbon nanotubes follow this trend,[123–126] and the strain hardening could only be observed when strong interactions were present due to surface modification of the nanotubes.[124]

Spherical nanoparticles, such as silica, also have similar effects,[127,128] and the several possibilities of surface modification can lead to very high strain hardening values, such as the ones observed for poly(lactic acid) (PLA) nanocomposites containing silica functionalized via two different grafting methods,[128] as shown in Figure 10.23. The PLA-grafted silica nanoparticles were prepared using

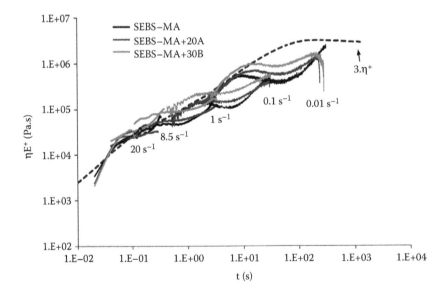

**FIGURE 10.22** Curves of steady elongation of SEBS–MA and its nanocomposites along the transverse (T) direction held at 200°C. The dashed line corresponds to three times the steady shear stress growth data for pure SEBS. (Reprinted from *European Polymer Journal*, 49, Carastan, D. J. et al. Morphological evolution of oriented clay-containing block copolymer nanocomposites under elongational flow, 1391–405. Copyright 2013, with permission from Elsevier.)

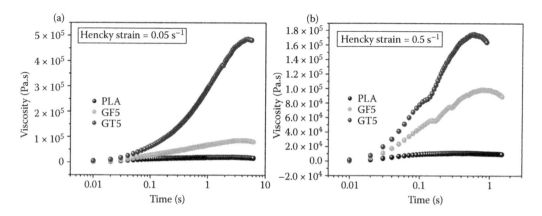

**FIGURE 10.23** **(See color insert.)** Time variation of elongational viscosity $\eta_E(\dot{\varepsilon},t)$ for molten PLA and its silica nanocomposites at 158.5°C. (a) $\dot{\varepsilon} = 0.05\,s^{-1}$ and (b) $\dot{\varepsilon} = 0.5\,s^{-1}$. GF5 contains 5 wt% PLA-treated silica prepared by the "grafting from" method, and GT5 contains 5 wt% PLA-treated silica prepared by the "grafting to" method. (Reprinted from *Polymer*, 55, Wu, F. et al. Inorganic silica functionalized with PLLA chains via grafting methods to enhance the melt strength of PLLA/silica nanocomposites, 5760–72, Copyright 2014, with permission from Elsevier.)

either a "grafting to" or a "grafting from" method. The nanoparticles modified using the "grafting to" method were covered by higher molecular weight PLA, although the grafting density was not so high. The "grafting from" method resulted in the opposite effect, producing nanoparticles covered with lower molecular weight PLA at a higher grafting density. The "grafting to" method proved more efficient in increasing the extensional viscosity in PLA nanocomposites, indicated by GT5 in Figure 10.23, in comparison to the "grafting from" (GF5) samples. This result indicates that the higher molecular weight of the polymer covering the nanoparticles in GT5 was the main reason for increased melt strength on the nanocomposite.

## 10.3.5 EFFECT OF ADDITION OF CLAY ON DIE SWELL AND EXTRUSION INSTABILITIES

Die swell is a phenomenon that is common in polymer extrusion. It occurs when the polymer is forced to pass through the die. Due to the elastic nature of the polymer, the polymer will expand at the exit of the die. In the die, the polymer molecules are elongated by shear and recoil when going out of the die. The extent of swell at the exit of the die depends on the rheological properties of the polymer, the temperature, the extrusion rate, and also the geometry of the die. It normally increases with decreasing temperature, increasing extrusion rate, and decreasing $L/D$ die (where $L$ is the length and $D$ the diameter of the die). The die swell is normally quantified in a circular die as

$$\alpha = \frac{D}{D_o} \tag{10.21}$$

where $D$ is the diameter after die swell and $D_o$ is the diameter of the die.

Several reports in the literature showed that die swell was lessened by the presence of clays.[66,129–132] Figure 10.24 shows the extrudate swell for PS clay nanocomposites obtained by Zhong et al.[131] measured using a Monsanto capillary rheometer. It can be seen that on addition of clay, the die swell is lessened. Kader et al.[130] attributed the reduced die swell to the decrease of elasticity and viscosity of fluoroelastomer on addition of clay at high shear rates.

When polymers are extruded at high shear rates, melt instability can occur leading to melt fracture, extrudate distortion, or sharkskin effects, which result in an extrudate with an irregular

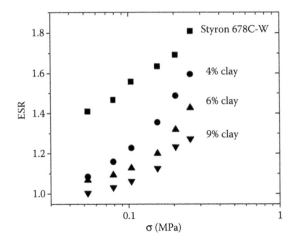

**FIGURE 10.24**   Die swell of PS/clay nanocomposites measured using a Monsanto Capillary rheometer at 180°C. (Reprinted from *Polymer,* 46, Zhang, Y.; Zhu, Y.; Wang, S. Q. Synthesis and rheological properties of polystyrene/layered silicate nanocomposite, 3006–13, Copyright 2005, with permission from Elsevier.)

surface. Kader et al.[130] showed that the addition of clay to fluoroelastomers resulted in the formation of smoother surfaces of the extrudates. This improvement of the extrudate appearance was attributed to the presence of surfactant of the clay that could decrease the melt elasticity and also to the decrease of melt viscosity and elasticity on addition of clay.

## 10.4   CONCLUSIONS

Rheology is a very powerful tool to study the structure and behavior of polymeric systems. The rheological properties of polymers are affected by the addition of other components, such as nanoparticles. The structure and processing characteristics of polymer nanocomposites, therefore, can be assessed by suitable rheological tests.

The microstructure of nanocomposites can be explored using a series of experiments within the range of the linear viscoelastic regime, being complementary to microscopy and other techniques of microstructural analysis. Some rheological properties are very sensitive to the dispersion, concentration, shape, and orientation of the nanoparticles. The linear viscoelastic behavior of nanocomposites can most of the time be explained by the formation of a percolating network that renders the polymer more solid-like, with a behavior closer to that of a gel.

Whereas the linear viscoelasticity is related to the study of materials in an almost undisturbed state near the equilibrium, the nonlinear viscoelastic regime appears when high strains or strain rates are applied to a sample, usually altering its original microstructure. Although much more complex to understand, the properties of nanocomposites in the nonlinear regime are very important, especially when studying the behavior of these materials in processing. The nanoparticles tend to align under strong shear or extensional flows, disrupting their percolating networks, an effect that sometimes is reversible, once the flow stops.

When dealing with phenomena in the nanoscale, it is very important to understand the interactions present in the systems. The rheological properties are strongly affected by polymer–nanoparticle and/or nanoparticle–nanoparticle interactions, which can be used to create mathematical models to predict the behavior of nanocomposites. However, despite research advances in the last years, there are still many phenomena not completely understood, as nanostructured materials do not usually behave according to classical rheological models. Anyhow, rheology will always be a very rich source of information to understand the properties and structures of materials.

## ACKNOWLEDGMENTS

The authors would like to thank FAPESP, CNPq, CAPES, and NSERC for financial support.

## REFERENCES

1. Okada, A.; Fukushima, Y.; Kawasumi, M.; Inagaki, S.; Usuki, A.; Sugiyami, S.; Kurauchi, T.; Kamigaito, O. US Patent Number 4739007, 1988 (assigned to Toyota Motor Co. Japan).
2. Kawasumi, M.; Kohazaki, M.; Kojima, Y.; Okada, A; Kamigaito, O.; US Patent Number 4810737, 1989 (assigned to Toyoto Motor Co. Japan).
3. Kojima, Y.; Usuki, A.; Kawasumi, M.; Okada, A.; Kurauchi, T.; Kamigaito, O. Synthesis of nylon 6-clay hybrid by montmorillonite intercalated with ε-caprolactam. *Journal of Polymer Science Part A: Polymer Chemistry* 31, 1993: 983–86.
4. Kim, H.; Abdala, A.; Macosko, C. W. Graphene/polymer nanocomposites. *Macromolecules* 43, 2010: 6515–30.
5. Dong, W. Z.; Han, Z. D.; Han, B. Z. Boron nitride filled immiscible blends of polyethylene and ethyl-ene-vinyl acetate copolymer: Morphology and dielectric properties. In *Properties and Applications of Dielectric Materials, IEEE, 10th International Conference on the Properties and Applications of Dielectric Materials*. Bangalore, India, 2012.
6. Cao, Y.; Irwin, P. C.; Younsi, K. The future of nanodielectrics in the electrical power industry. *IEEE Transactions on Dielectrics and Electrical Insulation* 11, 2004: 797–807.
7. Xingyi, H.; Pingkai, J.; Tanaka, T. A review of dielectric polymer composites with high thermal conduc-tivity. *Electrical Insulation Magazine IEEE* 27, 2011: 8–16.
8. Kojima, Y.; Usuki, A.; Kawasumi, M.; Okada, A.; Fukushima, Y.; Karauchi, T.; Kamigaito, O. Mechanical properties of nylon-6–clay hybrid. *Journal of Materials Research* 6, 1993: 1185–89.
9. Levchenko, V.; Mamnunya, Y.; Boiteux, G. et al. Influence of organo-clay on electrical and mechanical properties of PP/MWCNT/OC nanocomposites. *European Polymer Journal* 47, 2011: 1351–60.
10. David, E.; Fréchette, M. F. Polymer nanocomposites—Major conclusions and achievements reached so far. *Electrical Insulation Magazine IEEE* 29, 2013: 29–36.
11. Oliveira, C. F. P.; Carastan, D. J.; Demarquette, N.; Fechine, G. J. M. Photooxidative behavior of poly-styrene–montmorillonite nanocomposites. *Polymer Engineering and Science* 48, 2008: 1511–17.
12. Bourbigot, S.; Le Bras, M.; Dabrowski, F.; Gilman, J. W.; Kashiwagi, T. PA-6 clay nanocomposite hybrid as char forming agent in intumescent formulations. *Fire Materials* 24, 2000: 201–08.
13. Alexandre, M.; Dubois, P. Polymer-layered silicate nanocomposites: Preparation, properties and uses of a new class of materials. *Materials Science and Engineering* 28, 2000: 1–63.
14. Vermogen, A.; Masenelli-Varlot, K; Séguéla, R.; Duchet-Rumeau, J.; Boucard S.; Prele, P. Evaluation of the structure and dispersion in polymer-layered silicate nanocomposites. *Macromolecules* 38, 2005: 9661–69.
15. Vermogen, A.; Masenelli-Varlot, K.; Vigier, G.; Sixou, B.; Thollet, G.; Duchet-Rumeau, J. Clay disper-sion and aspect ratios in polymer–clay nanocomposites. *Journal of Nanoscience and Nanotechnology* 7, 2007: 3160–71.
16. Carastan, D. J.; Vermogen, A.; Masenelli-Varlot, K.; Demarquette, N. R. Quantification of clay disper-sion in nanocomposites of styrenic polymers. *Polymer Engineering and Science* 50, 2010: 257–67.
17. Fornes, T. D.; Yoon, P. J.; Keskkula, H.; Paul, D. R. Nylon 6 nanocomposites: The effect of matrix molecular weight. *Polymer* 42, 2001: 9929–40.
18. Dealy, J.; Larson, R. *Structure and Rheology of Molten Polymers—From Structure to Flow Behavior and Back Again.* Munich, Germany: Hanser Verlag, 2006.
19. He, C.; Wood-Adams, P.; Dealy, J. M. Broad frequency range characterization of molten polymers. *Journal of Rheology* 48, 2004: 711–14.
20. Palierne, J. F. Linear rheology of viscoelastic emulsions with interfacial tension. *Rheologica Acta* 29, 1990: 204–14.
21. Graebling, D.; Froelich, D.; Muller, R. Viscoelastic properties of polydimethylsiloxane–polyoxyethylene blends in the melt emulsion model. *Journal of Rheology* 33, 1989: 1283–91.
22. Graebling, D.; Muller, R.; Palierne, J. F. Linear viscoelastic behavior of some incompatible polymer blends in the melt. Interpretation of data with a model of emulsion of viscoelastic liquids. *Macromolecules* 26, 1993: 320–29.
23. Dealy, J. M.; Wang, J. *Melt Rheology and its Application in the Plastic Industry.* Netherlands: Springer Verlag, 2013.

24. Cogswell, F. N. *Polymer Melt Rheology: A Guide for Industrial Practice*. Cambridge, England: Woodhead Publishing, 1997.

25. Dintcheva, N. T.; Rosamaria, M.; La Mantia, F. P. Effect of the extensional flow on the properties of oriented nanocomposite films for twist wrapping. *Journal of Applied Polymer Science* 120, 2011: 2772–79.

26. Barnes, H. A.; Hutton, J. F.; Walters F. R. S. K. *An Introduction to Rheology*. Amsterdam: Elsevier Science Publishers, 1989.

27. Vermant, J.; Ceccia, S.; Dolgovskij, M. K.; Maffettone, P. L.; Macosko, C. W. Quantifying dispersion of layered nanocomposites via melt rheology. *Journal of Rheology* 51, 2007: 429–50.

28. Mobuchon, C. P; Carreau, P.; Heuzey, M. C. Effect of flow history on the structure of a non-polar polymer/clay nanocomposite model system. *Rheologica Acta* 46, 2007: 1045–56.

29. Treece, M. A.; Oberhauser, J. P. Soft glassy dynamics in polypropylene–clay nanocomposites. *Macromolecules* 40, 2007: 571–82.

30. Treece, M. A.; Oberhauser, J. P. Ubiquity of soft glassy dynamics in polypropylene–clay nanocomposites. *Polymer* 48, 2007: 1083–95.

31. Zouari, R.; Domenech, T.; Vergnes, B.; Peuvrel-Disdier, E. J. Time evolution of the structure of organoclay/polypropylene nanocomposites and application of the time-temperature superposition principle. *Journal of Rheology* 56, 2012: 725–42.

32. Ghanbari, A.; Heuzey, M. C.; Carreau, P. J.; Ton-That, M. T. Morphological and rheological properties of PET/clay nanocomposites. *Rheologica Acta* 52, 2013: 59–74.

33. Solomon, M. J.; Almusallam, A. S.; Seefeldt, K. F.; Smowangthanaroj, A.; Varadan, P. Rheology of polypropylene/clay hybrid materials. *Macromolecules* 34, 2001: 1864–72.

34. Li, Q.; Huang, Y.; Chen, G.; Lv, Y. Effect of compatibilizer content on the shear and extensional rheology of polypropylene/clay nanocomposites. *Journal of Macromolecular Science Part B: Physics* 51, 2012: 1776–93.

35. Singh, S.; Ghosh, A. K.; Maiti, S. N.; Raha, S.; Gupta, R. K.; Bhattacharya, S. Morphology and rheological behavior of polylactic acid/clay nanocomposites. *Polymer Engineering and Science* 52, 2012: 225–32.

36. Zhang, X.; Guo, J.; Zhang, L.; Yang, S.; Zhang, J.; He, Y. Rheological properties of polypropylene/attapulgite nanocomposites. *Journal of Nanoscience and Nanotechnology* 10, 2010: 5277–81.

37. Lertwimolnun, W.; Vergnes, B. Influence of compatibilizer and processing conditions on the dispersion of nanoclay in a polypropylene matrix. *Polymer* 46, 2005: 3462–71.

38. Lerwitmolnun, W.; Vergnes, B. Effect of processing conditions on the formation of polypropylene/organoclay nanocomposites in a twin screw extruder. *Polymer Engineering and Science* 46, 2006: 314–23.

39. Durmus, A.; Kasgoz, A.; Macosko, C. W. Linear low density polyethylene (LLDPE)/clay nanocomposites. Part I: Structural characterization and quantifying clay dispersion by melt rheology. *Polymer* 48, 2007: 4492–502.

40. Krishnamoorti, R.; Giannelis, E. P. Rheology of end-tethered polymer layered silicate nanocomposites. *Macromolecules* 30, 1997: 4097–102.

41. Krishnamoorti, R.; Vaia, R. A.; Giannelis, E. P. Structure and dynamics of polymer–layered silicate nanocomposites. *Chemistry of Materials* 8, 1996: 1728–34.

42. Médéric, P.; Le Pluart, L.; Aubry, T.; Madec, P. J. Structure and rheology of polyethylene/imidazolium-based montmorillonite nanocomposites. *Journal of Applied Polymer Science* 127, 2013: 879–87.

43. Galgali, G.; Ramesh, C.; Lele, A. A rheological study on the kinetics of hybrid formation in polypropylene nanocomposites. *Macromolecules* 34, 2001: 852–58.

44. Saikat, B.; Mangala, J.; Anup, G. Investigations on clay dispersion in polypropylene/clay nanocomposites using rheological and microscopic analysis. *Journal of Applied Polymer Science* 130, 2013: 4464–73.

45. Mattausch, H.; Laske, S.; Duretek, I.; Kreith, J.; Maier, G.; Holzer, C. Investigation of the influence of processing conditions on the thermal, rheological and mechanical behavior of polypropylene nanocomposites. *Polymer Engineering and Science* 53, 2013: 1001–10.

46. Solomon, M. J.; Almusallam, A. S.; Seefeldt, K. F.; Somwangthanaroj, A.; Varadan, P. *Macromolecules* 34, 2001: 1864–72.

47. Lele, A.; Mackley, M.; Galgali, G.; Ramesh, C. In situ rheo-x-ray investigation of flow-induced orientation in layered silicate–syndiotactic polypropylene nanocomposite melt. *Journal of Rheology* 46, 2002: 1091–110.

48. Treece, M. A.; Zhang, W.; Moffitt, R. D.; Oberhauser, J. P. Twin-screw extrusion of polypropylene–clay nanocomposites: Influence of masterbatch processing, screw rotation mode, and sequence. *Polymer Engineering and Science* 47, 2007: 898–911.

49. Incarnato, L.; Scarfato, P.; Scatteia, L.; Acierno, D. Rheological behavior of new melt compounded copolyamide nanocomposites. *Polymer* 45, 2004: 3487–96.
50. Alix, S.; Follain, N.; Tenn, N. et al. Effect of highly exfoliated and oriented organoclays on the barrier properties of polyamide based nanocomposites. *Journal of Physical Chemistry C* 116, 2012: 4937–47.
51. Utracki, L.; Lyngaar-Jørgensen, J. Dynamic melt flow of nanocomposites based on poly-ε-caprolactam. *Rheologica Acta* 41, 2002: 394–407.
52. Hoffmann, B.; Dietrich, C.; Thomann, R.; Friedrich, C.; Mülhaupt, R. Morphology and rheology of polystyrene nanocomposites based upon organoclay. *Macromolecular Rapid Communications* 21, 2000: 57–61.
53. Sohn, J. I.; Lee, C. H.; Lim, S. T.; Kim, T. H.; Choi, H. J.; Jhon, M. S. Viscoelasticity and relaxation characteristics of polystyrene/clay nanocomposites. *Journal of Materials Science* 38, 2003: 1849–52.
54. Kim, T. H.; Lim, S. T.; Lee, C. H.; Choi, H. J.; Jhon, M. S. Preparation and rheological characterization of intercalated polystyrene/organophilic montmorillonite nanocomposite. *Journal of Applied Polymer Science* 87, 2003: 2106–12.
55. Ren, J.; Casanueva, B. F.; Mitchell, C. A.; Krishnamoorti, R. Disorientation kinetics of aligned polymer layered polymer layered silicate nanocomposites. *Macromolecules* 36, 2003: 4188–94.
56. Lim, Y. T.; Park, O. O. Phase morphology and rheological behavior of polymer/layered silicate nanocomposite. *Rheologica Acta* 40, 2001: 220–29.
57. Zhao, J.; Morgan, A. B.; Harris, J. D. Rheological characterization of polystyrene–clay nanocomposites to compare the degree of exfoliation and dispersion. *Polymer* 46, 2005: 8641–60.
58. Ray, S. S.; Maiti, P.; Okamoto, M.; Yamada, K.; Ueda, K. New polylactide/layered silicate nanocomposites. 1. Preparation, characterization, and properties. *Macromolecules* 35, 2002: 3104–10.
59. Wang, B.; Wan, T.; Zeng, W. Dynamic rheology and morphology of polylactide/organic montmorillonite nanocomposites. *Journal of Applied Polymer Science* 121, 2011: 1032–39.
60. Singh, S.; Ghosh, A. K.; Maiti, S. N.; Raha, S.; Gupta, R. K.; Bhattacharya, S. Morphology and rheological behavior of polylactic acid/clay nanocomposites. *Polymer Engineering and Science* 52, 2012: 225–32.
61. Krishnamoorti, R.; Giannelis, E. P. Strain hardening in model polymer brushes under shear. *Langmuir* 17, 2001: 1448–52.
62. Maiti, P. Influence of miscibility on viscoelasticity, structure, and intercalation of oligo-poly(caprolactone)/ layered silicate nanocomposites. *Langmuir* 19, 2003: 5502–10.
63. Lepoittevin, B.; Devalckenaere, M.; Pantoustier, N. et al. Poly(ε-caprolactone)/clay nanocomposites prepared by melt intercalation: Mechanical, thermal and rheological properties. *Polymer* 43, 2002: 4017–23.
64. Lee, K. M.; Han, C. D. Effect of hydrogen bonding on the rheology of polycarbonate/organoclay nanocomposites. *Polymer* 44, 2003: 4573–88.
65. Bandyopadhyay, J.; Maity, A.; Khatua, B. B.; Ray, S. S. Thermal and rheological properties of biodegradable poly[(butylene succinate)-*co*-adipate]. *Journal of Nanoscience and Nanotechnology* 10, 2010: 4184–95.
66. Saadat, A.; Nazockdast, H.; Sepehr, F.; Mehranpour, M. Linear and nonlinear melt rheology and extrudate swell of acrylonitrile-butadiene-styrene and organoclay-filled acrylonitrile-butadiene-styrene nanocomposite. *Polymer Engineering and Science* 50, 2010: 2340–49.
67. Gupta, R. K.; Pasanovic-Zujo, V.; Bhattacharya, S. N. Shear and extensional rheology of EVA/layered silicate-nanocomposites. *Journal of Non-Newtonian Fluid Mechanics* 128, 2005: 116–25.
68. Labaume, I.; Huitric, J.; Médéric, P.; Aubry, T. Structural and rheological properties of different polyamide/polyethylene blends filled with clay nanoparticles: A comparative study. *Polymer* 54, 2013: 3671–79.
69. Ren, J.; Silva, A. S.; Krishnamoorti, R. Linear viscoelasticity of disordered polystyrene–polyisoprene block copolymer based layered-silicate nanocomposites. *Macromolecules* 33, 2000: 3739–46.
70. Carastan, D. J.; Vermogen, A.; Masenelli-Varlot, K.; Demarquette, N. R.; Quantification of clay dispersion in nanocomposites of styrenic polymers. *Polymer Engineering and Science* 50, 2010: 257–67.
71. Carastan, D. J.; Demarquette, N. R.; Vermogen, A.; Masenelli-Varlot, K. Linear viscoelasticity of styrenic block copolymers–clay nanocomposites. *Rheologica Acta* 47, 2008: 521–36.
72. Lee, K. M.; Han, C. D. Linear dynamic viscoelasticity properties of funcionalized block copolymer/ organoclay nanocomposites. *Macromolecules* 36, 2003: 804–15.
73. Mitchell, C. A.; Krishnamoorti, R. Rheological properties of diblock copolymer/layered silicate nanocomposites. *Journal of Polymer Science part B: Polymer Physics* 40, 2002: 1434–43.
74. Okamoto, M.; Nam, P. H.; Maiti, P.; Kotaka, T.; Hasegawa, N.; Usuki A. A house of cards structure in polypropylene/clay nanocomposites under elongational flow. *Nano Letters* 1, 2001: 295–98.

75. Suh, C. H.; White, J. L. Talc-thermoplastic compounds: Particle orientation in flow and rheological properties. *Journal of Non-Newtonian Fluid Mechanics* 62, 1996: 175–206.
76. Ray, S. S.; Okamoto, K.; Okamoto, M. Structure–property relationship in biodegradable poly(butylene succinate)/layered silicate nanocomposites. *Macromolecules* 36, 2003: 2355–67.
77. Carastan, D. J.; Demarquette, N. R. Microstructure of nanocomposites of styrenic polymers. *Macromolecular Symposia* 233, 2006: 152–60.
78. Spencer, M. W.; Henter, D. L.; Knesek, B. W.; Paul, D. R. Morphology and properties of polypropylene nanocomposites based on a silanized organoclay. *Polymer* 52, 2011: 5369–77.
79. Naveau, E.; Dominkovics, Z.; Detrembleur, C. et al. Effect of clay modification on the structure and mechanical properties of polyamide-6 nanocomposites. *European Polymer Journal* 47, 2011: 5–15.
80. Jancar, J.; Douglas, J. F.; Starr, F. W. et al. Current issues in research on structure–property relationships in polymer nanocomposites. *Polymer* 51, 2010: 3321–43.
81. Cassagnau, P. Melt rheology of organoclay and fumed silica nanocomposites. *Polymer* 49, 2008: 2183–96.
82. Ren, J.; Krishnamoorti, R. Nonlinear viscoelastic properties of layered-silicate-based intercalated nanocomposites. *Macromolecules* 36, 2003: 4443–51.
83. Du, F.; Scogna, R. C.; Zhou, W. et al. Nanotube networks in polymer nanocomposites: Rheology and electrical conductivity. *Macromolecules* 37, 2004: 9048–55.
84. Huang, Y. Y.; Ahir, S. V.; Terentjev, E. M. Dispersion rheology of carbon nanotubes in a polymer matrix. *Physical Review* B 73, 2006: 125422.
85. Cipiriano, B. H.; Kashiwagi, T.; Raghavan, S. R. et al. Effects of aspect ratio of MWNT on the flammability properties of polymer nanocomposites. *Polymer* 48, 2007: 6086–96.
86. Chatterjee, T.; Krishnamoorti, R. Rheology of polymer carbon nanotubes composites. *Soft Matter* 9, 2013: 9515–29.
87. Kim, H.; Macosko, C. W. Processing–property relationships of polycarbonate/graphene composites. *Polymer* 50, 2009: 3797–809.
88. El Achaby, M.; Arrakhiz, F.-E.; Vaudreuil, S.; Qaiss, A.; Bousmina, M.; Fassi-Feri, O. Mechanical, thermal, and rheological properties of graphene-based polypropylene nanocomposites prepared by melt mixing. *Polymer Composites* 33, 2012: 733–45.
89. Tan, Y.; Song, Y.; Zheng, Q. Hydrogen bonding-driven rheological modulation of chemically reduced graphene oxide/poly(vinyl alcohol) suspensions and its application in electrospinning. *Nanoscale* 4, 2012: 6997–7005.
90. Sadasivuni, K. K.; Saiter, A.; Gautier, N.; Thomas, S.; Grohens, Y. Effect of molecular interactions on the performance of poly(isobutylene-*co*-isoprene)/graphene and clay nanocomposites. *Colloid and Polymer Science* 291, 2013: 1729–40.
91. Akcora, P.; Kumar, S. K.; Moll, J. et al. 'Gel-like' mechanical reinforcement in polymer nanocomposite melts. *Macromolecules* 43, 2010: 1003–10.
92. Bartholome, C.; Beyou, E.; Bourgeat-Lami, E. et al. Viscoelastic properties and morphological characterization of silica/polystyrene nanocomposites synthesized by nitroxide-mediated polymerization. *Polymer* 46, 2005: 9965–73.
93. Park, S. H.; Lee, S. G.; Kim, S. H. The use of a nanocellulose-reinforced polyacrylonitrile precursor for the production of carbon fibers. *Journal of Materials Science* 48, 2013: 6952–59.
94. Hassanabadi, H. M.; Rodrigue, D. Linear and non-linear viscoelastic properties of ethylene vinyl acetate/nano-crystalline cellulose composites. *Rheologica Acta* 51, 2012: 127–42.
95. Joshi, M.; Butola, B. S.; Simon, G.; Kukaleva, N.; Rheological and viscoelastic behavior of HDPE/octamethyl–POSS nanocomposites. *Macromolecules* 39, 2006: 1839–49.
96. Hassanabadi, H. M.; Rodrigue, D. Relationships between linear and nonlinear shear response of polymer nano-composites. *Rheologica Acta* 51, 2012: 991–1005.
97. Hassanabadi, H. M.; Abbassi, M.; Wilhem, M.; Rodrigue, D. Validity of the modified molecular stress function theory to predict the rheological properties of polymer nanocomposites. *Journal of Rheology* 57, 2013: 881–99.
98. Hamley, I. W. *The Physics of Block Copolymers*. Oxford, New York, Tokyo: Oxford University Press, 1998.
99. Kossuth, M. B.; Morse, D. C.; Bates, F. S. Viscoelastic behavior of cubic phases in block copolymer melts. *Journal of Rheology* 43, 1999: 167–96.
100. Shih, W. H.; Shih, W. Y.; Kim, S. I.; Aksay, A. I. Scaling behavior of the elastic properties of colloidal gels. *Physical Review A* 42, 1990: 4772–79.
101. Lertwimolnun, W.; Vergnes, B. Influence of screw profile and extrusion conditions on the microstructure of polypropylene/organoclay nanocomposites. *Polymer Engineering and Science* 47, 2007: 2100–09.

102. Vergnes, B. The use of apparent yield stress to characterize exfoliation in polymer nanocomposites. *International Polymer Processing* 26, 2011: 229–32.

103. Carreau, P. J.; De Kee, D. C. R.; Chhabra, R. P. *Rheology of Polymeric Systems.* New York: Hanser Publishers, 1997.

104. J. Ferry. *Viscoelastic Properties of Polymers.* New York: John Wiley & Sons, 1980.

105. Demarquette, N. R.; Dealy, J. M.; Non linear viscoelastic properties of concentrated polystyrene solutions. *Journal of Rheology* 36, 1992: 1007–32.

106. Li, J.; Zhou, C.; Wang, G.; Zhao, D. Study on rheological behavior of polypropylene/clay nanocomposites. *Journal of Applied Polymer Science* 89, 2003: 3609–17.

107. Rajabian, M.; Naderi, G.; Dubois, C.; Lafleur, P. Measurements and model predictions of transient elongational rheology of polymeric nanocomposites. *Rheologica Acta* 49, 2010: 105–18.

108. Hermans, J. J.; Hermans, P. H.; Vermaas, D.; Weidinger, A. Quantitative evaluation of orientation in cellulose fibres from the x-ray fibre diagram. *Recueil des Travaux Chimiques des Pays-Bas* 65, 1946: 427–47.

109. Rajabian, M.; Naderi, G.; Carreau, P. J.; Dubois, C. Flow-induced particle orientation and rheological properties of suspensions of organoclays in thermoplastic resins. *Journal of Polymer Science Part B: Polymer Physics* 48, 2010: 2003–11.

110. Letwimolnum, W.; Vergnes, B.; Ausias, G.; Carreau, P. J. Stress overshoots of organoclay nanocomposites in transient shear flow. *Journal of Non-Newtonian Fluid Mechanics* 141, 2007: 167–79.

111. Doi, M.; Edwards; S. F. *Theory of Polymer Dynamics.* Oxford: Oxford University Press, 1986.

112. Botta, L.; Scaffaro, R.; La Mantia, P. F.; Dintcheva, N. T. Effect of different matrices and nanofillers on the rheological behavior of polymer–clay nanocomposites. *Journal of Polymer Science Part B: Polymer Physics* 48, 2010: 344–55.

113. Saadat, A.; Nazockdast, H.; Sepehr, F.; Mehranpour, M. Linear and nonlinear melt rheology and extrudate swell of acrylonitrile-butadiene-styrene and organoclay-filled acrylonitrile-butadiene-styrene nanocomposites. *Polymer Engineering and Science* 50, 2010: 2340–49.

114. Lee, S. H.; Cho, E.; Youn, J. R. Rheological behavior of polypropylene/layered silicate nanocomposites prepared by melt compounding in shear and elongational flows. *Journal of Applied Polymer Science* 103, 2007: 3506–15.

115. Krishnamoorti, R.; Ren, J.; Silva, A. Shear response of layered silicate nanocomposites. *Journal of Chemical Physics* 114, 2001: 4968–73.

116. Park, J. U.; Kim, J. L.; Kim, D. H.; Ahn, K. H.; Lee, S. J.; Cho, K. S. Rheological behavior of polymer/layered silicate nanocomposites under uniaxial extensional flow. *Macromolecular Research* 14, 2006: 318–23.

117. Kotsilkova, R. Rheology–structure relationship of polymer/layered silicate hybrids. *Mechanics of Time-Dependent Materials* 6, 2002: 283–300.

118. Wang, M.; Hu, J. H.; Hsieh, A. J.; Rutledge, G. C. Effect of tethering chemistry of cationic surfactants on clay exfoliation, electrospinning and diameter of PMMA/clay nanocomposite fibers. *Polymer* 51, 2010: 6295–302.

119. Prasad, R.; Gupta, R. K.; Cser, F.; Bhattacharya, S. N. Investigation of melt extensional deformation of ethylene-vinyl acetate nanocomposites using small-angle light scattering. *Polymer Engineering and Science* 49, 2009: 984–92.

120. Pasanovic-Zujo, V.; Gupta, R. K.; Bhattacharya, S. N. Effect of vinyl acetate content and silicate loading on EVA nanocomposites under shear and extensional flow. *Rheologica Acta* 43, 2004: 99–108.

121. Carastan, D. J.; Amurin, L. G.; Craievich, A. F.; Gonçalves, M. C.; Demarquette, N. R. Clay-containing block copolymer nanocomposites with aligned morphology prepared by extrusion. *Polymer International* 63, 2014: 184–94.

122. Carastan, D. J.; Amurin, L. G.; Craievich, A. F.; Gonçalves, M. C.; Demarquette, N. R. Morphological evolution of oriented clay-containing block copolymer nanocomposites under elongational flow. *European Polymer Journal* 49, 2013: 1391–405.

123. Pötschke, P.; Krause, B.; Stange, J.; Münstedt, H. Elongational viscosity and foaming behavior of pp modified by electron irradiation or nanotube addition. *Macromolecular Symposia* 254, 2007: 400–08.

124. Lee, S. H.; Cho, E.; Jeon, S. H.; Youn, J. R. Rheological and electrical properties of polypropylene composites containing functionalized multi-walled carbon nanotubes and compatibilizers. *Carbon* 45, 2007: 2810–22.

125. Handge, U. A.; Pötschke, P. Deformation and orientation during shear and elongation of a polycarbonate/carbon nanotubes composite in the melt. *Rheologica Acta* 46, 2007: 889–98.

126. Huegun, A.; Fernández, M.; Muñoz, M. E.; Santamaría, A. Rheological properties and electrical conductivity of irradiated MWCNT/PP nanocomposites. *Composites Science and Technology* 72, 2012: 1602–07.

127. Lim, H. T.; Ahn, K. H.; Lee, S. J.; Hong, J. S. Design of new HDPE/silica nanocomposite and its enhanced melt strength. *Rheologica Acta* 51, 2012: 143–50.

128. Wu, F.; Zhang, B.; Yang, W.; Liu, Z.; Yang, M. Inorganic silica functionalized with PLLA chains via grafting methods to enhance the melt strength of PLLA/silica nanocomposites. *Polymer* 55, 2014: 5760–72.

129. Sadhy, S.; Bhowmick, A. K.; Unique rheological behavior of rubber based nanocomposites. *Journal of Polymer Science Part B: Polymer Physics* 43, 2005: 1854–64.

130. Kader, M. A.; Lyu, M.-Y.; Nah, C. A study on melt processing and thermal properties of fluoroelastomer nanocomposites. *Composite Science and Technology* 66, 2006: 1431–43.

131. Zhang, Y.; Zhu, Y.; Wang, S. Q. Synthesis and rheological properties of polystyrene/layered silicate nanocomposite. *Polymer* 46, 2005: 3006–13.

132. Bhattacharya, A.; Mondal, S.; Banduppodhay, A. Maleic anhydride grafted atactic polypropylene as exciting new compatibilizer for poly(ethylene-*co*-octene) organically modified clay nanocomposites: Investigations on mechanical and rheological properties. *Industrial and Engineering Chemistry Research* 52, 2013: 14143–53.

# 11 Mechanical and Thermomechanical Properties of Nanocomposites

*Nektaria-Marianthi Barkoula and Athanasios K. Ladavos*

## CONTENTS

## 11.1 INTRODUCTION

Different types of nanoreinforcement in combination with polymer, metal, or ceramic matrices find application in various areas of engineering and technology. The use of an additional phase at the nanoscale aims in tailoring the properties of the system for specific applications. In terms of the thermomechanical properties, the target is to gain high stiffening, significant load transfer ability, increased fracture toughness, and enhancement of the glass transition temperature ($T_g$) with very small addition of nanoreinforcement. This can be achieved via the introduction of rigid fillers with large interfacial area within the matrix materials. Most commonly used nanofillers have high elastic modulus along with high aspect ratio and/or extremely large surface area leading to one order of magnitude greater interfacial area compared to that of conventional composite materials. The level of enhancement is generally dependent on the constituent materials (mechanical properties, volume fraction, shape, and size of the filler particles), degree of dispersion/exfoliation/impregnation/orientation, and the interfacial adhesion of the nanoreinforcement in the matrix.

The number of scientific papers discussing the thermomechanical properties of polymer-based nanocomposites is enormous. The list of review papers in the area is also quite extensive. One important family that has been comprehensively investigated is that of polymer layered silicate (PLS) nanocomposites. Apart from layered silicates, polymer nanocomposites with other inorganic particles (nanooxides like nanosilica ($SiO_2$), nanoalumina ($Al_2O_3$), titanium dioxide ($TiO_2$), etc.) also have been discussed. Carbon nanotube (CNT) reinforced polymers is another class of nanocomposites that has received tremendous attention in the last 2–3 decades. Introduction of graphene inclusions in polymers to produce multifunctional nanocomposites has been in focus recently. At the same time, the concept of hierarchical nanocomposites has lately received considerable attention.

The scope of this chapter is not to provide a comprehensive review of all possible combinations of matrix/reinforcement materials that have been reported thus far in the open literature. Instead an overview of the progress in polymer nanocomposites will be provided based on key review articles and critical issues for mechanical reinforcement will be highlighted. The focus of the chapter will be on the mechanical and thermomechanical properties of materials that fall in the area of expertise of the authors of the present chapter. These include nanocomposites based on nonpolar and biodegradable matrices. The materials are reinforced with layered silicates targeting in flexible packaging applications. Thus, results from the research effort of the authors of the present chapter will be discussed.

## 11.2   OVERVIEW OF THE PROGRESS IN POLYMER NANOCOMPOSITES

### 11.2.1   Layered Silicate and Inorganic Filler Polymer Nanocomposites

The interest in PLS nanocomposites was boosted after the report from the Toyota research group (Okada et al., 1987), which stated that very small amounts of layered silicate loadings resulted in pronounced improvements of thermal and mechanical properties of Nylon-6 matrix. Since then, the addition of layered silicates has been investigated thoroughly in combination with almost all types of polymer matrices. Giannelis (1996) reported that Nylon-6 (PA6) and epoxy (EP)-based layered silicate nanocomposites present stiffness, strength, and barrier properties improvements at low inorganic content. Alexandre and Dubois (2000) discussed the preparation, properties, and uses of PLS nanocomposites covering polymer matrices ranging among thermoplastics, thermosets, and elastomers. Another important review in the same area is that of Ray and Okamoto (2003). Several examples of poly(methyl methacrylate) (PMMA), polypropylene (PP), polylactic acid (PLA), and PA6-based nanocomposites were presented with emphasis on preparation, characterization, and processing. In the same direction, Ray and Bousmina (2005) presented a comprehensive review of PLS biodegradable nanocomposites with focus on PLA. In all cases, it was suggested that the PLS nanocomposites exhibit improved thermomechanical properties compared to conventional composites because reinforcement in the PLS nanocomposites occurs in multiple directions. Rhim and Ng (2007), Rhim et al. (2013), and Arora and Padua (2010) focused on nanocomposite films for packaging applications and discussed several examples of natural biopolymers, such as starch, cellulose, protein, and polybutylene succinate (PBS), with or without further modification, for the preparation of the films with nanoclays. It was suggested that biopolymer-based nanocomposites present improved mechanical properties and decreased water sensitivity. Manias et al. (2007) reported on the performance of polymer–inorganic nanocomposites, where the polymers were thermoplastics and the inorganic nanoscale filler had high aspect ratio. Improvements across multiple properties were documented (mechanical, thermal, thermomechanical, etc.). In 2008, Pavlidou and Papaspyrides presented the mechanical (tensile strength, elongation at break, flexural modulus and strength, and Izod strength) and thermomechanical properties of PA, ethylene-vinyl acetate (EVA), polyurethane (PU), high density polyethylene (HPDE), EP, polyethylene terephthalate (PET), and PLA layered silicate nanocomposites. The reinforcing mechanism of layered silicates was effectively reviewed with emphasis on the understanding of how superior nanocomposites are formed. Mittal (2009) reviewed several aspects of PLS nanocomposites. In terms of mechanical properties, this review focuses on PA6, PP, EVA, PU, and HDPE matrices reinforced with various types of layered silicates. It was stated that the intercalation of nonpolar chains inside the polar silicate interlayers is difficult and thus improvement of the modulus in the case of polyolefin nanocomposites was not as high as the other polymers like PA or other polar polymers. In such cases, the use of compatibilizers is required. Based on the reviewed data, it was shown that compatibilizer addition at low contents had a positive effect on the exfoliation structure and led to an increase of the modulus. Excessive amount of compatibilizer, on the other hand, resulted in plasticization of the matrix reducing the modulus even though the extent of delamination of the silicate was increased. It was also shown that the strength and elongation are more dependent on the morphology of the nanocomposites rather than the intercalation of the clays. The thermomechanical properties of PLS nanocomposites were reviewed suggesting broadening of the $\tan\delta$ peak and increase of $T_g$ which was generally attributed to restricted segmental motions near the polymer/silicate interface. It was also observed that the storage modulus increases on the addition of the filler and this increment is more significant above the $T_g$. Gatos and Karger-Kocsis (2010) surveyed the research works performed on rubber/layered silicate nanocomposites concluding that layered silicate can be considered a potential substitute of carbon black and $SiO_2$ for enhanced mechanical properties and durability (outstanding mechanical performance also after exposure to various chemical and thermal environments). The thermomechanical properties improved significantly, particularly above the $T_g$. Azeez et al. (2013) provided a review on processing, properties, and applications of EP clay nanocomposites suggesting enhanced thermomechanical properties. The use of modified micromechanical models for the prediction of Young's modulus was also reviewed. Finally, Ojijo and Ray (2014) reviewed nanobiocomposite systems based on synthetic aliphatic polyesters and nanoclay. An extensive literature review was provided on the thermomechanical properties of such systems. A general improvement of the dynamic mechanical properties of nanocomposites over those of neat polymers was documented. As in previous studies, it was concluded that the mechanical properties were dependent mostly on the degree of dispersion of the silicate layers in the polymer matrix, an effect of the silicates on the matrix crystallinity, and finally anisotropy of the resultant nanocomposite due to the orientation of the dispersed clay particles.

Picken et al. (2008) discussed the structure and mechanical properties of PA6 nanocomposites with rod- and plate-shaped nanoparticles. It was suggested that the modulus increase and the shift of the $T_g$ can be related to the particle aspect ratios and the nanoparticle concentration. Zhang et al. (2009) discussed the fabrication and mechanical properties of various matrix systems reinforced with $SiO_2$ nanoparticles. Kalfus (2009) reviewed the viscoelastic properties of amorphous polymer nanocomposites with individual nanoparticles. Jeon and Baek (2010) presented a review on recent progress in polymer-based inorganic nanoparticle composites. It was suggested that the mechanical properties of nanocomposites prepared from various polymers (PU, EP, PA6, polyimide (PI), PP, PET, and polystyrene (PS)) and inorganic particles (layered silicates, $SiO_2$, zinc sulfide (ZnS), and $TiO_2$) did not always increase due to aggregation of the inorganic particles in the polymer matrices. It was suggested to optimize the content of the inorganic particles or to use organic additives as functionalizers in order to avoid aggregation. Kango et al. (2013) provided a review on the effect of surface modification of inorganic nanoparticles for the development of organic–inorganic nanocomposites. A series of matrices (PA6, PA6.6, acrylonitrile butadiene styrene (ABS), polyester, EP, polyphenylene sulfide (PPS), PMMA, polycarbonate (PC), etc.) in combination with inorganic particles ($SiO_2$, calcium carbonate ($CaCO_3$), $Al_2O_3$, $TiO_2$, silicon carbide (SiC), zinc oxide (ZnO), etc.) was reported in terms of their mechanical and tribological properties. The main conclusion drawn was that surface modification improved the interfacial interactions between the inorganic nanofillers and polymer matrices, which resulted in unique properties, such as very high mechanical toughness (even at low loadings of inorganic reinforcements).

## 11.2.2 CARBON FILLER-BASED POLYMER NANOCOMPOSITES

Xie et al. (2005) conducted a review on the dispersion and alignment of CNTs in polymer matrix, discussing the mechanical properties of CNTs modified PS, PMMA, PE, and PP nanocomposites. It was concluded that enhanced dispersion and alignment of CNTs in polymer matrices greatly improved mechanical properties. Coleman et al. (2006) provided a comprehensive review of the mechanical properties of CNT–polymer composites with thermoplastic and thermosetting matrices. The authors calculated the Young's modulus reinforcement per CNT volume content of various systems in order to compare different studies. Thostenson and Chou (2006) discussed the processing–structure–property relationship in CNT/epoxy composites. It was shown that the nanocomposites exhibited significantly enhanced fracture toughness at low CNT concentrations. Agglomerated CNTs resulted in slightly higher overall fracture toughness. Chou et al. (2010) worked in the same direction including fatigue data of EP-based CNT composites. The mechanical properties of CNT modified PP has been reviewed by Bikiaris in 2010. In this report, the author provided information regarding the effect of CNT addition on stress–strain response, the tensile modulus and strength, flexural modulus, impact strength, essential work of fracture, creep resistance, and thermomechanical response. Apart from experimental data from various studies, model predictions were also provided. Sahoo et al. (2010) presented a review of polymer nanocomposites based on functionalized CNTs and discussed the effect of functionalization on the mechanical properties of PU/CNTs and PI/CNTs nanocomposites. It was concluded that there is a competition between carbon–carbon bond damage and increased CNT–polymer interaction due to CNT functionalization. In the same direction is the review of Ma et al. (2010) which focused on the effects of CNT dispersion and functionalization on the properties of CNT/polymer nanocomposites. A book chapter on thermoplastic nanocomposites with CNTs is that of Sathyanarayana and Hübner (2013). The author reviewed the mechanical properties of thermoplastic–CNT composites and concluded that the theoretical potential of CNTs has not been achieved due to dispersion/distribution, alignment/orientation, reduced aspect ratio, and poor interfacial properties.

The research group of Drzal studied the incorporation of exfoliated graphite nanoplatelets in PP (Kalaitzidou et al., 2007a,b) and HDPE (Jiang and Drzal, 2010) matrices. Based on mechanical and thermomechanical data, it was suggested that exfoliated graphite nanoplatelets provided the ability to simultaneously improve multiple physical, thermal, and mechanical properties of the studied systems. Kuilla et al. (2010) reviewed graphene-based polymer composites and explored the effect of graphene/graphene oxide modification. The study concluded that graphene-based polymer nanocomposites exhibit superior mechanical properties compared to the neat polymer or conventional graphite-based composites. In parallel, the solution mixing process was superior compared to the melt mixing method for mechanical reinforcement. A critical review on the mechanics of graphene nanocomposites is that of Young et al. (2012). The structure and mechanical properties of both graphene and graphene oxide were reviewed. A comparison between various rigid polymers and elastomers in terms of modulus reinforcement due to graphene addition was provided. Both graphene and its derivatives had demonstrated their potential as reinforcements for high performance nanocomposites with high levels of strength and stiffness and superior mechanical properties at lower loadings although the levels of reinforcement were significantly lower than those anticipated. The key issues were related to defects and holes in graphene oxide compared to pristine graphene, poor interfacial properties particularly in the case of pristine graphene or graphite nanoplatelets, not complete exfoliation/poor dispersion of the reinforcement (Young et al., 2012). Desai and Njuguna (2013) reviewed the use of graphite-based nanofillers to enhance the mechanical properties of different polymer matrices. Based on the available data, mechanical reinforcement was in general achieved. It was suggested that graphene appeared to bond better to the polymers. Very recently, Mittal et al. (2015) provided a review on CNTs and graphene as fillers in reinforced polymer nanocomposites. It was stated that the mechanical properties of these systems depend on the type/purity/length/aspect ratio of CNTs, layers of graphene, dispersion of CNTs and graphene into the matrix, alignment, anti-agglomeration of CNTs and graphene into the matrix, interaction between the fillers and the matrix, etc.

### 11.2.3 POLYMER NANOCOMPOSITES WITH VARIOUS FILLERS

Many articles present results obtained using various types of nanoreinforcement. Thostenson et al. (2005) discussed the influence of nanoparticle, nanoplatelet, nanofiber, and CNTs addition on the mechanical properties and fracture toughness of various polymers. Jordan et al. (2005) made an attempt to understand the synthesis, processing, and properties of polymer nanocomposites and highlighted some of the issues related to the preparation and mechanical behavior of composites with nanosized reinforcement ($Al_2O_3$ beads, glass beads, $CaCO_3$, $SiO_2$, and layered silicate) in comparison with composites with larger micron-sized inclusions. The effect of nanosized reinforcement addition on the elastic modulus, yield stress/strain, ultimate stress/strain, strain to failure, $T_g$, and viscoelastic properties of the resulted nanocomposites was presented. It was found that the relative crystalline or amorphous nature of the polymer matrix as well as the interaction between the filler and matrix are of key importance for the final properties. Tjong (2006) provided a comprehensive review on the effects of silicate clay, ceramic nanoparticle, and CNTs addition on the structure and mechanical properties of thermoplastics, elastomers, and EP resins. Based on this report, it has been concluded that the mechanical performance depends on the types of nanofillers and polymeric matrices used. While silicate clays and ceramic nanoparticles offer considerable stiffening and strengthening in thermoplastics, elastomers, and EP resins, it was suggested that CNTs are more effective due to the higher aspect ratio (over 1000), mechanical strength, and stiffness. In terms of toughness, CNTs and silicate clays were beneficial for EP resins and elastomers since they resulted in bridging (CNTs) and/or deflection (silicate clays) of microcracks. Semicrystalline polymers presented reduction in toughness while in glassy thermoplastics toughness depended on the exfoliation level of the silicates. Hussain et al. (2006) presented a comprehensive discussion on technology, modeling, characterization, processing, manufacturing, applications, and health/ safety concerns for polymer nanocomposites with layered silicates, graphite nanoplatelets, CNTs, nanoparticles, and nanofibers. Ciardelli et al. (2008) reviewed nanocomposites based on polyolefins and functional thermoplastic materials showing enhanced thermomechanical properties. The use of nanophase surface modification and/or functionalized polyolefins to allow the dispersion of clays, CNTs, and various metal nanoparticles was discussed. Kaminsky (2014) investigated the properties of metallocene-based polyolefin nanocomposites using different inorganic nanomaterials such as $SiO_2$ balls, magnesium oxide (MgO), $Al_2O_3$, other inorganic materials as well as CNTs or carbon nanofibers (CNFs). It was demonstrated that the use of catalysts in the in situ polymerization process facilitated the deagglomeration process and uniform distribution of the nanosized particles or fibers in the polyolefin matrix. This was accompanied with a tremendous boost in the stiffness and thermomechanical properties.

A very interesting approach to evaluate the surface mechanical properties on the nanoscale level was presented in the review article of Díez-Pascual et al. (2015) using advanced indentation techniques. In this article, special emphasis was placed on nanocomposites incorporating carbon-based (CNTs, CNFs, graphene, nanodiamond, fullerenes, and carbon black) or inorganic (layered silicates and spherical nanoparticles) nanofillers. Based on the nanoindentation experiments, evaluation of the modulus, hardness, and creep enhancements on incorporation of the filler was discussed. It was shown that thermoset, glassy, and semicrystalline matrices exhibited distinct reinforcing mechanisms while improvements of mechanical properties were dependant on the nature of the filler and the dispersion and interaction with the matrix. Another important aspect of this review paper was the comparison between nanoindentation results and macroscopic properties.

## 11.2.4 HIERARCHICAL POLYMER NANOCOMPOSITES

One of the first reviews in the area of hierarchical polymer nanocomposites was that of Njuguna et al. (2007), which discussed the available literature in EP-fiber-reinforced composites manufactured using CNTs, CNFs, and nanoclays for reinforcement. In this review, it was highlighted that there are open areas of investigation, that is, the damage resistance characteristics of three-phase nanocomposite laminates, and the effect of the addition of nanofillers on the crystallization kinetics and resulting morphology of polymer-based nanocomposites. Qian et al. (2010) presented two alternative strategies for forming CNT-based hierarchical composites, that is, the dispersion of

CNTs into the composite matrix and their direct attachment onto the fiber surface. Mechanical tests of hierarchical CNT/fiber/epoxy composites confirmed that the fiber-dominated in-plane properties were not significantly affected by the introduction of CNTs while the matrix-dominated properties, particularly the interlaminar shear strength, the fracture toughness, and the flexural properties were significantly improved. Improvements in the bulk mechanical properties of CNT-grafted fiber systems were also observed. Next to the experimental results the authors presented a review of the modeling studies of CNT-based hierarchical composites which are however quite limited. Thermoplastic polymer nanocomposites based on inorganic fullerene-like nanoparticles and inorganic nanotubes were reviewed by Naffakh and Díez-Pascual (2014). In this review, improved mechanical/thermomechanical properties were reported for thermoplastic polymers like isotactic PP, PPS, or PEEK and their fiber-reinforced composites on the addition of inorganic fullerene-like nanoparticles and inorganic nanotubes. Recently a book has been devoted to exploring the unique properties of CNT as an additive in the matrix of fiber-reinforced plastics, for producing structural composites with improved mechanical performance as well as sensing/actuating capabilities (Paipetis and Kostopoulos, 2013). Two chapters discussed the mechanical performance of these hierarchical nanocomposites. One focused on the improvements in damage tolerance of aerospace structures by the addition of CNTs (Karapappas and Tsotra 2013) and the other covered the effect of environmental degradation mainly on the thermomechanical properties of such systems (Barkoula, 2013). It was concluded that the addition of CNTs enhanced greatly the damage tolerance and the durability due to environmental loadings (water absorption and thermal shock cycles).

### 11.2.5 MODELING THE MECHANICAL PROPERTIES OF POLYMER NANOCOMPOSITES

In the area of modeling the mechanical properties of PLS nanocomposites, Sheng et al. (2004) concluded that continuum-based micromechanical models can offer robust predictions of the overall elastic properties of PLS nanocomposites, provided that the hierarchical morphology of intercalated nanoclay is taken into account, along with the matrix morphology and the properties adjacent to the particles. Valavala and Odegard (2005) provided a review of modeling techniques for predicting the mechanical behavior of polymer nanocomposites reinforced with clays, CNTs, and nanoparticles. The approach for modeling the elastic properties, stress–strain behavior, and interfacial bonding/load transfer was via molecular dynamics, analytical micromechanics, and computational micromechanics techniques. Results found in the literature for the various modeling tools were tabulated and compared for six polymer nanocomposite systems. The comparison emphasized the flexibility of the modeling approaches for different polymer nanocomposite geometries. Manias et al. (2007) attempted to provide some theoretical insight into the mechanical properties, discussed existing and modified models with their shortcomings and approximations, and suggested important design parameters for improved mechanical performance. The authors proposed important design parameters for the mechanical performance of polymer nanocomposites. It was suggested that the effective filler aspect ratio and effective filler volume fraction should be considered when incomplete dispersion is present (Shia et al., 1998; Brune and Bicerano, 2002). Next to that in order to better predict the nanocomposite response, the filler-specific mechanisms of deformation and fracture should be known (Brune and Bicerano, 2002). Finally, it should be kept in mind that the small interfacial strength compared to the modulus of the filler can dramatically limit a filler's reinforcing effectiveness (Shia et al., 1998; Wang and Pyrz 2004a,b). Kutvonen et al. (2012) studied the structural and dynamical mechanisms of reinforcement of a polymer nanocomposite via coarse-grained molecular dynamics simulations. It was suggested that if the polymer–filler interactions are strong, the stress at failure of the nanocomposite is clearly correlated to the filler loading, the surface area of the polymer–filler interface, and the network structure. It was also found that small fillers (size of polymer monomers) are the most effective at reinforcing the matrix by surrounding the polymer chains and maximizing the number of strong polymer–filler interactions.

## 11.3 POLYMER LAYERED SILICATE NANOCOMPOSITE FILMS FOR FOOD PACKAGING APPLICATIONS

### 11.3.1 INTRODUCTION

Petrochemical-based plastics such as PET, polyvinylchloride (PVC), PE, PP, and PS are widely used in food packaging applications because of their exceptional properties, such as low cost, good mechanical performance, good barrier to oxygen, and easy processibility. The addition of layered silicates in petrochemical-based plastics has received great attention as reviewed in Section 11.2.1. As discussed, the motivation behind the addition of layered silicates in polymer matrices is superior mechanical properties such as modulus, strength, toughness, and barrier far from those of conventional microcomposites and unreinforced plastics. Due to environmental concerns caused by the solid waste after use of petrochemical-based plastics, the research interest has focused in recent years on biodegradable and/or biobased packaging materials. Starch, a natural polymer, has been considered as one of the most promising alternatives especially due to its attractive combination of low price and availability. Chitosan, which is industrially produced from chitin, the second most abundant polysaccharide in nature, is also an attractive candidate for packaging applications owing to its biodegradability, biocompatibility, and absence of toxicity (Sorrentino et al., 2007). In the following paragraphs, the mechanical and/or thermomechanical response of four different types of PLS nanocomposites, two based on petrochemical plastics, that is, PS (Giannakas, 2009; Giannakas et al., 2012) and low density PE (LDPE) (Giannakas et al., 2009; Grigoriadi et al., 2013) and two made out of renewable sources, that is, starch (Katerinopoulou et al., 2014) and chitosan (Giannakas et al., 2014; Grigoriadi et al., 2015) will be explored. PS and LDPE were selected based on the fact that they are nonpolar and present some challenges as matrices for PLS nanocomposites. Starch and chitosan were chosen as very promising biodegradable/biobased substitutes of petrochemical matrices in food packaging, with issues that are expected to be resolved after the addition of layered silicates.

### 11.3.2 PLS NANOCOMPOSITES BASED ON NONPOLAR MATRICES

The intercalation of molecular chains into the galleries of layered silicates depends greatly on the polarity of polymer matrices. For nonpolar PE and PS, homogenous dispersion and delamination of the silicate layers is difficult to achieve. One way to improve the dispersability of the layered silicates is by applying organic surfactants to the clays. Despite the organic modification of the silicates, there are still difficulties to disperse them in polyolefins due to their very high hydrophobicity (Jeon et al., 1998). The branched structure of LDPE makes the dispersion of inorganic nanoparticles much more difficult than in linear HDPE. To achieve miscibility between the two phases, modification via grafting is quite common (Gopakumar et al., 2002). As discussed previously (Mittal, 2009), the use of compatibilizer facilitated the exfoliation process in polyolefin-based layered silicate nanocomposites with some side effects (plasticization of the matrix) when used in excessive amounts.

In the following paragraphs, the effect of organic modification of layered silicates on the mechanical properties of PS and LDPE layered silicate nanocomposites will be discussed. The effects of synthesis parameters, such as solvent's polarity, surfactant type, and concentration as well as processing route are of key importance for the obtained morphology and structure and in turn for the mechanical properties of the nanocomposites. For the synthesis of the PS-based nanocomposites, the parameters that were considered are (a) solvent's polarity and (b) surfactant type and concentration. In the case of LDPE, two synthesis routes were compared: (a) melt mixing and (b) solvent mixing with the aim to prepare layered silicate nanocomposites based on unmodified LDPE. In both PS and LDPE, two types of layered silicates were applied: one with high aspect ratio (100–200) and one with low aspect ratio (20–30). The effect of synthesis and clay type on the mechanical is presented next. In order to facilitate this discussion, we will frequently refer to the morphology of the films based on published XRD data.

### 11.3.2.1 Polystyrene-Based Nanocomposites

The modification of the montmorillonite (MMT) clay surface from hydrophilic to organophilic (OMt) took place using hexadecyltrimethylammonium bromide (CTAB) as a surfactant while di(hydrogenated tallow) dimethylammonium chloride (2HT) was used for the modification of laponite (Lp) to organophilic (OLp). Various surfactant concentrations in the ion-exchange solution corresponding to 0.8, 1.5, and 3.0 times the CEC of the clay and two different solvents, that is, chloroform ($CHCl_3$) and carbon tetrachloride ($CCl_4$), were used for the preparation of the films with the solution casting method. Details on the materials and the preparation methodology can be found in Giannakas et al. (2012) and Giannakas (2009).

Figure 11.1 compares the Young's modulus of PS nanocomposites with (a) 2 and (b) 10 wt% OMt prepared using different surfactant concentrations (CTAB) and solvent types. Based on the obtained results, it is obvious that the addition of OMt results in enhanced modulus values, and this enhancement is higher at lower OMt contents (2 wt%) using $CCl_4$ as solvent. The observed modulus increase (up to 30%) is much higher than those documented in the past for materials with low OMt content

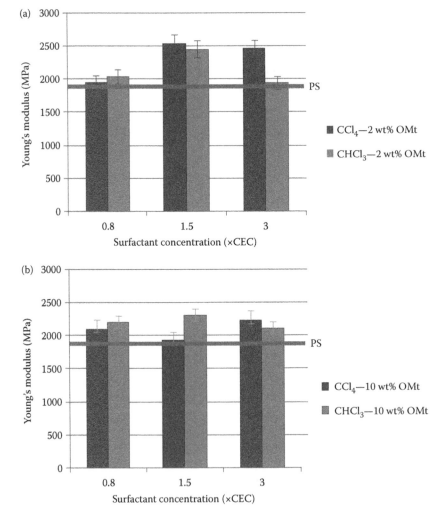

**FIGURE 11.1** Effect of surfactant concentration (CTAB) and solvent type on Young's modulus of PS-based nanocomposites with (a) 2 wt% and (b) 10 wt% OMt. (Data from Giannakas, A. et al., *Journal of Applied Polymer Science*, 114(1), 83–89, 2009.)

(2 wt%). More specifically, Uthirakumar et al. (2005) prepared exfoliated PS/clay nanocomposites via in situ polymerization and reported up to 25% increase in Young's modulus values. Noh and Lee (1999) prepared PS/NaMMT nanocomposites by emulsion polymerization and reported 7.4% respective improvement with 5 wt% MMT while Park et al. (2004) reported a marginal increase of 4%, with OMt addition of 3 wt%, using a melt mixing process. With further OMt addition (10 wt%), a considerable modulus increase is being obtained (up to 14%) which is also twice as high as that reported in the past for similar OMt contents (7% with 9 wt% OMt in Park et al., 2004). The use of $CHCl_3$ as solvent resulted in slightly lower modulus enhancement, which is up to 26% for 2 wt% and 18% for 10 wt% OMt addition. Intermediate surfactant concentrations (1.5× CEC) seem to work well with both $CCl_4$ and $CHCl_3$.

Based on Figure 11.2, it is obvious that the nanocomposites obtained using $CCl_4$ present significantly increased tensile strength with up to 40% reinforcement with the addition of 2 wt% OMt which supports the conclusion that $CCl_4$ is effective in the dispersion of a low amount of OMt in the PS matrix. The use of $CHCl_3$ as solvent results in strength values close to those of the PS matrix.

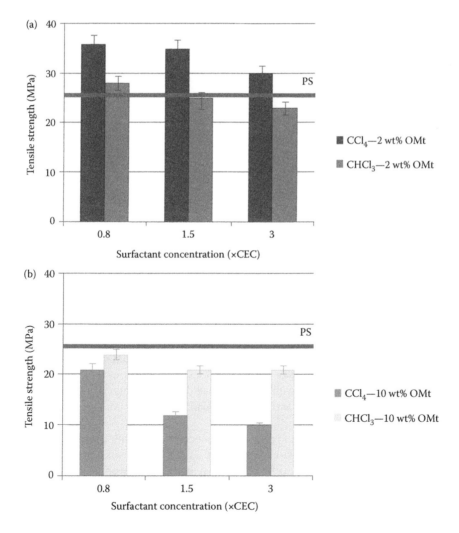

**FIGURE 11.2**   Effect of surfactant concentration and solvent type on the tensile strength of PS-based nanocomposites with (a) 2 wt% and (b) 10 wt% OMt. (Data from Giannakas, A. et al., *Journal of Applied Polymer Science*, 114(1), 83–89, 2009.)

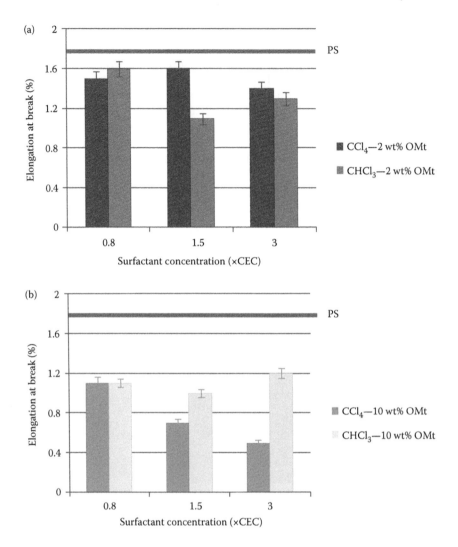

**FIGURE 11.3** Effect of surfactant concentration and solvent type on the elongation at break of PS-based nanocomposites with (a) 2 wt% and (b) 10 wt% OMt. (Data from Giannakas, A. et al., *Journal of Applied Polymer Science*, 114(1), 83–89, 2009.)

The tensile strength is slightly higher than that reported by Uthirakumar et al. (2005) (47%) and much greater than the reported value in the literature (21%) by Park et al. (2004). Note that the organoclay content does not coincide with that of silicate due to the amount of organic modification incorporated in the clay. To account for this modification, organoclay content translates to a lower percentage of silicate.

In terms of elongation at break (see Figure 11.3), nanocomposites with 2 wt% present a marginal reduction while those with 10 wt% OMt a clear reduction. Again $CCl_4$ seems to work better at low OMt contents, while $CHCl_3$ leads to lower reduction of strain at high OMt contents. The reduction of elongation at break in nanocomposite with 4.4 wt% MMT was reported to be 26% (Hasegawa et al., 1999) while Uthirakumar et al. (2005) reported an almost 50% reduction with the addition of 3 wt% MMT.

Based on the modulus, tensile strength, and elongation at break, it can be concluded that the systems prepared with 1.5 × CEC surfactant concentration and $CCl_4$ as solvent containing 2 wt% OMt presented the best performance, which is linked possibly with an exfoliated structure confirmed with the absence of clay-related deflection in the XRD results (Giannakas, 2009). It seems that phase separated microcomposites were obtained at 10 wt% OMt suggesting that it is difficult to achieve exfoliation/intercalation of high amounts of high aspect ratio fillers such as OMt.

Next an attempt was made to incorporate 10 wt% of OLp, which has significantly lower aspect ratio with the aim to improve the dispersing ability of the clays in the PS matrix and in turn the mechanical reinforcement. In Figure 11.4 is a comparison of Young's modulus (a), tensile strength (b), and elongation at break (c) of PS nanocomposites with 10 wt% OMt and OLp (Giannakas, 2009; Giannakas et al., 2012).

As can be seen in Figure 11.4, the addition of 10 wt% OLp resulted in up to 28% increase in the Young's modulus, leading to similar values as those obtained with 2 wt% OMt when $CCl_4$ was used as solvent. The respective increase with $CHCl_3$ as solvent was as low as 6%, which is probably due to better dispersion of the OLp layers in $CCl_4$. In terms of tensile strength and elongation at break, the best results were obtained with $CHCl_3$ as solvent. However, it was observed that the suspensions with $CHCl_3$ were not stable because they presented phase separation within a week. On the other hand, when $CCl_4$ was used as a solvent no sedimentation was observed for months, which was linked to the nature of $CCl_4$ being totally nonpolar (Giannakas et al., 2012).

### 11.3.2.2 Polyethylene-Based Nanocomposites

LDPE-based nanocomposites, without any polymer modification and with two kinds of clays, that is, high aspect ratio OMt and low aspect ratio OLp, were prepared using the solvent mixing and the melt mixing methodologies. Details on the materials and the preparation methodology can be found in Giannakas et al. (2009). The mechanical properties of the LDPE nanocomposites are presented in Table 11.1.

Based on the results of Table 11.1, it can be seen that the addition of OMt resulted in significant increase of Young's modulus (up to 45%) and the tensile stress at break (up to 225%) while the elongation at yield was less affected. The obtained results were linked to the intercalated or partially exfoliated structure, which was found in such systems. The addition of low OLp content (2 wt%) did not result in any increase of Young's modulus irrespective of the preparation method, indicating that the small addition of low aspect ratio filler does not induce any reinforcing effects. As the OLp content increased, the results presented significant differentiation depending on the preparation method. Particularly, in the case of 10 wt% OLp content, the modulus increased by 30% for the exfoliated nanocomposite prepared by the melt process whereas for the conventional composite received by solution blending method, the value of E decreased by 25% when compared with the neat LDPE polymer. The modulus increase obtained using the melt processing method is higher than that documented by Morawiec et al. (2005) (10% increase with 6 wt% OMt). Impressive modulus increase (83%) was presented from the research group of Malucelli et al. (2007) with the addition of 3 wt% OMt.

The materials prepared via the melt processing method were further tested in order to determine their thermomechanical response since the dispersion of layered silicates is of key importance for the linear viscoelastic properties PE-based nanocomposites (Ren et al., 2000; Krishnamoorti and Yurekli, 2001; La Mantia et al., 2006). Specifically when studying the thermomechanical properties of LDPE with compatibilizer and clays, Yang and Zheng (2004) concluded that the dynamic viscoelastic properties are strongly dependent on the intercalation of the clays. Krishnamoorti and Yurekli (2001) reported that not only the dispersion but also the orientation of the clay platelets could affect the viscoelastic properties of the nanocomposites. As discussed in previous paragraphs, the aspect ratio of the clays is also crucial for both static and dynamic mechanical properties. As seen in Figure 11.5a and b, the thermomechanical response of the LDPE/silicate nanocomposites reveal a significant enhancement of the storage modulus (40%–50%) at temperatures below ambient, which is more pronounced up to filler contents of 5 wt% in the case of OMt. Further increase of the filler content led to more conventional composites, which hindered the reinforcing ability of the silicates. On the other hand, for low aspect ratio OLp there was almost no effect at contents up to 5 wt% while a significant enhancement was achieved with 10 wt% addition.

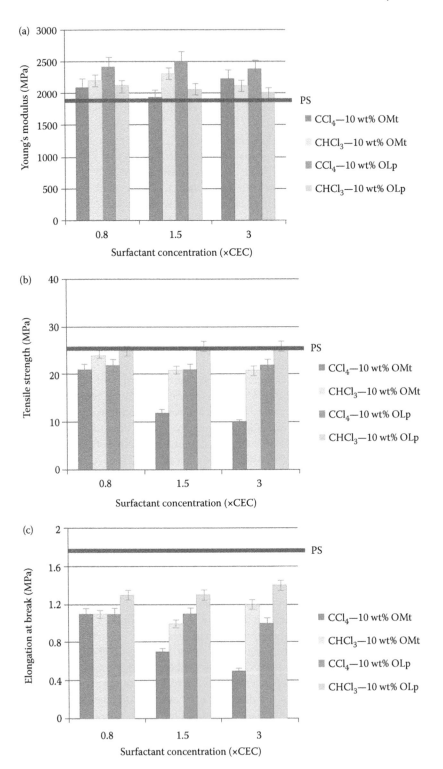

**FIGURE 11.4** Effect of layered silicate type, surfactant concentration, and solvent type on (a) Young's modulus, (b) tensile strength, and (c) elongation at break of PS-based nanocomposites with 10 wt% OMt and OLp. (Data from Giannakas, A. et al., *Journal of Applied Polymer Science*, 114(1), 83–89, 2009; Giannakas, A., Giannakas, A., Ladavos, A., *Polymer – Plastics Technology and Engineering*, 51(14), 1411–1415, 2012.)

## TABLE 11.1

## Mechanical Properties of LDPE Nanocomposites Prepared with Various Amounts of OMt and OLp Silicates via the Solvent Mixing (Solution) and Melt Mixing (Melt) Methodologies

| Material | Young's Modulus (MPa)[a] | Tensile Stress at Break (MPa)[b] | Elongation at Yield (%)[c] |
|---|---|---|---|
| LDPE (neat) | 288 | 6.0 | 10.2 |
| LDPE-2% OMt (Solution) | 311 | 7.4 | 9.9 |
| LDPE-10% OMt (Solution) | 350 | 19.6 | 8.9 |
| LDPE-2% OMt (Melt) | 303 | 12.1 | 10.4 |
| LDPE-10% OMt (Melt) | 423 | 17.5 | 9.9 |
| LDPE-2% OLp (Solution) | 286 | 10.2 | 10.1 |
| LDPE-5% OLp (Solution) | 295 | 12.2 | 8.8 |
| LDPE-10% OLp (Solution) | 215 | 11.1 | 7.3 |
| LDPE-2% OLp (Melt) | 269 | 12.7 | 9.4 |
| LDPE-5% OLp (Melt) | 305 | 14.4 | 9.9 |
| LDPE-10% OLp (Melt) | 345 | 11.6 | 10.8 |

*Source:* Adapted from Giannakas, A. et al., *Journal of Applied Polymer Science*, 114(1), 83–89, 2009. Copyright © 2009, Wiley Periodicals, Inc., with permission.

[a] Relative probable error 5%.

[b] Relative probable error 10%.

[c] Relative probable error 5%.

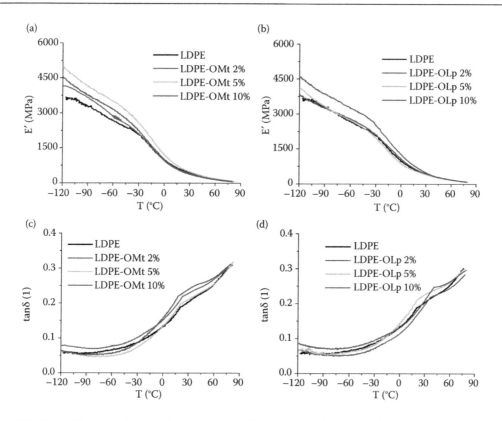

**FIGURE 11.5 (See color insert.)** Storage modulus E′ (a and b) and loss modulus E″ (c and d) as a function of temperature of LDPE modified nanocomposites with various loadings of (a) and (c) OMt and (b) and (d) OLp. (Adapted from Grigoriadi, K. et al., *Polymer Engineering and Science*, 53(2), 301–308, 2013. Copyright © 2012, Society of Plastics Engineers, with permission.)

**TABLE 11.2**

**Comparison of the Calculated E′ of the LDPE-OLp and LDPE-OMt Composites with the Measured ones at −120°C, where $E_c'$: Storage Modulus of the Composite**

| $W_f$ (%) | $W_m$ (%) | LDPE-OLp $E_c'$ (GPa) (Calculated) | LDPE-OLp $E_c'$ (GPa) (Measured) | LDPE-OMt $E_c'$ (GPa) (Calculated) | LDPE-OMt $E_c'$ (GPa) (Measured) |
|---|---|---|---|---|---|
| 2 | 98 | 3.64 | 3.82 | 4.42 | 4.17 |
| 5 | 95 | 3.86 | 4.46 | 5.86 | 5.00 |
| 10 | 90 | 4.25 | 4.64 | 8.40 | 4.58 |

*Source:* Adapted from Grigoriadi, K. et al., 2013, copyright © 2012, Society of Plastics Engineers. With permission.

It is interesting to note that LDPE-OMt 5% had better performance in the whole temperature range than LDPE-OLp 10% nanocomposites, which is in agreement with the static results presented above and can be linked to the higher aspect ratio, that is, higher active surface area of the OMt fillers. At temperatures higher than −20°C, the reinforcing efficiency was not too significant due to the increased mobility of the polymer chains. This temperature is related to the β-relaxation of the LDPE as confirmed by the peak of the E″ curves (Grigoriadi et al., 2013). The addition of layered silicates did not bring any significant alteration on β-relaxation since only a very small shift of the peak to higher values was found in the case of 10 wt% OMt (Grigoriadi et al., 2013). It is believed that β-relaxation is due to motion in the amorphous phase near branch points of LDPE (Ohta and Yasuda, 1994) where high content of high aspect ratio OMt might have resulted in some restriction. Based on the tanδ results (Figure 11.5c and d), the addition of OLp/OMt led to lower damping at temperatures below ambient and higher damping at high temperatures. This reflects the increasing reinforcement ability of the clay layers at lower temperatures where polymer chains present restricted mobility.

In order to link the reinforcing ability of low aspect ratio OLp and high aspect ratio OMt with the dispersion state of the silicates in the LDPE nanocomposites, we attempted to run back-calculations of the expected reinforcement using the modified Halpin-Tsai equation (Affdl and Kardos, 1976). The idea was to compare the calculated modulus with the measured one in order to get an indication of the degree of exfoliation/intercalation (for details, see Grigoriadi et al., 2013). The aspect ratio values used for the calculations are OMt (200) and OLp (30).

The data presented in Table 11.2 suggest a good comparison between the measured and calculated E′ in the case of the OLp nanocomposites where a partially exfoliated/exfoliated structure was found based on XRD data (Grigoriadi et al., 2013, Giannakas et al., 2009). A relatively good agreement was found in the case of OMt nanocomposites with 2 and 5 wt% contents again in line with the XRD observations. The overall conclusion was that the relatively high aspect ratio OMt clay platelets can induce superior mechanical properties to the LDPE polymer compared to lower aspect ratio clay (OLp) even if they are not fully exfoliated/intercalated (Grigoriadi et al., 2013).

### 11.3.3 PLS Nanocomposites Based on Biodegradable Matrices

As discussed in Section 11.3.1, starch and chitosan are two very interesting alternatives to replace synthetic plastics in the production of biodegradable-based nanocomposite films. The main limitation for the widespread use of these polymers is linked with their strongly hydrophilic character, which makes them not technologically useful for food packaging. One promising way to effectively control the moisture transfer is through the introduction of reinforcement at the nanoscale. The most intensive research thus far is on layered silicates. Due to the unique platelet-like structure of layered silicates, when oriented in a particular direction in the composite, they have the ability to alter the barrier properties creating a tortuous path as reviewed by Ray and Okamoto (2003). In order to facilitate the intercalation/exfoliation of the layered silicates in the matrix, and to effectively control the mechanical and barrier properties and the processability of biobased films, various methodologies have been followed.

In the following paragraphs, the effect of synthesis parameters, processing routes, and blending with other polymers/plasticizers will be discussed. For the preparation of the starch-based nanocomposites, the parameters that were considered are (a) plasticization and (b) blending with PVOH (Katerinopoulou et al., 2014). In the case of chitosan-based nanocomposites, we investigated the effect of (a) solution preparation methodology, (b) chitosan dilution level, (c) plasticization, and (d) pre- and postprocessing (Giannakas et al., 2014; Grigoriadi et al., 2015). The mechanical and thermomechanical data of the films prepared using the aforementioned parameters will be briefly reviewed next and key conclusions will be highlighted. In all cases, sodium MMT (NaMMT) was used as reinforcement.

### 11.3.3.1 Starch-Based Layered Silicate Nanocomposites

Thermoplastic acetylated corn starch (ACS)-based nanocomposite films, with or without addition of polyvinyl alcohol (PVOH), were prepared by casting using glycerol as a plasticizer. Two types of PVOH with different molecular weights were used in order to assess the effect of the chain length on the structure and performance of the obtained nanocomposites. Details on the materials and the preparation methodology can be found in Katerinopoulou et al. (2014).

The effect of the aforementioned parameters on Young's modulus, yield strength, tensile strength, elongation at yield, and elongation at break is presented in Figure 11.6a and e. The films contained different amounts of glycerol (20 and 30 wt%, designated as 20 G and 30 G) and two types of PVOH (PVOH1: lower molecular weight, PVOH2: higher molecular weight; the films contained different weight fractions of PVOH: 10 and 30 PVOH denotes 10 and 30 wt%, respectively).

Based on the obtained results it can be observed that the incorporation of NaMMT fillers resulted in significant increase in stiffness and strength of ACS systems with 20 and 30 wt% glycerol and slight decrease of the elongation at break. This was attributed to the presence of clays in an intercalated/exfoliated structure which resulted in constrained regions enhancing stiffness and strength and reducing the elongation properties (Ali et al., 2011). It was also observed that the reinforcing efficiency of the NaMMT filler was lower at higher glycerol contents (30 wt%), which was attributed to higher interactions between the glycerol and ACS chains instead of NaMMT filler and the ACS (Katerinopoulou et al., 2014).

In line with other studies, it was seen that replacement of glycerol with PVOH in the ACS/NaMMT system resulted in superior mechanical stiffness and strength due to the creation of hydrogen bonds between the ACS and the PVOH chains (Majdzadeh-Ardakani and Nazari, 2010). The higher the molecular weight of PVOH, the higher the enhancement of stiffness and strength.

The addition of NaMMT in the ACS–PVOH systems had a smaller effect on stiffness and strength, which was more obvious at high NaMMT contents (5 wt% NaMMT). The reinforcing phenomena in ACS–PVOH nanocomposite systems were promoted by the PVOH–ACS interactions rather than the clay–polymer interactions as documented in previous studies (Dean et al., 2008; Majdzadeh-Ardakani and Nazari, 2010).

Concurrent use of PVOH and glycerol reduced the stiffness and strength of the obtained films. The lower amount of PVOH (10 vs. 30 wt%) used in the combined systems and the presence of glycerol resulted in a lower amount of hydrogen bonds between the PVOH and ACS chains, which explains the deterioration in the mechanical performance (Katerinopoulou et al., 2014). Glycerol, on the other hand, effectively plasticized the films leading into the highest elongation at yield and break values. Systems with PVOH were less ductile with elongation at break values below 3%.

For the thermomechanical response presented in Figure 11.7, it is important to note that the $\tan\delta$ curve presents two relaxation peaks, one at approximately −45°C and a second one at approximately 30°C. The first ($T_\beta$) was attributed to the $T_g$ of glycerol, while the second ($T_\alpha$) was linked with the $T_g$ of

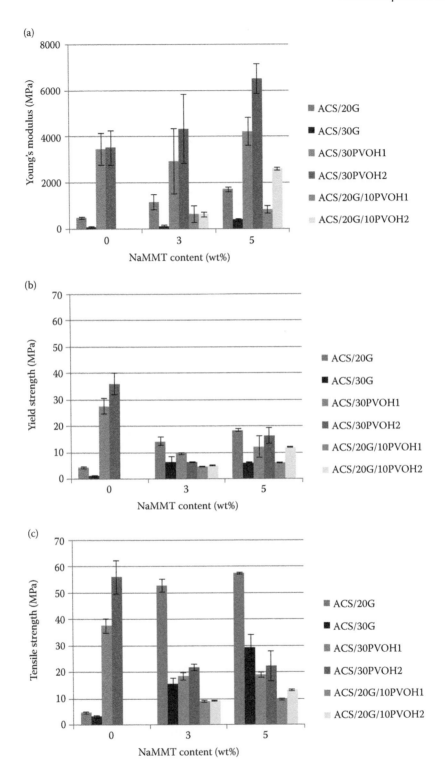

**FIGURE 11.6** Effect of glycerol and/or PVOH addition on the mechanical response of ACS-based films with 0, 3, and 5 wt% NaMMT: (a) Young's modulus, (b) yield strength, (c) tensile strength. (Data from Katerinopoulou, K. et al., *Carbohydrate Polymers*, 102(1), 216–222, 2014.)                 (*Continued*)

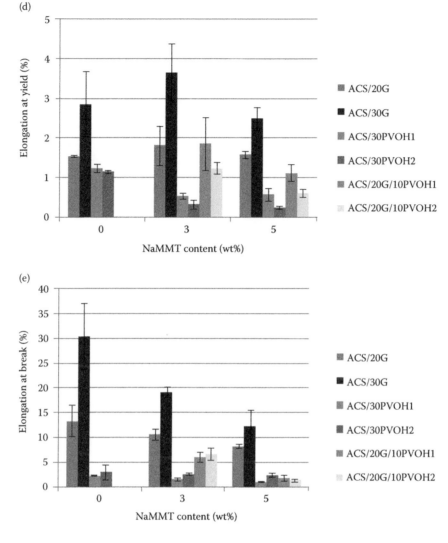

**FIGURE 11.6 (*Continued*)** Effect of glycerol and/or PVOH addition on the mechanical response of ACS-based films with 0, 3, and 5 wt% NaMMT: (d) elongation at yield, and (e) elongation at break. (Data from Katerinopoulou, K. et al., *Carbohydrate Polymers*, 102(1), 216–222, 2014.)

the ACS (Katerinopoulou et al., 2014). As discussed by Chivrac et al. (2010), this indicates the existence of phase separation between domains rich in carbohydrate chains and domains rich in plasticizer.

Based on Figure 11.7a, it is seen that systems with 5 wt% NaMMT presented slightly higher $T_\beta$ values and storage modulus compared to those with 3 wt% up to $T_\beta$ relaxation. Above this temperature due to the increased mobility of glycerol the interactions between NaMMT and glycerol chains were less significant. Systems with 5 wt% NaMMT presented broader peak of the tanδ curve around the $T_\alpha$ transition linked to the existence of interactions between the ACS chains and the NaMMT particles.

After blending ACS with glycerol and PVOH (Figure 11.7b), it is seen that the $T_\alpha$ transition became broader and was shifted by approximately 15°C to higher temperatures. The single $T_\alpha$ transition indicated that there was no phase separation between the two polymers ACS and PVOH due to the effective interaction of the PVOH with the ACS and the formation of hydrogen bonds between the two polymers. The peak of this transition did not change with the NaMMT content suggesting

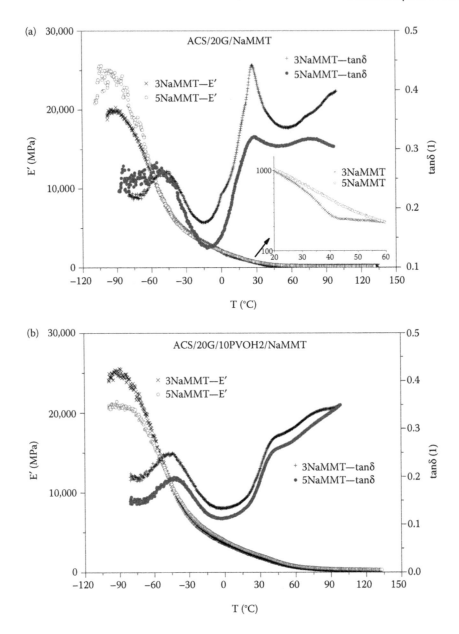

**FIGURE 11.7** Effect of glycerol and/or PVOH addition on the storage modulus (E′) and loss factor (tanδ) of (a) ACS/20G and (b) ACS/20G/10PVOH2-based films with 3 and 5 wt% NaMMT. (Reprinted from *Carbohydrate Polymers*, 102(1), Katerinopoulou, K. et al., Preparation and characterization of acetylated corn starch-(PVOH)/clay nanocomposite films, 216–222, Copyright 2014, with permission from Elsevier.)

that NaMMT interacted effectively with glycerol rather than PVOH and/or ACS, in line with the tensile results (Katerinopoulou et al., 2014).

### 11.3.3.2 Chitosan-Based Layered Silicate Nanocomposites

It has been widely acknowledged that chiotosan's solution temperature and acidity as well as the amount of acetic acid residues are of key importance for the degree of intercalation, the amount of hydrogen bonding, and the crystallinity level of the final films in chitosan-based nanocomposites (Günister et al., 2007; Wang et al., 2005; Potarniche et al., 2012). For this reason, we have recently prepared chitosan-based films using a newly introduced reflux-solution methodology on two different chitosan acetate dilution levels (2 w/v% and 1 w/v% in chitosan) in order to define the required acetate concentration for effective charge of the chitosan's chains with limited residues (Giannakas et al., 2014). Selected films were prepared with the conventional solution cast method for comparison. The use of glycerol as a plasticizer was also investigated (Giannakas et al., 2014). Finally, the effect of pre- and postprocessing was considered applying sonication or heat pressing prior/post-casting, respectively (Grigoriadi et al., 2015).

As seen in Figure 11.8, the reflux process facilitated the incorporation of NaMMT into the chitosan matrix resulting in greater enhancement of the stiffness (up to 100%) and strength (up to 70%) and respective reduction in the elongation at break (up to 75%) compared to that obtained when the no-reflux (conventional) methodology was followed (2 w/v% chitosan solutions). Best results were obtained after the addition of 3 wt% NaMMT, in agreement with the XRD results, which indicated an exfoliated structure for that system (Giannakas et al., 2014). A smaller enhancement in stiffness (up to 35%) and strength (up to 18%) and a relative lower decrease in the elongation at break (up to 65%) were found after the addition of NaMMT in 1 w/v% chitosan solutions. This was explained on the basis that dilution resulted in less charged amine groups leading to weaker electrostatic and hydrogen bond interactions among chitosan, NaMMT, and water (Giannakas et al., 2014). The mechanical properties applying the reflux process on undiluted solutions were comparable with the work of Petrova et al. (2012) and superior with most literature findings (Wang et al., 2005; Lavorgna et al., 2010; Hong et al., 2011). This was related to the effective interaction between the different constituents of the nanocomposite films since the reflux condenser does not allow any water or acetic acid to evaporate during processing (Giannakas et al., 2014). The concurrent addition of glycerol and NaMMT resulted in a direct reduction of the stiffness and the strength and in most cases in an increase in the elongation at break. This phenomenon was explained on the basis of more homogenous distribution of water and glycerol across the system in the presence of NaMMT, resulting in better plasticization effect (Xie et al., 2013).

The effect of plasticizer/NaMMT addition on the thermomechanical response of chitosan-based films is illustrated in Figure 11.9. Based on these data, the addition of plasticizer resulted in a significant decrease in the storage modulus. It is interesting to note that while the tanδ curve of the chitosan film presents on relaxation peak at approximately 170°C, which is associated with the $T_g$ of chitosan, the system containing glycerol presents two relaxation peaks: one well defined at approximately 90°C and a very broad one at approximately 130°C. As discussed in Section 11.3.3.1 for starch-based nanocomposites, this indicates the existence of phase separation between domains rich in carbohydrate chains and domains rich in plasticizer. The addition of NaMMT resulted in more homogenous thermomechanical response with one very broad transition at approximately 130°C. In line with the static results, the addition of NaMMT in chitosan/plasticizer films did not result in any reinforcing ability of the clay in the glassy region; it was, however, associated with a more homogenous distribution of water and plasticizer across the system, resulting in better plasticization effect.

The effect of the pre- and postprocessing on the mechanical properties of chitosan/NaMMT nanocomposites is illustrated in Figure 11.10. As discussed thoroughly in Grigoriadi et al. (2015) the greatest stiffness and strength enhancement was achieved at 3 wt% NaMMT for "unpressed" and "sonicated" films while "pressed" films presented the highest enhancement at 10 wt%. The overall level of enhancement was similar for all three processing routes. Based on the XRD results presented in Grigoriadi et al. (2015) it was concluded that there is a competition among the degree of dispersion, exfoliation, or intercalation of the NaMMT nanoparticles in the polymer matrix, and the morphology of the matrix (crystalline/amorphous) with pressed films presenting partially exfoliated structure and higher degree of amorphous phase and unpressed/sonicated films presenting mostly intercalated structures and a higher amount of crystalline phase.

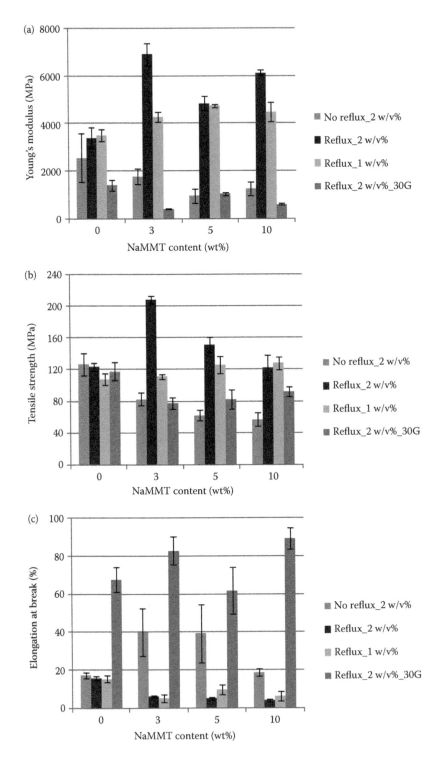

**FIGURE 11.8** Effect of the reflux processing, dilution, and glycerol addition on (a) Young's modulus, (b) tensile strength, and (c) elongation at break of chitosan/NaMMT films with 3, 5, and 10 wt% NaMMT (designation: 2 w/v%, chitosan solution with 2 w/v% in 1 v/v% acetic acid; 1 w/v%, chitosan solution 1 w/v% in 0.5 v/v% acetic acid; and 30G, 30 wt% glycerol). (Data from Giannakas, A. et al., *Carbohydrate Polymers*, 108(1), 103–111, 2014.)

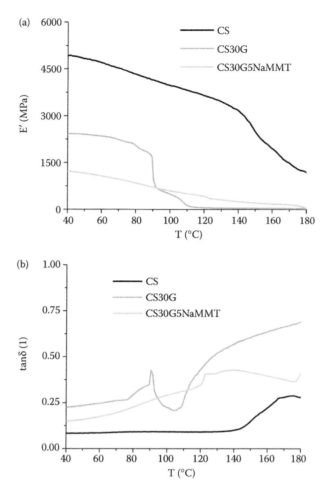

**FIGURE 11.9**  Effect of glycerol addition on (a) storage modulus (E′) and (b) loss factor (tanδ) chitosan-based films (designation: CS, chitosan films; CS30G, chitosan with 30 wt% glycerol; and CS30G5NaMMT, chitosan with 30 wt% glycerol and 5 wt% NaMMT).

## 11.4  CONCLUDING REMARKS

As discussed in the previous paragraphs, both experimental and modeling works have been conducted in order to determine key influencing parameters for the mechanical and thermomechanical performance of polymer-based nanocomposites. These are

- *Interfacial strength*: As in the case of microcomposites, there is a strong effect of the interfacial bond on the mechanical properties; therefore, many studies are devoted to improving the load transfer between the nanoscale reinforcement and the polymer matrix in order to benefit from the tremendous mechanical properties of the nanoscale reinforcements. For instance, complete exfoliation and/or higher aspect ratio nanoreinforcement results in higher stiffness, provided that there is a strong interface.
- *Structure/size*: Properties of nanostructured composites are highly structure/size dependent and there is a need of fundamental understanding of the properties and their interactions across various length scales. Jordan et al. (2005) established that with reduction in particle size the interface layers from adjacent particles overlap, altering the bulk properties significantly.

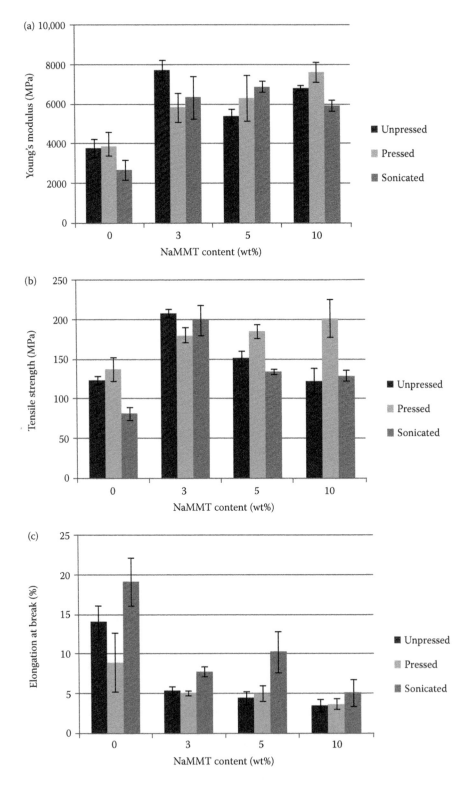

**FIGURE 11.10** Effect of processing on (a) Young's modulus, (b) tensile strength, and (c) elongation at break of chitosan/NaMMT films with 3, 5, and 10 wt% NaMMT. (Data from Grigoriadi, K. et al., *Polymer Bulletin*, 72(5), 1145–1161, 2015.)

- *Dispersion and alignment*: Good dispersion and alignment of the nanoscaled reinforcement are two more requirements for achieving mechanical reinforcement.
- *Matrix polarity*: The polarity of the matrix defines the polymer/clay interfacial interactions and the degree of exfoliation and is crucial for the strength and elongation of the obtained nanocomposites. For instance, better intercalated or exfoliated nanostructure of natural biopolymers is formed with natural NaMMT rather than organically modified MMT because hydrophilic NaMMT is compatible with most hydrophilic biopolymers (Rhim and Ng, 2007). On the other hand, organic modification is facilitating the intercalation/exfoliation process in nonpolar matrices.

In terms of fracture toughness, the mechanisms that have been recognized as critical are

- *Size/shape dependent*: Nanoreinforced composites present similar fracture mechanisms as traditional fiber-reinforced composites where crack bridging, interfacial debonding, and pullout are mechanisms of energy dissipation. When the interface is very strong, fracture toughness may improve due to crack deflection mechanisms. The theory of Chan et al. (2002) suggested that in the case where the size of the nanofiller is of the order of 50 nm or less, these fillers may not be able to resist the propagation of the crack.
- *Matrix ductility*: From J-integral tests, Chan et al. (2002) suggested that small nanoparticles would trigger large-scale plastic deformation of the matrix, which consumes tremendous fracture energy.
- *Dispersion/deagglomeration of particles*: The mobility of the nanofillers in a polymer controls their ability to dissipate energy, which would increase toughness of the polymer nanocomposites in the case of proper thermodynamic state of the matrix (Gersappe, 2002). Prerequisite for this mobility is the weakening of the interaction between the nanoparticles and the enhancement of the nanofiller/matrix interaction.

## LIST OF ABBREVIATIONS

Acetylated corn starch (ACS)
Alumina ($Al_2O_3$)
Calcium carbonate ($CaCO_3$)
Carbon nanotube (CNT)
Carbon nanofibers (CNFs)
Carbon tetrachloride ($CCl_4$)
Chloroform ($CHCl_3$)
Di(hydrogenated tallow) dimethylammonium chloride (2HT)
Epoxy (EP)
Ethylene-vinyl acetate (EVA)
Glass transition temperature ($T_g$)
Hexadecyltrimethylammonium bromide (CTAB)
High density polyethylene (HDPE)
Laponite (Lp)
Magnesium oxide (MgO)
Montmorillonite (MMT)
Nylon-6 (PA6)
Organophilic laponite (OLp)
Organophilic montmorillonite (OMt)
Polybutylene succinate (PBS)
Polycarbonate (PC)
Polyethylene terephthalate (PET)

Polyimide (PI)
Polylactic acid (PLA)
Polymer layered silicate (PLS)
Poly(methyl methacrylate) (PMMA)
Polyphenylene sulfide (PPS)
Polypropylene (PP)
Polystyrene (PS)
Polyurethane (PU)
Polyvinyl alcohol (PVOH)
Polyvinylchloride (PVC)
Silica ($SiO_2$)
Silicon carbide (SiC)
Titanium dioxide ($TiO_2$)
Zinc oxide (ZnO)
Zinc sulfide (ZnS)

## REFERENCES

Affdl, J.C.H., Kardos, J.L. 1976. The Halpin-Tsai equations: A review. *Polymer Engineering and Science*, 16(5), 344–352.

Alexandre, M., Dubois, P. 2000. Polymer-layered silicate nanocomposites: Preparation, properties and uses of a new class of materials. *Materials Science and Engineering R: Reports*, 28(1), 1–63.

Ali, S.S., Tang, X., Alavi, S., Faubion, J. 2011. Structure and physical properties of starch/poly vinyl alcohol/sodium montmorillonite nanocomposite films. *Journal of Agricultural and Food Chemistry*, 59(23), 12384–12395.

Arora, A., Padua, G.W. 2010. Review: Nanocomposites in food packaging. *Journal of Food Science*, 75(1), R43–R49.

Azeez, A.A., Rhee, K.Y., Park, S.J., Hui, D. 2013. Epoxy clay nanocomposites – Processing, properties and applications: A review. *Composites Part B: Engineering*, 45(1), 308–320.

Barkoula, N.-M. 2013. Environmental degradation of carbon nanotube hybrid aerospace composites. *Solid Mechanics and its Applications*, 188, 337–376.

Bikiaris, D. 2010. Microstructure and properties of polypropylene/carbon nanotube nanocomposites. *Materials*, 3(4), 2884–2946.

Brune, D.A., Bicerano, J. 2002. Micromechanics of nanocomposites: Comparison of tensile and compressive elastic moduli, and prediction of effects of incomplete exfoliation and imperfect alignment on modulus. *Polymer*, 43, 369–337.

Chan, C.-M., Wu, J., Li, J.-X., Cheung, Y.-K. 2002. Polypropylene/calcium carbonate nanocomposites. *Polymer*, 43(10), 2981–2992.

Chou, T.-W., Gao, L., Thostenson, E.T., Zhang, Z., Byun, J.-H. 2010. An assessment of the science and technology of carbon nanotube-based fibers and composites. *Composites Science and Technology*, 70(1), 1–19.

Chivrac, F., Pollet, E., Dole, P., Avérous, L. 2010. Starch-based nano-biocomposites: Plasticizer impact on the montmorillonite exfoliation process. *Carbohydrate Polymers*, 79(4), 941–947.

Ciardelli, F., Coiai, S., Passaglia, E., Pucci, A., Ruggeri, G. 2008. Nanocomposites based on polyolefins and functional thermoplastic materials. *Polymer International*, 57(6), 805–836.

Coleman, J.N., Khan, U., Blau, W.J., Gun'ko, Y.K. 2006. Small but strong: A review of the mechanical properties of carbon nanotube–polymer composites. *Carbon*, 44(9), 1624–1652.

Dean, K.M., Do, M.D., Petinakis, E., Yu, L. 2008. Key interactions in biodegradable thermoplastic starch/poly(vinyl alcohol)/montmorillonite micro- and nanocomposites. *Composites Science and Technology*, 68(6), 1453–1462.

Desai, S., Njuguna, J. 2013. Graphite-based nanocomposites to enhance mechanical properties. In: Njuguna, J. (ed.), *Structural Nanocomposites, Engineering Materials*, Springer-Verlag, Berlin, Heidelberg.

Díez-Pascual, A.M., Gómez-Fatou, M.A., Ania, F., Flores, A. 2015. Nanoindentation in polymer nanocomposites. *Progress in Materials Science*, 67, 1–94.

Gatos, K.G., Karger-Kocsis, J. 2010. Rubber/clay nanocomposites: Preparation, properties and applications. In: Thomas, S. and Stephen, R. (eds.), *Rubber Nanocomposites: Preparation, Properties, and Applications*, John Wiley & Sons, Ltd, Chichester, UK.

Gersappe, D. 2002. Molecular mechanisms of failure in polymer nanocomposites. *Physical Review Letters*, 89(5), 058301/1–058301/4.

Giannakas, A. 2009. PhD Thesis Synthesis, characterization and study of properties of innovative nanocomposite materials based on PS, PE and silicates.

Giannakas, A., Xidas, P., Triantafyllidis, K.S., Katsoulidis, A., Ladavos, A. 2009. Preparation and characterization of polymer/organosilicate nanocomposites based on unmodified LDPE. *Journal of Applied Polymer Science*, 114(1), 83–89.

Giannakas, A., Giannakas, A., Ladavos, A. 2012. Preparation and characterization of polystyrene/organolaponite nanocomposites. *Polymer – Plastics Technology and Engineering*, 51(14), 1411–1415.

Giannakas, A., Grigoriadi, K., Leontiou, A., Barkoula, N.-M., Ladavos, A. 2014. Preparation, characterization, mechanical and barrier properties investigation of chitosan–clay nanocomposites. *Carbohydrate Polymers*, 108(1), 103–111.

Giannelis, E.P. 1996. Polymer layered silicate nanocomposites. *Advanced Materials*, 8(1), 29–35.

Grigoriadi, K., Giannakas, A., Ladavos, A., Barkoula, N.-M. 2013. Thermomechanical behavior of polymer/layered silicate clay nanocomposites based on unmodified low density polyethylene. *Polymer Engineering and Science*, 53(2), 301–308.

Grigoriadi, K., Giannakas, A., Ladavos, A.K., Barkoula, N.-M. 2015. Interplay between processing and performance in chitosan-based clay nanocomposite films. *Polymer Bulletin*, 72(5), 1145–1161.

Gopakumar, T.G., Lee, J.A., Kontopoulou, M., Parent, J.S. 2002. Influence of clay exfoliation on the physical properties of montmorillonite/polyethylene composites. *Polymer*, 43(20), 5483–5491.

Günister, E., Pestreli, D., Ünlü, C.H., Atici, O., Güngör, N. 2007. Synthesis and characterization of chitosan–MMT biocomposite systems. *Carbohydrate Polymers*, 67(3), 358–365.

Hasegawa, N., Okamoto, H., Kawasumi, M., Usuki, A. 1999. Preparation and mechanical properties of polystyrene–clay hybrids. *Journal of Applied Polymer Science*, 74(14), 3359–3364.

Hong, S.I., Lee, J.H., Bae, H.J., Koo, S.Y., Lee, H.S., Choi, J.H., Kim, D.H., Park, S.-H., Park, H.J. 2011. Effect of shear rate on structural, mechanical, and barrier properties of chitosan/montmorillonite nanocomposite film. *Journal of Applied Polymer Science*, 119(5), 2742–2749.

Hussain, F., Hojjati, M., Okamoto, M., Gorga, R.E. 2006. Review article: Polymer–matrix nanocomposites, processing, manufacturing, and application: An overview. *Journal of Composite Materials*, 40(17), 1511–1575.

Jeon, H.G., Jung, H.-T., Lee, S.W., Hudson, S.D. 1998. Morphology of polymer/silicate nanocomposites: High density polyethylene and a nitrile copolymer. *Polymer Bulletin*, 41(1), 107–113.

Jeon, I.-Y., Baek, J.-B. 2010. Nanocomposites derived from polymers and inorganic nanoparticles. *Materials*, 3(6), 3654–3674.

Jiang, X., Drzal, L.T. 2010. Multifunctional high density polyethylene nanocomposites produced by incorporation of exfoliated graphite nanoplatelets 1: Morphology and mechanical properties. *Polymer Composites*, 31(6), 1091–1098.

Jordan, J., Jacob, K.I., Tannenbaum, R., Sharaf, M.A., Jasiuk, I. 2005. Experimental trends in polymer nanocomposites – A review. *Materials Science and Engineering A*, 393(1–2), 1–11.

Kalfus, J. 2009. Viscoelasticity of amorphous polymer nanocomposites with individual nanoparticles In: Karger-Kocsis, J. and Fakirov, St. (eds.), *Nano- and Micromechanics of Polymer Blends and Composites*, Carl Hanser Verlag GmbH & Co. KG, München, pp. 207–240.

Kalaitzidou, K., Fukushima, H., Drzal, L.T. 2007a. Multifunctional polypropylene composites produced by incorporation of exfoliated graphite nanoplatelets. *Carbon*, 45(7), 1446–1452.

Kalaitzidou, K., Fukushima, H., Drzal, L.T. 2007b. Mechanical properties and morphological characterization of exfoliated graphite–polypropylene nanocomposites. *Composites Part A: Applied Science and Manufacturing*, 38(7), 1675–1682.

Kango, S., Kalia, S., Celli, A., Njuguna, J., Habibi, Y., Kumar, R. 2013. Surface modification of inorganic nanoparticles for development of organic–inorganic nanocomposites – A review. *Progress in Polymer Science*, 38(8), 1232–1261.

Kaminsky, W. 2014. Metallocene based polyolefin nanocomposites. *Materials*, 7(3), 1995–2013.

Karapappas, P., Tsotra, P. 2013. Improved damage tolerance properties of aerospace structures by the addition of carbon nanotubes. *Solid Mechanics and its Applications*, 188, 267–336.

Katerinopoulou, K., Giannakas, A., Grigoriadi, K., Barkoula, N.M., Ladavos, A. 2014. Preparation and characterization of acetylated corn starch-(PVOH)/clay nanocomposite films. *Carbohydrate Polymers*, 102(1), 216–222.

Krishnamoorti, R., Yurekli, K. 2001. Rheology of polymer layered silicate nanocomposites. *Current Opinion in Colloid and Interface Science*, 6(5–6), 464–470.

Kuilla, T., Bhadra, S., Yao, D., Kim, N.H., Bose, S., Lee, J.H. 2010. Recent advances in graphene based poly-mer composites. *Progress in Polymer Science* (Oxford), 35(11), 1350–1375.

Kutvonen, A., Rossi, G., Ala-Nissila, T. 2012. Correlations between mechanical, structural, and dynamical properties of polymer nanocomposites. *Physical Review E – Statistical, Nonlinear, and Soft Matter Physics*, 85(4), 041803.

La Mantia, F.P., Dintcheva, N.T., Filippone, G., Acierno, D. 2006. Structure and dynamics of polyethylene/clay films. *Journal of Applied Polymer Science*, 102(5), 4749–4758.

Lavorgna, M., Piscitelli, F., Mangiacapra, P., Buonocore, G.G. 2010. Study of the combined effect of both clay and glycerol plasticizer on the properties of chitosan films. *Carbohydrate Polymers*, 82(2), 291–298.

Ma, P.-C., Siddiqui, N.A., Marom, G., Kim, J.-K. 2010. Dispersion and functionalization of carbon nanotubes for polymer-based nanocomposites: A review. *Composites Part A: Applied Science and Manufacturing*, 41(10), 1345–1367.

Majdzadeh-Ardakani, K., Nazari, B. 2010. Improving the mechanical properties of thermoplastic starch/poly(vinyl alcohol)/clay nanocomposites. *Composites Science and Technology*, 70(10), 1557–1563.

Malucelli, G., Ronchetti, S., Lak, N., Priola, A., Dintcheva, N.T., La Mantia, F.P. 2007. Intercalation effects in LDPE/o-montmorillonites nanocomposites. *European Polymer Journal*, 43(2), 328–335.

Manias, E., Polizos, G., Nakajima, H., Heidecker, M.J. 2007. Fundamentals of polymer nanocomposite technology. In: Morgan, A.B. and Wilkie, C.A. (eds.), *Flame Retardant Polymer Nanocomposites*, John Wiley & Sons, Inc., Hoboken, NJ, USA.

Mittal, G., Dhand, V., Rhee, K.Y., Park, S.-J., Lee, W.R. 2015. A review on carbon nanotubes and graphene as fillers in reinforced polymer nanocomposites. *Journal of Industrial and Engineering Chemistry*, 21, 11–25.

Mittal, V. 2009. Polymer layered silicate nanocomposites: A review. *Materials*, 2(3), 992–1057.

Morawiec, J., Pawlak, A., Slouf, M., Galeski, A., Piorkowska, E., Krasnikowa, N. 2005. Preparation and prop-erties of compatibilized LDPE/organo-modified montmorillonite nanocomposites. *European Polymer Journal*, 41(5), 1115–1122.

Naffakh, M., Díez-Pascual, A.M. 2014. Thermoplastic polymer nanocomposites based on inorganic fullerene-like nanoparticles and inorganic nanotubes. *Inorganics*, 2, 291–312.

Njuguna, J., Pielichowski, K., Alcock, J.R. 2007. Epoxy-based fibre reinforced nanocomposites. *Advanced Engineering Materials*, 9(10), 835–847.

Noh, M.W., Lee, D.C. 1999. Synthesis and characterization of PS-clay nanocomposite by emulsion polymer-ization. *Polymer Bulletin*, 42(5), 619–626.

Ohta, Y., Yasuda, H. 1994. Influence of short branches on the $\alpha$, $\beta$ and $\gamma$-relaxation processes of ultra-high strength polyethylene fibers. *Journal of Polymer Science, Part B: Polymer Physics*, 32(13), 2241–2249.

Ojijo, V., Ray, S.S. 2014. Nano-biocomposites based on synthetic aliphatic polyesters and nanoclay. *Progress in Materials Science*, 62, 1–57.

Okada, A., Kawasumi, M., Usuki, A., Kurauchi, T., Kamigaito, O. 1987. Synthesis and characterization of nylon 6-clay hybrid. *American Chemical Society, Polymer Preprints, Division of Polymer Chemistry*, 28(2), 447–448.

Paipetis, A.S., Kostopoulos, V. 2013. Carbon nanotubes for novel hybrid structural composites with enhanced damage tolerance and self-sensing/actuating abilities. *Solid Mechanics and its Applications*, 188, 1–20.

Park, C.I.L., Choi, W.M., Kim, M.H.O., Park, O.O.K. 2004. Thermal and mechanical properties of syndiotac-tic polystyrene/organoclay nanocomposites with different microstructures. *Journal of Polymer Science Part B: Polymer Physics*, 42(9), 1685–1693.

Pavlidou, S., Papaspyrides, C.D. 2008. A review on polymer-layered silicate nanocomposites. *Progress in Polymer Science* (Oxford), 33(12), 1119–1198.

Petrova, V.A., Nud'Ga, L.A., Bochek, A.M., Yudin, V.E., Gofman, I.V., Elokhovskii, V. Yu., Dobrovol'Skaya, I.P. 2012. Specific features of chitosan–montmorillonite interaction in an aqueous acid solution and properties of related composite films. *Polymer Science – Series A*, 54(3), 224–230.

Picken, S.J., Vlasveld, D.P.N., Bersee, H.E.N., Özdilek, C., Mendes, E. 2008. Structure and mechanical prop-erties of nanocomposites with rod- and plate-shaped nanoparticles. In: Knauth, P. and Schoonman, J. (eds.), *Nanocomposites: Ionic Conducting Materials and Structural Spectroscopies*, Springer, Boston, MA.

Potarniche, C.G., Vuluga, Z., Donescu, D., Christiansen, J.D.C., Eugeniu, V., Radovici, C., Serban, S., Ghiurea, M., Somoghi, R., Beckmann, S. 2012. Morphology study of layered silicate/chitosan nanohy-brids. *Surface and Interface Analysis*, 44(2), 200–207.

Qian, H., Greenhalgh, E.S., Shaffer, M.S.P., Bismarck, A. 2010. Carbon nanotube-based hierarchical compos-ites: A review. *Journal of Materials Chemistry*, 20(23), 4751–4762.

Ray, S.S., Okamoto, M. 2003. Polymer/layered silicate nanocomposites: A review from preparation to processing. *Progress in Polymer Science* (Oxford), 28(11), 1539–1641.

Ray, S.S., Bousmina, M. 2005. Biodegradable polymers and their layered silicate nanocomposites: In greening the 21st century materials world. *Progress in Materials Science*, 50(8), 962–1079.

Ren, J., Silva, A.S., Krishnamoorti, R. 2000. Linear viscoelasticity of disordered polystyrene–polyisoprene block copolymer based layered-silicate nanocomposites. *Macromolecules*, 33(10), 3739–3746.

Rhim, J.-W., Ng, P.K.W. 2007. Natural biopolymer-based nanocomposite films for packaging applications. *Critical Reviews in Food Science and Nutrition*, 47(4), 411–433.

Rhim, J.-W., Park, H.-M., Ha, C.-S. 2013. Bio-nanocomposites for food packaging applications. *Progress in Polymer Science*, 38(10–11), 1629–1652.

Sahoo, N.G., Rana, S., Cho, J.W., Li, L., Chan, S.H. 2010. Polymer nanocomposites based on functionalized carbon nanotubes. *Progress in Polymer Science* (Oxford), 35(7), 837–867.

Sathyanarayana, S., Hübner, C. 2013. Thermoplastic nanocomposites with carbon nanotubes. In: Njuguna, J. (ed.), *Structural Nanocomposites, Engineering Materials*, Springer-Verlag, Berlin, Heidelberg.

Sheng, N., Boyce, M.C., Parks, D.M., Rutledge, G.C., Abes, J.I., Cohen, R.E. 2004. Multiscale micromechanical modeling of polymer/clay nanocomposites and the effective clay particle. *Polymer*, 45(2), 487–506.

Shia, D., Hui, C.Y., Burnside, S.D., Giannelis, E.P. 1998. An interface model for the prediction of Young's modulus of layered silicate–elastomer nanocomposites. *Polymer Composites*, 19(5), 608–617.

Sorrentino, A., Gorrasi, G., Vittoria, V. 2007. Potential perspectives of bio-nanocomposites for food packaging applications. *Trends in Food Science and Technology*, 18(2), 84–95.

Thostenson, E.T., Li, C., Chou, T.-W. 2005. Nanocomposites in context. *Composites Science and Technology*, 65(3–4), 491–516.

Thostenson, E.T., Chou, T.-W. 2006. Processing–structure–multi-functional property relationship in carbon nanotube/epoxy composites. *Carbon*, 44(14), 3022–3029.

Tjong, S.C. 2006. Structural and mechanical properties of polymer nanocomposites. *Materials Science and Engineering R: Reports*, 53(3–4), 73–197.

Uthirakumar, P., Song, M.-K., Nah, C., Lee, Y.-S. 2005. Preparation and characterization of exfoliated polystyrene/clay nanocomposites using a cationic radical initiator–MMT hybrid. *European Polymer Journal*, 41(2), 211–217.

Valavala, P.K., Odegard, G.M. 2005. Modeling techniques for determination of mechanical properties of polymer nanocomposites. *Reviews on Advanced Materials Science*, 9(1), 34–44.

Wang, J., Pyrz, R. 2004a. Prediction of the overall moduli of layered silicate-reinforced nanocomposites—Part I: Basic theory and formulas. *Composites Science and Technology*, 64(7–8), 925–934.

Wang, J., Pyrz, R. 2004b. Prediction of the overall moduli of layered silicate-reinforced nanocomposites—Part II: Analyses. *Composites Science and Technology*, 64(7–8), 935–944.

Wang, S.F., Shen, L., Tong, Y.J., Chen, L., Phang, I.Y., Lim, P.Q., Liu, T.X. 2005. Biopolymer chitosan/montmorillonite nanocomposites: Preparation and characterization. *Polymer Degradation and Stability*, 90(1), 123–131.

Xie, X.-L., Mai, Y.-W., Zhou, X.-P. 2005. Dispersion and alignment of carbon nanotubes in polymer matrix: A review. *Materials Science and Engineering R: Reports*, 49(4), 89–112.

Xie, D.F., Martino, V.P., Sangwan, P., Way, C., Cash, G.A., Pollet, E., Dean, K.M., Halley, P.J., Avérous, L. 2013. Elaboration and properties of plasticised chitosan-based exfoliated nano-biocomposites. *Polymer* (United Kingdom), 54(14), 3654–3662.

Yang, H.M., Zheng, Q. 2004. The dynamic viscoelasticity of polyethylene based montmorillonite intercalated nanocomposites. *Chinese Chemical Letters*, 15(1), 74–76.

Young, R.J., Kinloch, I.A., Gong, L., Novoselov, K.S. 2012. The mechanics of graphene nanocomposites: A review. *Composites Science and Technology*, 72(12), 1459–1476.

Zhang, M.Q., Rong, M.Z., Ruan, W.H. 2009. Nanoparticles/polymer composites: Fabrication and mechanical properties In: Karger-Kocsis, J. and Fakirov, S. (eds.), *Nano- and Micromechanics of Polymer Blends and Composites*, Carl Hanser Verlag GmbH & Co. KG, München, pp. 91–140.

# Index

Printed and bound by CPI Group (UK) Ltd, Croydon, CR0 4YY

01/11/2024

01782601-0005